Microreactor Technology

This book offers a comprehensive and forward-looking exploration of microreactor systems as transformative tools for modern chemical, materials, and environmental research. Designed to connect fundamental principles with cutting-edge applications, it presents how miniaturized, precisely controlled reaction platforms are reshaping synthesis, remediation, and sustainable process development. It overviews design principles, operational mechanisms, and the transformative impact of microreactors across various scientific domains.

Features:

- Provides readers with a thorough understanding of microreactor technology, covering fundamental principles, design, and a wide range of applications.
- Aligns with global efforts toward sustainability by showcasing how microreactors can reduce waste and enhance energy efficiency.
- Emphasizes that microreactors are primarily valuable for generating insights and refining designs before larger-scale pilot plant installations.
- Balances the focus on both the "how" and the "why" of microreactor technology, addressing both technical aspects and broader industrial implications.
- Includes practical examples, case studies, and pedagogical content.

Intended for researchers, postgraduate students, and professionals in chemical engineering, materials science, nanotechnology, and environmental engineering, this book serves as both a foundational reference and a guide to emerging trends.

Microreactor Technology

Edited by Satyabrata Si, Jayato Nayak,
Suraj K. Tripathy, and Sankha Chakrabortty

CRC Press
Taylor & Francis Group
Boca Raton London New York

CRC Press is an imprint of the
Taylor & Francis Group, an **informa** business

First edition published 2027
by CRC Press
2385 NW Executive Center Drive, Suite 320, Boca Raton FL 33431

and by CRC Press
4 Park Square, Milton Park, Abingdon, Oxon, OX14 4RN

CRC Press is an imprint of Taylor & Francis Group, LLC

ISBN: 978-1-032-99992-0 (hbk)
ISBN: 978-1-041-06100-7 (pbk)
ISBN: 978-1-003-63376-1 (ebk)

DOI: 10.1201/9781003633761

Typeset in Times
by SPi Technologies India Pvt Ltd (Straive)

Contents

About the Editors vii
Contributors ix

Chapter 1 Overview of Microreactor Technology 1
Satyabrata Si and Sankha Chakrabortty

Chapter 2 Design and Fabrication of Microreactors 17
Sushanta Kumar Behera, Sriya Naik, Naresh Kumar, and Subrata Biswas

Chapter 3 Organic Chemical Synthesis Using Microreactor 38
Thangamani Suppan, Kiruthika Paramasivam, Moorthi Pichumani, M. Habibur Rahman, and Rama Ranjan Bhattacharjee

Chapter 4 Polymer-Based Material Synthesis Using Microreactors 59
Avik Mukherjee, Papita Das, and Saswata Bose

Chapter 5 Carbon-Based Nanomaterial Synthesis Using Microreactors 88
Madhubonti Pal, Rahul S. Tanpure, and Sankha Chakrabortty

Chapter 6 Metal and Metal Oxide Nanoparticle Synthesis Using Microreactors 108
Mugdha Sinha, Taisia Miroshnichenko, and Priti Sundar Mohanty

Chapter 7 Biomaterial Synthesis and Biofunctionalization Using Microreactors 136
Jayato Nayak and Sourab Paul

Chapter 8 Quantum Dot Synthesis Using Microreactors 148
Sarojini Nanthakumar, Bhargav Midathani, Priyadharshini Sivalingam, Subash Baskaran, and Subhendu K. Panda

Chapter 9 Degradation of Organic Pollutants Using Microreactors 170

Arunava Sengupta, Mekhla Banerjee, Arindam Banerjee, and Sankha Chakrabortty

Chapter 10 Use of Microphotobioreactor in Algal Bioremediation: A Hypothetical Discussion for a Potential Future 195

Soumya Banerjee

Chapter 11 Advanced Oxidation and Photocatalytic Remediation Using Microreactors 210

Rashmi Acharya, Reema Nayak, Anubhav Naik, and Subhasish Mishra

Chapter 12 Hybrid and Integrated Approaches to Microreactor-Based Remediation 240

Akshay Kumar Dey, Sinchon Roy, Anik Das, Monisankar Roy, Sankha Chakrabortty, Arindam Mandal, and Biswajit Kamila

Chapter 13 Microreactors for Renewable Energy Production 262

Prasenjit Chakraborty and Ramesh Kumar

Chapter 14 Quantum Dot Synthesis Using Microreactors and Its Applications in Energy Storage 274

Akshay Kumar Dey, Monisankar Roy, Anik Das, Sinchon Roy, Sankha Chakrabortty, Biswajit Kamila, and Arindam Mandal

Chapter 15 Bridging of Microreactor Technology with CFD-Based Approaches in Sustainable Process Development............................ 294

Krishanu Roy, Sabyasachi Paitandi, Darpan Borah, Supriya Jena, and Shirsendu Banerjee

Index 310

About the Editors

Satyabrata Si is a faculty member in the School of Chemical Engineering at KIIT Deemed to be University, India. His research expertise spans nanomaterials, advanced catalysis, polymeric composites, and functional biomaterials for biomedical, environmental, and energy applications. His work particularly focuses on gold and silver nanoparticles, silica-coated nanostructures, graphene quantum dots, hydrogel-based nanomaterials, and catalytic process development for sustainable technologies. Dr. Si has authored and edited several book chapters and review articles in the fields of nanotechnology, catalysis, green chemistry, and process engineering. His research integrates material synthesis, characterization, and application-driven design, with an emphasis on translational outcomes in healthcare, sensing, antimicrobial systems, and environmental remediation. He is actively involved in developing interdisciplinary research proposals, including projects related to healthcare, advanced catalytic systems, energy materials, etc. Beyond research, Dr. Si contributes to curriculum design, mentoring of undergraduate and postgraduate students, and academic administration.

Jayato Nayak, Assistant Professor at Mahindra University & Center for Life Sciences, is a leading researcher with over a decade of experience in novel design and applications focused on process intensification. His expertise spans water and wastewater treatment, graphene-integrated photocatalyst development, sludge treatment, and bio-synthesis of value-added products. He collaborates nationally with IITs, NITs, CSIR labs, and internationally with professionals in Poland, Brazil, Malaysia, Vietnam, among others. With 35+ international peer-reviewed journal papers, 5 edited books, 35+ book chapters, and 3 filed patents, his research has garnered over 700 citations, achieving an h-index of 14 and i10-index of 15. Dr. Nayak serves as a resource person for webinars, holds editorial roles in 30+ international journals, and has received 5 distinguished awards.

Suraj K. Tripathy, Head of the School of Chemical Engineering, KIIT, also leads the Chemical & Bioprocess Engineering Lab (CBEL) at KIIT, which focuses on achieving sustainability in materials processing and utilization. CBEL explores opportunities in the valorization of waste materials (secondary resources) and investigates their applications in catalysis, water treatment and biomedical systems. CBEL also works closely with industries to develop suitable waste management and resource recycling strategies to optimize the potential of the circular economy model. Dr. Suraj has published more than 70 papers in several good SCI/SCOPUS indexed journals, and he has experience in handling projects worth over 10 crore rupees in core chemical process engineering.

Sankha Chakrabortty is an Associate Professor at the School of Chemical Engineering & Biotechnology, KIIT Deemed to be University, Bhubaneswar. A gold medalist Ph.D. from NIT Durgapur, he has over eight years of postdoctoral experience in teaching, research, and technology development. His research focuses on membrane technology, water and wastewater treatment, CO_2 capture and conversion, biomass valorization, and sustainable chemical processes. He has led and contributed to projects funded by DST, UGC, Ministry of Mines, NALCO, and NMDC, etc. Dr. Chakrabortty has published over 85 peer-reviewed articles, edited books, and holds several patents (14), including a US patent. He serves as Associate Editor for Scientific Reports, South African Journal of Chemical Engineering, and PLOS Water. An active proposal evaluator for BIRAC and other agencies, he has received multiple awards, including the Hiyoshi Young Leaf Award and Scientist of the Year.

Contributors

Rashmi Acharya
Department of Chemistry
I.T.E.R, Siksha O Anusandhan Deemed to be University
Bhubaneswar, Odisha, India

Arindam Banerjee
Gelatex Technologies OÜ
Mahtra 30a, 13811, Tallinn, Estonia, EU

Mekhla Banerjee
Department of Environment Science
The Burdwan University
Burdwan, West Bengal, India

Shirsendu Banerjee
School of Chemical Engineering
KIIT Deemed to be University
Bhubaneswar, Odisha, India

Soumya Banerjee
Hooghly Engineering & Technology College
Chinsurah, West Bengal, India

Subash Baskaran
Solar Cells Laboratory, EPEM Division
CSIR-Central Electrochemical Research Institute
Karaikudi, Tamil Nadu, India

Sushanta Kumar Behera
Department of Chemical Engineering
National Institute of Technology (NIT) Andhra Pradesh
Tadepalligudem, India

Rama Ranjan Bhattacharjee
Department of Chemistry
Sister Nivedita University
New Town, Kolkata, West Bengal, India

Subrata Biswas
Department of Paper & Packaging Technology
Indian Institute of Technology (IIT) Roorkee
Roorkee 247667, India

Darpan Borah
School of Chemical Engineering
KIIT Deemed to be University
Bhubaneswar, Odisha, India

Saswata Bose
Chemical Engineering Department
Jadavpur University
Kolkata, India

Prasenjit Chakraborty
Central Institute of Petrochemicals Engineering & Technology (CIPET): IPT
Bihta
Patna-801103, Bihar, India

Sankha Chakrabortty
School of Chemical Engineering
KIIT Deemed to be University
Bhubaneswar, Odisha, India

Anik Das
Department of Chemical Engineering
University of Calcutta
Kolkata, India

Papita Das
School of Advanced Studies in Industrial Pollution Control Engineering
Jadavpur University
Kolkata, India

and

Chemical Engineering Department
Jadavpur University
Kolkata, India

Akshay Kumar Dey
Department of Chemical Engineering
University of Calcutta
Kolkata, India

and

Department of Chemistry
Brainware University
Kolkata, India

Supriya Jena
School of Chemical Engineering
KIIT Deemed to be University
Bhubaneswar, Odisha, India

Biswajit Kamila
Department of Chemical Engineering
University of Calcutta
Kolkata, India

Naresh Kumar
Department of Chemical Engineering
National Institute of Technology (NIT)
Andhra Pradesh, Tadepalligudem, India

Ramesh Kumar
Department of Earth Resources & Environmental Engineering
Hanyang University
Seoul, Republic of Korea

Arindam Mandal
Department of Chemical Engineering
University of Calcutta
Kolkata, India

Bhargav Midathani
Solar Cells Laboratory, EPEM Division
CSIR-Central Electrochemical Research Institute
Karaikudi, Tamil Nadu, India

Taisia Miroshnichenko
School of Computing Engineering & the Built Environment
Edinburgh EH10 5DT, UK

Subhasish Mishra
Department of Chemistry
I.T.E.R, Siksha O Anusandhan Deemed to be University
Bhubaneswar, Odisha, India

Priti Sundar Mohanty
School of Chemical Engineering
Kalinga Institute of Industrial Technology (KIIT) Deemed to be University
Bhubaneswar - 751024, India

and

School of Biotechnology
Kalinga Institute of Industrial Technology (KIIT) Deemed to be University
Bhubaneswar - 751024, India

and

Centre for Soft Matter & Bio-Engineering (CSMBE)
Kalinga Institute of Industrial Technology (KIIT)
Deemed to be University
Bhubaneswar 751024, India

Avik Mukherjee
School of Advanced Studies in Industrial Pollution Control Engineering
Jadavpur University
Kolkata, India

and

Chemical Engineering Department
Jadavpur University
Kolkata, India

Anubhav Naik
Department of Chemistry
I.T.E.R, Siksha O Anusandhan Deemed to be University
Bhubaneswar, Odisha, India

Sriya Naik
Department of Chemical Engineering
Indian Institute of Technology (IIT) Kharagpur
West Bengal, India

Sarojini Nanthakumar
Solar Cells Laboratory, EPEM Division
CSIR-Central Electrochemical Research Institute
Karaikudi, Tamil Nadu, India

and

Academy of Scientific and Innovative Research (AcSIR)
Ghaziabad, India

Jayato Nayak
Centre for Life Science, Mahindra University
Bahadurpally, Jeedimetla, Hyderabad, Telangana, India

Reema Nayak
Department of Chemistry
I.T.E.R, Siksha O Anusandhan Deemed to be University
Bhubaneswar, Odisha, India

Sabyasachi Paitandi
Birbhum Institute of Engineering and Technology
Suri, Birbhum, West Bengal, India

Madhubonti Pal
School of Chemical Sciences
Association for the Cultivation of Science
Jadavpur, Kolkata, India

Kiruthika Paramasivam
SparkVy Advanced Research & Innovations
Coimbatore, Tamil Nadu, India

Subhendu K. Panda
Solar Cells Laboratory, EPEM Division, CSIR-Central Electrochemical Research Institute
Karaikudi, Tamil Nadu, India

and

Academy of Scientific and Innovative Research (AcSIR)
Ghaziabad, India

Sourab Paul
Centre for Life Science
Mahindra University
Bahadurpally, Jeedimetla, Hyderabad, Telangana, India

Moorthi Pichumani
Department of Nano Science and Technology, Sri Ramakrishna Engineering College
Vattamalaipalayam, NGGO Colony post, Coimbatore, Tamil Nadu, India

M. Habibur Rahman
Department of Chemistry,
University of Rajshahi,
Rajshahi 6205, Bangladesh

Krishanu Roy
Dr. B. C. Roy Engineering College
Durgapur, West Bengal, India

Monisankar Roy
Department of Chemical Engineering
University of Calcutta, Kolkata, India

Sinchon Roy
Department of Chemical Engineering
University of Calcutta,
Kolkata, India

Arunava Sengupta
Department of Environment Science
The Burdwan University
Burdwan, West Bengal

Satyabrata Si
School of Chemical Engineering
Institute of Industrial Technology (KIIT) Deemed to be University
Bhubaneswar - 751024, INDIA

and

School of Biotechnology
Kalinga Institute of Industrial Technology (KIIT) Deemed to be University
Bhubaneswar - 751024, India

and

Centre for Soft Matter & Bio-Engineering (CSMBE)
Kalinga Institute of Industrial Technology (KIIT)
Deemed to be University
Bhubaneswar 751024, India

Mugdha Sinha
School of Biotechnology
KIIT Deemed to be University
Bhubaneswar, India

Priyadharshini Sivalingam
Solar Cells Laboratory, EPEM Division
CSIR-Central Electrochemical Research Institute
Karaikudi, Tamil Nadu, India

and

Academy of Scientific and Innovative Research (AcSIR)
Ghaziabad, India

Thangamani Suppan
Department of Chemistry
KPR Institute of Engineering and Technology
Coimbatore, Tamilnadu, India

Rahul S. Tanpure
Department of Earth Resources & Environmental Engineering
Hanyang University
Seoul, Republic of Korea

1 Overview of Microreactor Technology

Satyabrata Si and Sankha Chakrabortty

1.1 OVERVIEW OF MICROREACTOR TECHNOLOGY

The miniaturization of chemical processes has revolutionized modern science and engineering. Microreactor technology represents a paradigm shift from traditional batch reactors to continuous-flow systems operating in microscale environments. A microreactor is essentially a compact reaction system, typically characterized by channels with dimensions in the sub-millimeter range, that allows precise control over mass transfer, heat transfer, and reaction kinetics. The unique combination of microscale geometry, controlled flow patterns, and high surface-to-volume ratios has opened new avenues for material synthesis, chemical processing, and environmental applications.

The advent of microreactor technology (also called microreaction technology) in the 1990s corresponded well with the ultimate development of process intensification (PI) that entails the use of smaller reaction volumes as a result of the enhanced reaction kinetics through its evolutionary development cycle (Stankiewicz and Moulijn, 2004). As its name suggests, microreactor technology relies on the use of microchannel structures (with their lateral characteristic dimension usually in the range of 0.01–1 mm) to perform chemical reactions as well as other unit operations in a continuous-flow fashion (Figure 1.1) (Marre et al., 2010). The miniaturization of channel sizes to typically sub-millimeter or micrometer scale in microreactors brings about distinct advantages in reactor operation and significant potential for PI. The prevailing laminar flow conditions, and the predominance of surface tension forces or viscous forces (e.g., over buoyant or inertial forces), under wide operational conditions in microreactors facilitates the formation of regular flow patterns (Figure 1.2). For example, a precise manipulation of bubbles, droplets, and (single-phase, miscible, or immiscible) fluid streams can be easily achieved in microchannel structures where fluid(–fluid) hydrodynamics and fluid–solid interactions are addressed properly. The confinement of such single- or multiphase flow in microchannels renders a very high fluid–fluid or fluid–solid contact area (typically on the order of 10,000 m^2/m^3) (Kashid et al., 2011). This, combined with the reduced path for heat conduction and molecular diffusion in microchannels, greatly improves the temperature/concentration uniformity. Such significant heat and mass-transfer intensification enables the operation of microreactors (largely) under isothermal conditions and kinetic regime toward realizing maximized reaction performance. This makes microreactors well suited for processes that are sensitive to mixing, temperature, or interfacial heat-/mass-transfer

DOI: 10.1201/9781003633761-1

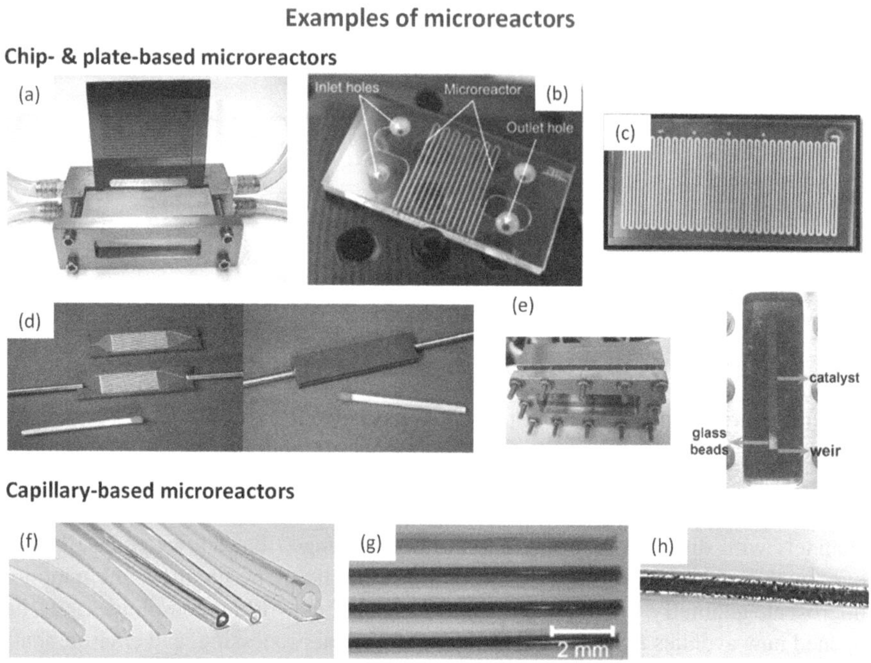

FIGURE 1.1 Examples of microreactor configurations include: (a) a high-pressure silicon–Pyrex microreactor system, (b) a glass microreactor chip used for esterification reactions, and (c) a glass chip coated with ZSM-5 for Pd-catalyzed methylisobutylketone synthesis. Additional designs shown are (d) a stainless-steel microreactor with Pt coating for methane combustion, (e) a packed-bed silicon–glass chip with Pt/Al_2O_3 catalyst for aldehyde oxidation, and (f) capillary microreactors made from PTFE (Polytetrafluoroethylene), polyphenylsulfone, glass, and PFA (Perfluoroalkoxy alkanes) with different internal diameters. Also illustrated are (g) fused-silica capillaries coated with increasing $Pd/\gamma\text{-}Al_2O_3$ catalyst loadings for hydrogenation applications and (h) a PFA capillary reactor packed with Ru/C catalyst for levulinic acid hydrogenation.

(Extracted with permission from Yue, 2022.)

limitations as frequently encountered in multiphase systems and conventional macroscale (batch or continuous-flow) reactors (Yue, 2018).

The high degree of control over these transport phenomena makes it possible to have not only a precise fine-tuning of the associated process behavior in microreactors (e.g., in terms of the desired mixing or separation efficiency, reaction conversion or selectivity, target product yield), but also a reliable microreactor performance prediction or interpretation. The small reagent inventories and tight temperature control in microreactors further lead to their inherent safety. Microreactors can also safely handle explosive reaction mixtures (e.g., gaseous hydrogen and oxygen) because the flame propagation is suppressed as the characteristic dimension of microchannels is often below the flame quenching distance. The above features and merits of microreactor technology have spurred enormous research interest over the recent two decades

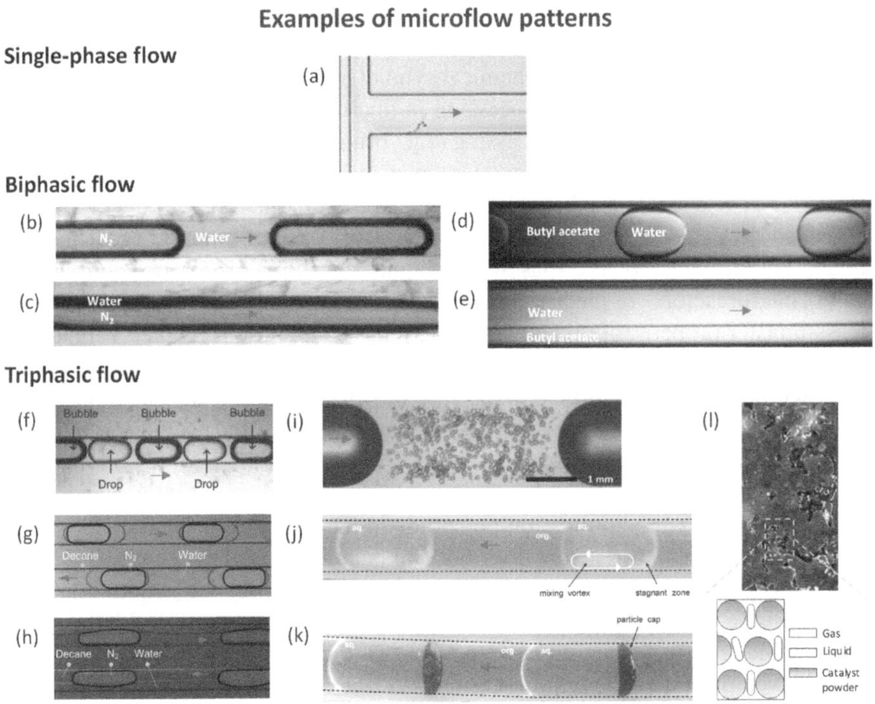

FIGURE 1.2 Diverse flow regimes in microreactors. Panel (a) shows the side-by-side laminar flow of two miscible liquids in a silicon–glass chip. Panels (b–c) display N_2–water slug and annular flow in a polymethyl methacrylate (PMMA) microchannel. Panels (d–e) present water–butyl acetate slug and parallel flow in a glass device. Panel (f) depicts gas–liquid–liquid slug flow in a PMMA chip, while panels (g–h) show N_2–water–decane slug and parallel-slug flow in glass channels. Panel (i) highlights gas–liquid–solid slurry slug flow in a PFA capillary reactor. Panels (j–k) demonstrate liquid–liquid–solid slurry slug flow in PTFE channels with suspended catalyst particles. Finally, panel (l) represents oxygen–benzyl alcohol slug flow packed with Au-Pd/TiO_2 catalyst for benzaldehyde production, along with its simplified flow schematic. Red arrows indicate flow direction.

(Extracted with permission from Yue, 2022.)

in the field of chemical synthesis (Jähnisch et al., 2004). Various materials (typically glass, silicon, stainless steel, and perfluorinated polymers) have been used to fabricate or assemble microreactors in a chip-, plate-, or capillary-based configuration (Figure 1.1). Chip- or plate-based microreactors are usually custom-made by tailored microfabrication techniques, which allow one to integrate different functionalities, e.g., additional channels for heat exchanging (Rebrov et al., 2001), sensors, or other mechanisms for the online measurement of reaction parameters within the same device for assisting reaction diagnosis, optimization, or automation. In comparison, capillary-based microreactors are easily assembled using commercially available polymeric (e.g., made of PTFE, PFA) or metal (e.g., made of stainless steel) tubings, which are relatively cheap and flexible, though with less room to accommodate high functionality. Especially, the latter microreactor type has been widely adopted in the rapidly growing field of flow chemistry for accelerated chemistry discovery and optimization in, among others, fine chemical synthesis (Plutschack et al., 2017), and the prominent advantages of flow processing in microreactors have been demonstrated well over conventional batch processing (e.g., bringing new chemical routes, process simplification, fast and effective telescoped synthesis) (Baumann et al., 2021). A myriad of reactions have benefited from operation in microreactors, which can be classified according to phases present, such as gas–solid, biphasic (gas–liquid or liquid–liquid), and triphasic (gas–liquid–liquid, gas–liquid–solid, or liquid–liquid–solid) microreactors. A precise manipulation of fluid flow patterns with superior mass-transfer properties (e.g., biphasic or triphasic slug flow), coupled with an efficient incorporation of solid catalysts (e.g., as wall coatings, or powders contained in a packed bed or within slurries) in the case of heterogeneous catalysis (Figure 1.2), has been realized toward significant transport intensification and reaction performance improvement in microreactors.

So far, microreactors are commonly considered as an important means of PI with broad promising application potentials in the chemical industry. Herein, the small volume of microreactors does not hamper PI toward larger industrial scale, given the promise of facile production capacity increase with continuous-flow processing (e.g., via the flow rate increase) and, more importantly, with the modular, flexible, and fast scale-up concept. That is, the scale-up of microreactors can be achieved primarily via numbering-up (i.e., the replication of functional microreactor units) and, if necessary, combined with the scale-out approach (i.e., by enlarging selectively the reactor dimension while maintaining PI merits) (Zhang et al., 2017).

1.2 DESIGN ASPECT OF MICROREACTOR

The engineering and design of microreactors are intrinsically governed by fundamental microscale transport phenomena, which differ remarkably from those encountered in conventional batch or continuous-flow reactors. Owing to their miniature channel dimensions, microreactors predominantly operate in the laminar flow regime, where Reynolds numbers typically remain far below the turbulent threshold ($Re < 2000$). In such low-Re conditions, natural turbulence is practically absent, significantly restricting convective mixing. Consequently, transport of reactants and products relies mainly on molecular diffusion, making well-engineered channel geometries and mixing promoters indispensable for enhancing interfacial contact and achieving uniform species distribution throughout the reactor. Techniques such as incorporating bends, geometric constrictions, staggered herringbone grooves, and split-and-recombine pathways help strengthen secondary flows and interfacial renewal, thereby overcoming poor mixing behavior at the microscale (Zhang et al., 2021). Beyond fluid dynamics, the compact dimensions of microchannels enable superior thermal control; the high surface-to-volume ratio facilitates rapid heat exchange, preventing thermal runaway in strongly exothermic reactions and eliminating temperature gradients that are common in large-scale vessels. Such efficient thermal management ensures stable reaction environments and is especially advantageous for catalytic, photochemical, and polymerization systems that require precise thermal regulation (Rodríguez-Guerra et al., 2016).

Several design fundamentals play a central role in optimizing the operational performance of microreactor systems. High surface-area exposure accelerates heat and mass transfer, resulting in intensified reaction rates and improved selectivity. Precise control over residence-time distribution (RTD) is essential, as deviations can lead to over-processing, incomplete conversion, or formation of undesired by-products. Microreactors allow tight residence-time regulation due to their well-defined flow channels, enabling narrow RTD profiles that directly translate into uniform product quality. Additionally, maintaining low pressure drop is a critical aspect of microreactor design, especially when dealing with viscous media or multiphase mixtures. Streamlined channel structures and optimized flow paths help uphold energy-efficient operation while supporting high throughput. To broaden application scope, reactor designs often incorporate flexibility for handling gas–liquid, liquid–liquid, or gas–solid catalytic reactions, ensuring compatibility with multiphase chemistries commonly used in pharmaceutical synthesis, renewable fuel production, and fine chemical manufacturing (El Zanati et al., 2018; Hessel, 2015).

Geometry and material selection further define microreactor performance. Channels typically range from 50 to 500 μm in width and depth, with lengths extending up to several meters when folded or spiraled into compact footprints. The hydraulic diameter, expressed as $D_H = 2\text{wh}/(w + h)$, serves as a key parameter for estimating frictional resistance, mixing behavior, and mass-transfer characteristics. For chemical reactions driven by fast kinetics or involving hazardous intermediates, microreactors offer exceptional safety benefits—small reaction volumes inherently limit risk, enabling safe handling of energetic reagents, strong oxidants, and unstable intermediates. Since laminar flow restricts natural mixing, PI strategies such as micromixers, reactant pre-dispersion, or pulsating flow operation are often used to enhance reagent contact. Residence time can be precisely tuned by adjusting the channel length or flow rates, allowing millisecond- to hour-scale control depending on reaction demands. Thermal gradients are eliminated rapidly, enabling tight temperature control even for highly exothermic processes such as nitrations, hydrogenations, and photochemical oxidations.

Industrial scalability of microreactors is realized through numbering-up approaches—operating many identical modules in parallel rather than enlarging individual reactor dimensions. This strategy preserves microlevel transport advantages while meeting pilot- and commercial-scale production demands. Modular manufacturing, digital flow control, and integrated diagnostics further support automation and continuous-manufacturing paradigms that are increasingly adopted in chemical and pharmaceutical sectors. When implemented strategically, microreactor systems offer enhanced resource efficiency, greater process safety, reduced chemical waste, and robust scalability compared to traditional batch synthesis.

Microreactor technology combines advanced transport behavior, modular scalability, precise reaction control, and enhanced thermal and mixing performance. By harnessing laminar-flow physics, efficient heat transfer, and tailored microchannel geometries, microreactors provide a transformative platform for high-efficiency chemical production, greener synthesis, and next-generation continuous manufacturing.

Their principles allow consistent and high-quality product formation while supporting safe and flexible scale-up, establishing microreactors as a cornerstone of modern PI strategies and sustainable industrial chemistry.

1.3 MICROREACTOR TECHNOLOGY APPLICATIONS AT A GLANCE

1.3.1 Microreactor Applications in Material Synthesis

Microreactor systems have emerged as transformative platforms in modern chemical processing owing to their exceptionally high surface-area-to-volume ratios and microscale channel architectures (Fletcher et al., 2002; Ekinci et al., 2024). These structural attributes fundamentally enhance thermal and mass transport phenomena beyond what is achievable in conventional batch reactors. As a result, microreactors can dissipate heat almost instantaneously, enabling tight thermal control during highly exothermic reactions such as nitrations, peroxidations, and other energetically intense transformations (Balakrishnan et al., 2023). This rapid heat handling capability allows reactions to be maintained close to isothermal conditions, drastically mitigating the risk of thermal runaways and improving operational safety. Additionally, the pressurizable microscale channels make it feasible to operate at temperatures exceeding the boiling points of common solvents, thereby enhancing reaction kinetics and throughput while maintaining precise temperature regulation (Fanelli et al., 2017).

Beyond heat management, microreactors also offer unparalleled control over mass-transfer processes. Micromixing phenomena and segmented-flow regimes within microchannels significantly accelerate molecular diffusion, particularly in gas–liquid or liquid–liquid biphasic systems (Feng et al., 2022). This makes microreactors highly attractive for reactions that are otherwise diffusion-limited, including catalytic hydrogenations and aerobic oxidations. Under these confined flow conditions, nearly ideal plug-flow behavior can be attained, ultimately improving conversion efficiencies, selectivity, and reaction uniformity (Fanelli et al., 2017). Moreover, residence time can be finely tuned by adjusting flow rates, allowing chemists to intercept fleeting intermediates and harness short-lived reactive species that are difficult to manipulate in batch settings. For instance, highly reactive organolithium intermediates whose stability is measured in milliseconds at cryogenic temperatures can be safely produced, handled, and consumed in continuous-flow environments (Fanelli et al., 2017).

Safety is another central advantage driving the adoption of microreactors. Their inherently small internal volumes drastically reduce the quantity of hazardous material present at any given moment, thereby minimizing the risk in handling toxic, explosive, or thermally unstable chemicals (Fletcher et al., 2002). This feature enables the generation and immediate consumption of dangerous intermediates such as diazomethane, eliminating the need for storage and significantly improving laboratory safety protocols. Likewise, reactions historically considered dangerous in batch systems, such as energetic aromatic nitrations, can be executed with enhanced reproducibility and reduced hazard in microreactor environments (Willms et al., 2020).

Vimalanathan et al. (2022) demonstrated how multiple parameters like oxidant, activation method, flow rate/mixing, and laser power were tuned during a continuous-flow fabrication of "green GO" in a vortex fluidic device (VFD) and how it affects the properties of the resultant material. In this work, they used 30% H_2O_2, a greener oxidant compared to permanganate/acids. Initially, they applied UV LED irradiation to generate hydroxyl radicals from H_2O_2 and then pulsed near-IR (1064 nm) laser irradiation to oxidize basal plane, combined with high shear from VFD. They varied the rotational speed of the VFD (4000–9000 rpm), the inclination angle (0°–90°), and the flow rate under continuous flow. They found optimal condition: 8000 rpm, 45° inclination, and flow rate 0.5 mL/min. They also varied NIR laser pulse energy (260, 450, 650 mJ), which impacted the degree of oxidation in the basal plane. At higher power, oxidation approached levels of the conventional Hummers' method; at lower power, oxidation was lower but still stable colloids were obtained. By varying these parameters, they could control where oxidation occurs (edges vs basal plane), the oxidation degree (atomic % oxygen), colloidal stability (lifetime of dispersion), emission properties (optical), and maintenance of large intact sp^2 domains for electrical or optical applications. Another study, Prestowitz et al. (2021), explained reduction/exfoliation and graphene gel formation in a glycol–thermal continuous-flow system. Here, they showed how flow rate, temperature, solvent environment, and the addition of chemical additives (e.g., base) influence the properties of the gel/reduced GO network. The ethylene glycol solvent was heated to about 150 °C to effect thermal deoxygenation and gel formation. They used PTFE tubing with flow rate ~45 mL/h, giving ~22 min residence time in the heated tubing for the reaction. The addition of NaOH to sequester CO_2 evolved during deoxygenation prevents blockages (because CO_2 bubbles can disrupt flow or eject the gel prematurely). This chemical additive is a kind of "catalyst"/process enhancer in a broad sense. The resulting gel (or xerogel) had retained moderate oxidation and, after further annealing, could be converted to higher-conductivity materials. Density, volumetric capacitance, and mechanical stability depend heavily on deoxygenation extent, residence time, and temperature.

Nanoscale materials (1–100 nm) have emerged as transformative components across biotechnology, catalysis, electronics, and environmental remediation due to their unique physicochemical attributes. Their exceptionally high surface-area-to-volume ratio, modified thermal and electronic properties, tunable band-gap behavior, and quantum-size effects impart performance advantages unattainable in bulk materials. However, conventional batch-based nanoparticle synthesis approaches often suffer from inherent drawbacks such as poor reproducibility, limited control over nucleation and growth kinetics, and difficulties associated with scaling while maintaining homogeneous product quality. Variations in heat and mass transfer, inhomogeneous mixing, and operator-dependent variability frequently result in inconsistent particle size, morphology, and surface chemistry.

In contrast, microreactor-assisted continuous-flow synthesis presents a robust and modern alternative, offering precise manipulation of reaction parameters and superior spatiotemporal control. Microreactors enable enhanced heat dissipation, rapid mass transfer, and plug-flow behavior, which collectively facilitate uniform supersaturation and tightly regulated nucleation–growth pathways. These characteristics translate into narrow particle-size distributions, reproducible crystallinity, and higher throughput under intensified reaction conditions. Furthermore, continuous-flow platforms significantly reduce reaction time, energy consumption, and solvent usage, making them attractive for both laboratory-scale optimization and industrial deployment through modular "numbering-up" strategies.

Critical parameters such as pH, temperature, precursor concentration, pressure, residence time, and stabilizer dosage greatly influence nanoparticle formation, particularly in biological and green-synthesis systems. Eco-benign synthesis routes utilizing plant extracts, microorganisms, biopolymers, and deep eutectic solvents are gaining strong momentum owing to their biocompatibility, reduction in hazardous waste, and alignment with sustainable chemistry practices. Nonetheless, challenges such as particle aggregation, toxicity assessment, and the translation of nanomaterial properties from controlled laboratory conditions to industrial-scale reactors continue to require attention. Modern flow-assisted reactors address many of these constraints by enabling real-time monitoring, inline purification, and continuous product harvesting.

Microreactor platforms are increasingly emerging as versatile tools for the synthesis and processing of nanomaterials in biomedical engineering. Their precise control over reaction environments, enhanced heat and mass transfer, and ability to incorporate external stimuli make them ideal for producing advanced functional materials. For example, Wang et al. (2024) demonstrated the ultrasound-assisted fabrication of superparamagnetic Fe_3O_4 nanoparticles using a vortex microreactor system. Traditional nanoparticle synthesis approaches often suffer from limitations such as particle aggregation, broad size distributions, and channel clogging due to inadequate mixing. In contrast, the integration of ultrasound within the microreactor significantly intensified micromixing, suppressed channel fouling, and enabled uniform nucleation and growth of Fe_3O_4 nanocrystals. The nanoparticles obtained exhibited highly uniform size distribution and improved magnetic sensitivity, showcasing the potential of microreactors to combine flow processing with external physical fields (ultrasound, magnetic, or electric) for enhanced nanomaterial performance. Such control is particularly valuable for biomedical domains—including targeted drug delivery, MRI contrast agents, and magnetically responsive biosensing platforms—where small changes in particle morphology and surface chemistry can dramatically influence biological behavior and therapeutic outcomes.

In addition to engineering inorganic nanostructures, microreactors are proving equally transformative for biological and hybrid bio-chemo production strategies. Ding et al. (2025) reported the use of water-in-water Pickering emulsion microreactors to encapsulate engineered *Bacillus amyloliquefaciens* for biocatalytic conversion of precursors into 4-methyl-5-thiazoleethanol (4M5TE), an important thiazole derivative. The compartmentalized microenvironments offered by the emulsion microreactors minimized product loss via volatilization, enhanced substrate–cell interactions, and supported sustained microbial viability and metabolism. This led to higher 4M5TE yields compared to conventional batch systems, underscoring the promise of microreactors in facilitating greener, more efficient biomanufacturing processes for high-value biomolecules. These findings highlight the broader potential of microfluidic bioreactors for sustainable biomaterial and bio-pharmaceutical production, leveraging microbial engineering within controlled microscale environments.

Beyond synthesis, microreactor technology plays a pivotal role in biofunctionalization—the modification of biomaterial surfaces to impart targeted biological functions. Biofunctionalization strategies are essential in biomedical devices, controlled drug delivery systems, and antimicrobial coatings, as they govern biocompatibility, targeting specificity, and cellular interaction. Microreactors enable uniform surface treatment through precise reagent delivery and continuous-flow operation, often resulting in superior functionalization outcomes compared to batch methods. Surface stabilization of drug nanocrystals represents one such application. Similarly, advances in photocatalytic microreactor engineering demonstrate the applicability of these systems beyond biomedical science. Tuntithavornwat et al. (2025) fabricated a low-cost PMMA-based photocatalytic microreactor coated with immobilized TiO_2 or ZnO nanoparticles. To further tailor the surface properties, cetyltrimethylammonium bromide (CTAB) was introduced, reducing the water contact angle and improving surface wettability—factors that directly translated to enhanced photocatalytic performance. Such systems offer dual advantages for environmental and biological sectors, facilitating the degradation of pollutants, disabling microbial contaminants, and serving as models for antibacterial or antiviral surface technologies.

Carbon dots (CDs) have emerged as a versatile class of nanomaterials, finding utility across biomedical, catalytic, and optoelectronic platforms. Their low cytotoxicity and excellent biocompatibility have positioned CDs as promising substitutes for conventional semiconductor quantum dots (QDs) in biomedical imaging and sensing applications (Molaei, 2019; Lin et al., 2021). Beyond the biomedical domain, CDs have demonstrated strong performance in photocatalytic systems. A compelling illustration of their process-intensified synthesis is presented by Lu et al. (2014), who systematically investigated solvent and precursor combinations using a continuous-flow microreactor platform. Starting from solubility-based screening, fifteen precursor-solvent systems were shortlisted, followed by optimization across more than 300 reaction conditions. Using a simple capillary-based microreactor (1 mm ID tube heated in an oil bath and fed via syringe pumps), the authors achieved precise control over residence time and thermal exposure, enabling high-throughput screening where each condition required only ~15 minutes. This methodology produced CDs with tunable photoluminescence (330–550 nm) and superior quantum yields, demonstrating the power of microflow synthesis for rapid material optimization.

Although nucleation and growth kinetics of CDs are not yet fully elucidated, earlier mechanistic frameworks—such as the Watzke–Finke (W–F) model, historically applied to noble-metal nanoparticle formation (Amirjani and Haghshenas, 2018)—provide foundational insight. The prevailing hypothesis suggests that carbonaceous nuclei first form individually and subsequently aggregate into nanoclusters, ultimately evolving into fluorescent carbon nanoparticles or QDs. Continuous-flow reactors facilitate controlled nucleation–growth separation, resulting in narrower particle-size distributions relative to conventional batch reactors.

More recently, microreactor-based synthesis methods have gained prominence for producing carbon QDs, owing to their exceptional heat-transfer capability and ability to sustain high-temperature, high-pressure reaction environments—conditions often required to pyrolyze organic precursors into carbon nanostructures. Microreactor systems can safely operate above 10 bar, with PTFE units tolerating up to ~300 °C and 50 bar, and glass systems exceeding 15 bar depending on configuration. Such operating windows accelerate reaction rates, enhance monodispersity, and provide fine control over optical properties, particularly emission color, by manipulating flow rate and residence time. The ability to tune emission characteristics through PI rather than post-synthetic modification represents a unique advantage of microflow systems over batch synthesis routes.

The integration of microreactors into nanomaterial and biomaterial processing offers unprecedented precision and flexibility, enabling scalable synthesis, controlled biofunctionalization, and innovative biocatalytic pathways. The studies discussed collectively reflect a growing trend toward microreactor-assisted material innovation, where continuous-flow design, external stimuli, and micro-compartmentalized biological systems converge to push the boundaries of next-generation healthcare and environmental technologies.

1.3.2 Microreactor Applications for Pollutant Remediation

Microreactor systems have emerged as a transformative platform for pollutant degradation due to their unparalleled efficiency, accuracy, and controllability compared to conventional bulk reactors. These compact reaction devices, typically fabricated with internal channels on the order of tens to hundreds of micrometers, create distinctive physicochemical environments that dramatically intensify reaction performance. A defining characteristic of microreactors is their exceptionally high surface-area-to-volume ratio—frequently exceeding 10,000 m^2/m^3—which promotes superior heat and mass transfer and drastically shortens diffusion distances to micrometer scales. This intensified transfer behavior is especially advantageous for contaminant-removal reactions where mass-transfer limitations are dominant. Additionally, flow inside microchannels generally remains laminar ($Re < 100$), ensuring highly predictable residence-time profiles and precise manipulation of fluid behavior. Operating with such small reaction volumes also reduces reagent use, minimizes risks when treating hazardous waste streams, and enables safe handling of toxic pollutants. These fundamental benefits allow microreactors to provide tight spatiotemporal regulation over reaction conditions, supporting reaction pathways and degradation mechanisms that are difficult to achieve in larger-scale units.

Among microreactor-enabled treatment routes, advanced oxidation processes (AOPs) represent a central strategy owing to their ability to produce highly reactive hydroxyl radicals (•OH) and other oxidative intermediates capable of mineralizing resistant organic contaminants. Microscale confinement significantly improves AOP performance by boosting interphase mass transfer, promoting rapid mixing, and maintaining precise operational parameters. Fenton-based systems are widely implemented, wherein ferrous ions (Fe^{2+}) react with hydrogen peroxide to yield hydroxyl radicals ($Fe^{2+} + H_2O_2 \rightarrow Fe^{3+} + \bullet OH + OH^-$). Microreactor designs ensure accurate dosage control, preventing excess iron precipitation and maximizing radical formation efficiency. When coupled with optical excitation, photo-Fenton configurations exhibit augmented performance through regenerative cycling of Fe^{3+} back to Fe^{2+} and photolytic radical generation. Microchannels also overcome irradiation constraints common in conventional photoreactors by ensuring uniform photon distribution and minimal light attenuation.

Ozonation processes benefit similarly when integrated into microreactor formats. Gas–liquid contacting efficiency is greatly improved by the narrow channel dimensions, which enable exceptionally high interfacial areas and prevent bypassing or poor gas distribution. Tailored microstructured surfaces and microporous interfaces further intensify ozone dissolution and transfer rates, overcoming bottlenecks associated with large-volume ozonation systems. These enhancements allow rapid oxidation of persistent organic pollutants, dyes, pharmaceuticals, and microbial contaminants.

In addition to physicochemical pathways, biological catalytic systems have also been successfully miniaturized. Enzymatic degradation microreactors leverage biocatalysts—including oxidoreductases and hydrolases—to break down pollutants under mild temperatures and near-neutral pH. This approach provides remarkable substrate specificity and eliminates the need for high chemical doses or energy-intensive conditions. The microscale format boosts enzyme efficiency by reducing transport resistance, stabilizing enzyme configurations, and ensuring a constant supply of substrate to active sites. Laccases, peroxidases, and cutinases are among the most studied biocatalysts for microreactor-assisted pollutant degradation. Peroxidases utilize hydrogen peroxide to oxidize phenolic pollutants and dyes, whereas laccases utilize molecular oxygen to achieve similar oxidative removal. Cutinases, in contrast, preferentially hydrolyze ester bonds and have demonstrated strong potential for degrading polymeric micro- and nanoplastics, particularly polyethylene terephthalate (PET).

Enzyme immobilization is a key design factor in enzymatic microreactors. Common immobilization approaches include physical adsorption, covalent bonding, entrapment, and cross-linking. Recently introduced asymmetric Janus microspheres, featuring dual functional regions and tailored porosity, have allowed improved catalyst loading and minimized mass-transfer resistance. Studies have shown that cutinase supported on functional Janus carriers retained more than 78% of its catalytic activity after eight reuse cycles, while soybean peroxidase immobilized in microchannels displayed superior reusability and decreased leaching relative to dissolved enzyme systems.

Microenvironment modulation is critical for enzymatic microreactor performance. Parameters such as pH, ionic strength, and cofactor availability can be continuously regulated, eliminating fluctuations that compromise enzyme stability in traditional batch reactors. Moreover, microreactors facilitate rapid product removal and prevent inhibitory product accumulation, thus preserving enzyme turnover efficiency. Computational Fluid Dynamics (CFD) simulations confirm that despite laminar flow, optimized microchannel geometries can achieve near-ideal radial mixing, maintaining uniform substrate availability across immobilized catalytic surfaces.

The results demonstrated by these systems are highly promising. For instance, immobilized soybean peroxidase in a microreactor configuration achieved greater than 95% degradation of 2,4,6-trichlorophenol at residence times of only a few minutes. Similarly, cutinase-based microfluidic systems enabled efficient PET nanoplastic breakdown with stable activity across multiple operational cycles. Due to their rapidly tunable chemistry, reduced footprints, and ability to treat trace-level persistent contaminants—including endocrine disruptors, pharmaceuticals, and microplastics—enzymatic microreactors represent a compelling advancement for next-generation wastewater purification.

Micro-photobioreactors (μPBRs) represent an innovative platform for algae-based bioremediation, offering a compact and efficient system for treating wastewater and generating value-added biomass. Unlike conventional photobioreactors, μPBRs operate at microscale volumes, making them particularly advantageous for laboratory research and controlled experimental studies. Microalgae cultivated in μPBRs can assimilate a wide spectrum of contaminants, including organic pollutants, excess nutrients such as nitrogen and phosphorus, and trace heavy metals, thereby contributing significantly to wastewater purification. The microscale architecture of μPBRs provides a high surface-area-to-volume ratio, enhancing light utilization and mass-transfer efficiency, which are critical for rapid algal growth and pollutant removal.

In addition, μPBR systems can be integrated with advanced monitoring tools—such as pH, temperature, dissolved oxygen, and nutrient sensors—allowing precise regulation of environmental conditions and real-time operational control. This level of automation enables fine-tuning of algal metabolic performance to maximize removal efficiency and biomass productivity, often surpassing the treatment efficacy of conventional open or less-regulated systems. Beyond wastewater remediation, the biomass generated from μPBR cultivation can be leveraged for downstream applications, including biofuel synthesis, animal feed formulation, and bioproduct development, supporting circular-economy and resource-recovery principles. Thus, μPBR technology not only improves pollutant mitigation strategies but also facilitates sustainable valorization of algal biomass for industrial and environmental benefit.

AOPs are increasingly recognized as powerful water treatment strategies due to their capacity to produce highly reactive oxygen species (ROS), including hydroxyl (•OH), ozone (O_3), superoxide ($•O_2^-$), and sulfate ($SO_4•^-$) radicals. These oxidizing agents possess exceptional redox potential and broad-spectrum reactivity, enabling the breakdown and complete mineralization of persistent organic contaminants. Despite their proven effectiveness, conventional AOP configurations often face limitations related to energy demand, reaction efficiency, and scale-up practicality. In recent years, microreactor technology has emerged as a promising solution to overcome these constraints. The unique microscale architecture offers superior heat and mass transfer, controlled laminar flow, and significantly enhanced surface-area-to-volume ratios, thereby improving radical generation, accelerating reaction kinetics, and minimizing transport resistance. When integrated with AOPs, microreactors facilitate intensified degradation pathways with reduced chemical consumption, improved catalyst utilization, and lower environmental footprint. This chapter presents a detailed overview of microreactor-assisted AOPs, with emphasis on photocatalytic systems for the elimination of heavy metals and recalcitrant organic pollutants from water. Mechanistic aspects of photocatalytic oxidation and reduction are elaborated, and recent innovations in microstructured water treatment reactors are highlighted, along with future directions for scalable, energy-efficient purification platforms aligned with sustainable environmental engineering objectives.

1.3.3 Microreactor Applications for Energy Application and Others

Microreactors have emerged as an innovative direction in advanced energy and chemical-processing systems, with particular promise in the next generation of nuclear energy solutions. Their compact size and modular architecture enable deployment in settings where conventional, large-scale power systems are impractical or economically unsuitable (Black et al., 2023). Typically operating within a power range of 1–20 MWe, and potentially up to 50 MWe at maximum design capacity, these systems can integrate both advanced non-light-water and light-water reactor technologies to deliver electricity and/or high-grade thermal energy (Ingersoll, 2009; Dittmeyer et al., 2017). If challenges related to manufacturing, operational costs, policy approval, and institutional readiness are successfully addressed, microreactors could play a crucial role in energy transition strategies, particularly in remote regions, defense applications, off-grid industrial zones, and areas facing fuel transportation barriers or grid limitations (Ekinci et al., 2024).

Microreactors hold unique strategic advantages in areas lacking reliable electrical infrastructure, facing harsh geographical or climatic constraints, or requiring highly mobile and flexible power solutions where conventional renewable systems face deployment limitations (Yadav et al., 2023; Yan et al., 2020). Technological progress in reactor miniaturization, heat-transfer efficiency, and system modularity presents both engineering opportunities and integration challenges in modern power systems. Their contribution is especially relevant in the context of global objectives for carbon-neutral energy production, climate resilience, and national preparedness for extreme events, where power reliability is linked to public safety, industrial continuity, and economic stability.

Deployment of microreactors can improve national resilience by ensuring uninterrupted power for essential infrastructures such as healthcare units, critical government facilities, industrial clusters, and emergency response networks (Mehla et al. 2022). Although economic competitiveness with other low-carbon technologies remains under evaluation, modeling studies indicate that microreactors may offer cost benefits where large nuclear plants cannot be justified due to scale or logistical constraints (Yadav et al., 2023). Beyond nuclear applications, microreactors are gaining traction in renewable chemical manufacturing pathways, including biofuel, bio-oil, and hydrogen synthesis. In biodiesel production, for instance, microreactor-based transesterification accelerates reaction kinetics, promotes rapid mixing, reduces residence time, and enables continuous, high-throughput conversion systems (Madhawan et al., 2018). Early biodiesel microreactors relied on capillary-type architectures; since then, numerous advanced designs have been fabricated using diverse fabrication strategies and materials, enabling higher efficiency, automation potential, and integration into decentralized biorefineries.

Pandey et al. (2021) demonstrated that monodisperse CdTe QDs can be efficiently synthesized using a continuous-flow active microreactor, where key operating variables were translated into mathematical process models for precise control.

The resulting nanocrystals exhibited an impressive photoluminescence quantum yield (PLQY) of 61.5%, along with strong photostability, retaining 44% of their emission intensity after 96 minutes of UV irradiation. These high-quality QDs are highly suitable for clean solar-energy applications, offering several attractive features:

- Tunable band gap (emission range: 500–630 nm), allowing spectral matching with solar irradiation
- Narrow particle-size distribution (FWHM ≈ 40 nm), minimizing energetic disorder within QD thin films
- PDMS (polydimethylsiloxane) encapsulation, improving resistance to environmental degradation

When integrated into QD-sensitized solar cells (QDSSCs), the microreactor-generated CdTe QDs achieved remarkable performance gains—~35% enhancement in power conversion efficiency compared to their batch-synthesized counterparts. Moreover, they exhibited excellent stability, maintaining>90% of their initial efficiency after 1000 hours of operation, and delivered a broad spectral harvesting range spanning the visible to near-infrared region.

1.4 CONCLUSION

Microreactor technology has emerged as a cornerstone of modern PI, offering a compact and efficient alternative to conventional reactors. The unique microscale geometry enables tight control over reaction parameters, rapid heat removal, and enhanced mass transfer, resulting in superior reaction performance and safety. These advantages make microreactors particularly effective for fast, exothermic, or hazardous reactions, as well as for multiphase and catalytic systems. Their compatibility with inline sensors and automation further enhances reaction monitoring and optimization. Importantly, microreactors support scalable production through numbering-up strategies, allowing industrial-level output without compromising process precision. As research advances, integration with smart control systems, nanomaterial catalysts, and renewable-energy-driven processes will expand their applicability across pharmaceuticals, fine chemicals, energy, and environmental technologies. In summary, microreactor technology provides a powerful platform for sustainable, precise, and flexible chemical manufacturing, positioning it as a key enabler of future continuous-flow and modular chemical processing systems.

REFERENCES

Amirjani, A., & Haghshenas, D. F. (2018). Modified Finke–Watzky mechanisms for the two-step nucleation and growth of silver nanoparticles. *Nanotechnology*, *29*(50), 505602. https://doi.org/10.1088/1361-6528/aae3dd

Balakrishnan, D., El Maiss, J., Olthuis, W., & García, C. P. (2023). Miniaturized control of acidity in multiplexed microreactors. *ACS Omega*, *8*, 7587–7594. https://doi.org/10.1021/acsomega.2c06897

Baumann, M., Moody, T. S., Smyth, M., & Wharry, S. (2021). Evaluating the green credentials of flow chemistry towards industrial applications. *Synthesis*, *53*, 3963–3976.

Black, G., Shropshire, D., Araújo, K., & van Heek, A. (2023). Prospects for nuclear microreactors: A review of the technology, economics, and regulatory considerations. *Nuclear Technology*, *209*(S1), S1–S20.

Ding, X., Huang, W., Cao, H., Ao, S., Zou, D., Wei, X., & Liu, S. (2025). Microreactors derived from Pickering emulsions for microbial synthesis. *Bioresource Technology*, *438*, 133231.

Dittmeyer, R., et al. (2017). Micro and micro membrane reactors for advanced applications in chemical energy conversion. *Current Opinion in Chemical Engineering*, *17*, 108–125.

Ekinci, F., Guzel, M. S., Acici, K., & Asuroglu, T. (2024). The future of microreactors: Technological advantages, economic challenges, and blockchain licensing. *Applied Sciences*, *14*(15), 6673. https://doi.org/10.3390/app14156673

El Zanati, E., et al. (2018). Design and scale-up of continuous flow micro-reactor. *Journal of Engineering Science and Technology*, *13*(8), 2481–2494.

Fanelli, F., Parisi, G., Degennaro, L., & Luisi, R. (2017). Contribution of microreactor technology and flow chemistry to sustainable synthesis. *Beilstein Journal of Organic Chemistry*, *13*, 520–542.

Feng, H., Zhang, Y., Liu, J., & Liu, D. (2022). Nanocatalyst deposition in continuous-flow microreactors. *Molecules*, *27*(22), 8052. https://doi.org/10.3390/molecules27228052

Fletcher, P. D. I., Haswell, S. J., Pombo-Villar, E., Warrington, B. H., Watts, P., Wong, S. Y. F., & Zhang, X. (2002). Microreactors: Principles and applications in organic synthesis. *Tetrahedron*, *58*, 4735–4757.

Hessel, V. (Ed.). (2015). *Design and engineering of microreactor and smart-scaled flow processes*. MDPI.

Ingersoll, D. T. (2009). Deliberately small reactors and the second nuclear era. *Progress in Nuclear Energy*, *51*, 589–603.

Jähnisch, K., Hessel, V., Löwe, H., & Baerns, M. (2004). Chemistry in microstructured reactors. *Angewandte Chemie International Edition*, *43*, 406–446.

Kashid, M. N., Renken, A., & Kiwi-Minsker, L. (2011). Gas–liquid and liquid–liquid mass transfer in microreactors. *Chemical Engineering Science*, *66*, 3876–3897.

Lin, L., et al. (2021). Microfluidic fabrication of fluorescent nanomaterials: A review. *Chemical Engineering Journal*, *425*, 131511.

Lu, Y., Zhang, L., & Lin, H. (2014). Microreactor for rapid reaction condition screening for carbon dots. *Chemistry – A European Journal*, *20*, 4246–4250.

Madhawan, A., Arora, A., Das, J., Kuila, A., & Sharma, V. (2018). Microreactor technology for biodiesel production: A review. *Biomass Conversion and Biorefinery*, *8*(2), 485–496.

Marre, S., Adamo, A., Basak, S., Aymonier, C., & Jensen, K. F. (2010). High-pressure and high-temperature microreactors. *Industrial & Engineering Chemistry Research*, *49*, 11310–11320.

Mehla, S., et al. (2022). Additive manufacturing for chemical engineering. In S. Thomas, N. Kalarikkal, R. Abraham, & S. V. Nair (Eds.), *Additive manufacturing for chemical sciences and engineering* (pp. 239–307). Springer.

Molaei, M. J. (2019). Carbon quantum dots: Biomedical applications. *RSC Advances*, *9*, 6460–6481.

Pandey, S., Mukherjee, D., Kshirsagar, P., Patra, C., & Bodas, D. (2021). Multiplexed bioimaging using CdTe QDs synthesized in active microreactor. *Materials Today Bio*, *11*, 100123. https://doi.org/10.1016/j.mtbio.2021.100123

Plutschack, M. B., Pieber, B., Gilmore, K., & Seeberger, P. H. (2017). Flow chemistry review. *Chemical Reviews*, *117*, 11796–11893.

Prestowitz, L. C. O., McElroy, C. R., Gun'ko, Y. K., & Coleman, J. N. (2021). Glycol-thermal continuous flow synthesis of graphene gel. *ACS Omega*, *6*(28), 17903–17910.

Rebrov, E., de Croon, M. H. J. M., & Schouten, J. C. (2001). Microstructured reactor with integrated heat exchanger. *Catalysis Today*, *69*, 183–192.

Rodríguez-Guerra, J., et al. (2016). Micro-/milli-channel heat exchanger reactors. *Industrial & Engineering Chemistry Research*, *55*(22), 6435–6442.

Stankiewicz, A. I., & Moulijn, J. A. (2004). *Process intensification*. Marcel Dekker.

Tuntithavornwat, S., et al. (2025). PMMA photocatalytic microreactor. *Chemical Engineering Journal Advances*, *23*, 100815.

Vimalanathan, K., Scott, J., Pan, X., Luo, X., Rahpeima, S., Sun, Q., Zou, J., Bansal, N., Prabawati, E., Zhang, W., Darwish, N., Andersson, M. R., Li, Q., & Raston, C. L. (2022). Continuous flow fabrication of green graphene oxide in aqueous hydrogen peroxide. *Nanoscale Advances*, *4*, 3121–3130.

Wang, S., et al. (2024). Vortex microreactors for Fe_3O_4 nanoparticle synthesis. *Chemical Engineering Journal*, *501*, 157672.

Willms, T., Kryk, H., & Hampel, U. (2020). Microreactor oxidation processes. *Catalysis Today*, *346*, 3–9.

Yadav, D., Lu, X., Vishwakarma, C. B., & Jing, D. (2023). Microreactors for hydrogen production via steam reforming. *Journal of Power Sources*, *585*, 233621.

Yan, B. H., Wang, C., & Li, L. G. (2020). Micro heat-pipe cooled reactor technology. *Annals of Nuclear Energy*, *135*, 106948.

Yue, J. (2018). Multiphase catalytic processing in microreactors. *Catalysis Today*, *308*, 3–19.

Yue, J. (2022). Green process intensification using microreactor technology for the synthesis of biobased chemicals and fuels. *Chemical Engineering and Processing Process Intensification*, *177*, 109002.

Zhang, J., Chen, X., & Li, Z. (2021). Microreactors for nanomaterial synthesis: A review. *Nanomaterials*, *11*(1), 121.

Zhang, J., Wang, K., Teixeira, A. R., Jensen, K. F., & Luo, G. (2017). Design and scale-up of microchemical systems. *Annual Review of Chemical and Biomolecular Engineering*, *8*, 285–305.

2 Design and Fabrication of Microreactors

Sushanta Kumar Behera, Sriya Naik, Naresh Kumar, and Subrata Biswas

2.1 INTRODUCTION

Microreactors, also called microstructured reactors or microchannel reactors, have reshaped chemical engineering through microscale platforms that enable chemical reactions in channels with dimensions typically below 1 millimeter. Their portable size and unique architecture provide clear advantages over traditional reactors, such as enhanced energy efficiency, faster reaction rates, greater safety, and improved process control. In recent years, microreactor technology has progressed markedly in both academic and industrial research, based on its potential for cost-effective, safer, and sustainable chemical production methods.

The design and fabrication of a microreactor is characterized by the fusion of microfluidic principles, materials science, and engineering innovation, executing a vast range of chemical processes from nanoparticle synthesis to pharmaceutical production. Microreactors are produced through several fabrication techniques, such as micromachining, wet etching, and advanced molding, and utilize diverse materials, such as glass, silicon, polymers, and metals, which are selected for their compatibility with specific reaction systems and operational demands. It is crucial to select appropriate fabrication strategies and material combinations to achieve optimal reactor performance, chemical resistance, and operational safety.

Furthermore, innovative solutions are demanded for reactor packaging, fluid handling, and individual reactor control in the move from lab-scale microreactors to industrially viable systems, particularly when multiple reactors are integrated. These engineering challenges represent the next frontier in microreactor technology development, with ongoing research increasingly focusing on overcoming these barriers to extend the impact of microreactors in chemical, pharmaceutical, and biotechnological industries.

This chapter focuses on a minute examination of recent developments in microreactor design and fabrication. Material selection, innovative fabrication techniques, and integrated manufacturing strategies are highlighted, with a specific emphasis placed on evolving methods that enable precise microscale structuring, high-resolution feature control, and scalable production.

The design and fabrication of microreactors have achieved remarkable heights over the past 30 years, transforming steadily from laboratory-scale concepts to scalable, high-performance platforms for chemical production. The first microreactors were

DOI: 10.1201/9781003633761-2

heavily influenced by the semiconductor industry, such as injection molding, wet etching, lithography, and micromachining. These methods facilitated the construction of microchannel networks in a wide range of materials such as silicon, ceramics, glass, polymers, and metals. The field prefers glass and silicon for their extraordinary chemical robustness and thermal conductivity, metals and ceramics provide durability under corrosive or high-temperature conditions, while polymers such as polydimethylsiloxane (PDMS) are valued for low cost and flexibility (Suryawanshi et al., 2018).

The integration of modern technologies like machine learning, automation, and additive manufacturing is shaping recent advances in microreactor fabrication. The coupling of machine learning with microfluidic systems facilitates the creation of intelligent cyber-physical systems that autonomously accelerate chemical synthesis and process optimization. Likewise, additive manufacturing (3D printing) has altered the prototyping paradigm for microreactor design. It enables the rapid and economical fabrication of intricate three-dimensional microreactor architectures in both metals and plastics. These approaches reduce development time and support modular and disposable designs tailored for specialized chemical processes. Conventional techniques such as wire electrical discharge machining (EDM), high-precision milling, and imprint molding are widely employed for fabricating microfluidic structures with high accuracy. Multi-material combinations like silicon-Pyrex reactors offer excellent control over flow and temperature. Nanoscale synthesis, like quantum dot production, is leading to extensive research on microreactor technology. Modern microreactor design prefers embedding sensors and an advanced process control system within reactor platforms. The remote monitoring and control of the microreactor through cloud-based platforms have been permitted by advances in Internet of Things (IoT) technology, greatly enhancing the flexibility and responsiveness of chemical production processes. The implementation of embedded sensor networks to monitor key process variables, including pressure, flow, and temperature, yields a continuous telemetry stream of real-time feedback data, which is significant for upholding optimal reaction setpoints and rendering optimum product quality. Together, these advances give rise to a new era of autonomous microreactors that are capable of self-optimization and adapt rapidly to fluctuating process parameters. Sustainability and material selection are impacting the future of microreactor technology, where economic viability and environmental impact are fundamental criteria in the fabrication process. The advancement of microreactors using recyclable polymers, additively manufactured metal components, and innovative catalyst coating techniques is offered as a pathway to reduce material cost, advance circular lifecycle principles, and enhance operational durability.

The framework of a microreactor is the keystone of its serviceability. The structural design, particularly the convoluted patterning of its microchannels, directly imposes key considerations of the reaction kinetics by supervising the residence time, flow regime, and crucial interfacial area between reactants. Therefore, the essential supremacy of laminar, well-defined flow makes up a highly consistent and conventional reaction environment. The specific geometry and configuration of these channels must be painstakingly customized to the nature of the intended chemical transformations. For example, parallel channel arrays are frequently employed for quantifying feed by a strategy known as numbering-up, while serpentine or split-and-recombine geometries are selected to optimize mixing and residence time distribution (RTD). A significant advantage of this design is that the vicinity of the reactor wall substantially reduces the characteristic heat and mass transfer length scales. This special feature ensures quick and homogeneous temperature distribution, a vital capability for managing highly exothermic reactions and boosting overall process safety, selectivity, and efficiency. In microreactor design, material selection directly influences chemical compatibility, mechanical integrity, thermal management, and cost-effectiveness. Inorganic materials, namely silicon, glass, and metals (namely stainless steel and titanium alloys), are continually employed due to their high thermal conductivity, structural robustness, and broad chemical inertness, which make them suitable for strict reaction environments and raised operating pressure. On the other hand, polymeric substrates are produced through some techniques such as hot embossing, injection molding, or additive manufacturing, which provide advantages of low production cost, quick prototyping, and manageable customization, making them particularly attractive for biological appraisals and liquid-phase reactions where hazardous chemicals are not involved. Up-and-coming trends are driven by the use of sustainable and recyclable material platforms in conjunction with engineered surface coatings and advanced functional films to reduce fouling, improve chemical resistance, and enhance environmental performance, thus reinforcing the green testimonials of microreactor innovation and facing key issues such as waste minimization and circular manufacturing.

The relationship between design and synthesis is heavily mediated by the interaction between channel geometry, wall structure, and reactor material. For instance, microreactors with a high aspect ratio and good thermal conductivity are suitable for faster heat removal and promoting the use of microreactors in high-energy or hazardous reactions. Droplet formation in a multiphase system is regulated, fluid flow profiles are defined, and fouling or clogging is minimized by microreactor wall properties, such as smoothness, wettability, and surface treatments. Furthermore, advanced microreactors can couple with built-in sensors that continuously monitor real-time parameters such as temperature, flow rate, and infrared or ultraviolet absorption during the reaction. The combination of a synthesis platform and an analytical instrument is a unique synergy for microreactors (Engelbrecht et al., 2020).

Fabrication methodologies for microreactors extend a progression from ordinary hierarchical lithographic processes to pioneering advanced manufacturing techniques. Some fabrication methodologies or microfabrication processes, namely photolithography and micromachining, are underlying techniques, commencing from the microelectronics industry, in which light is used to transfer a precise geometric pattern of silicon and glass (light-sensitive photoresist materials) from a photomask at submicron levels. These methods encourage both planar and three-dimensional architectures and persist imperative in research contexts, specifically when assimilation with microfluidic or advanced sensing is prescribed. Meanwhile, technologies like electroforming and laser ablation have been designed to create metallic microreactors with complex, multi-level geometries suitable for high-temperature or corrosive chemicals (Reid et al., 2023).

The widespread accessibility of quick prototyping capabilities has been catalyzed by the improvement in additive manufacturing (3D printing), which enables heretofore unsurpassed channel designs, functional gradients, and embedded features to be interpreted directly from digital files. Modern techniques like direct-ink-writing (DIM), two-photon polymerization (2PP), and selective laser sintering (SLS) can now yield microreactors from polymers, ceramics, and even metals, noticeably diminishing chronology and cultivating design innovation. Regardless, several fabrication routes present a balance of quid pro quo; traditional photolithographic ways excel in dimensional accuracy and surface finish, but have less volumetric feed, while 3D printing offers multifacetedness and scalability, but often struggles with material selection and post-processing requirements (Yadav et al., 2023).

Multiple empirical expositions are an example of these designs and fabrication innovations in operative chemical concern. In pharmaceutical synthesis, microreactorial architectures have facilitated the continuous, autonomized production of bioactive moieties with decisively mobilized reaction times, elongated productivities, and inherent safety interlocks. Commercialized platforms are now capable of pliant "chemical synthesis on demand," rearrangeable for different drugs, and apposite to point-of-care or diffusely distributed manufacturing environments. Within the realms of fine chemicals and green chemistry, the ability to function with attenuated hazard solvents, recycle process streams, and minimize waste has led to manifest advances in sustainability, cost efficiency, and regulatory conformity. For energy and materials synthesis, encompassing the use of heat pipe structures in next-generation nuclear microreactorial systems, large-scale monolithic fabrication and modern capillary wick configurations facilitate deployment in demanding, high-temperature environments with attenuated material expenditure and elevated operational dependability (Hessel et al., 2024).

Hydrodynamic fluxes and heat/mass transfer constitute the foundational determinants of any microchannel, undergirding their transformative impact. Due to high surface-area-to-volume proportions, heat dissipation transpires with assiduity and uniformity, effectively quelling localized thermal maxima and thermal runaway even during highly energetic synthesis. Mass transfer, dictated by short diffusion distances and laminar flow regimes, is similarly magnified, permitting reaction kinetics and discriminative selectivity unsurpassed at macroscopic scales. Such demonstrated regulation empowers microreactors to destine flash chemistry, highly exothermic alteration, multiphasic catalytic operations, and even hazardous or explosive protocols with augmented safety margins and elevated product quality. Computational design tools encompassing computational fluid dynamics analyses have become *de rigueur* for prognosticating reactor performance and refining geometries prior to manufacturing (Gateway for Accelerated Innovation in Nuclear, 2025).

2.2 DESIGN OF MICROREACTORS

2.2.1 Design Principles

Microreactor design is governed by microscale transport phenomena. At small channel dimensions, laminar flow dominates (Reynolds number typically <2000), resulting in limited turbulent mixing. Therefore, mixing must be achieved either through diffusion or by implementing geometrical strategies that enhance interfacial area (Zhang et al., 2021). Additionally, microreactors must ensure rapid heat dissipation to prevent hot spots in exothermic reactions, while also offering sufficient residence time for complete reactions (Rodríguez-Guerra et al., 2016).

Microreactor design relies on several fundamental principles that enhance both process efficiency and product quality in chemical synthesis. A high surface-to-volume ratio is crucial, as it supports rapid mass and heat transfer, ensuring that temperature and concentration gradients are minimized within the reaction channels. Controlling the RTD with precision is essential for maintaining product selectivity, as it directly influences conversion rates and reaction outcomes. Additionally, effective design must achieve low pressure drop, enabling high throughput without compromising operational stability. A flexible approach to channel design helps facilitate multiphase flows, which are common in many chemical processes and vital for broader reactor applicability (El Zanati et al., 2018; Hessel, 2015).

Key aspects in microreactor engineering include selecting appropriate channel geometry—such as straight, serpentine, spiral, or split-and-recombine configurations—based on the reaction's requirements. Channels are typically fabricated with widths and depths between 50 and 500 micrometers, and lengths that can extend up to several meters if arranged in complex, folded patterns. Hydraulic diameter, calculated as $D_h = \frac{2wh}{w+h}$ (where w is the channel width and h the height), helps in quantifying flow dynamics. Mixing efficiency presents another challenge, as laminar flow dominates at the microscale; hence, microreactors often incorporate intricate features like bends or herringbone structures to optimize mixing. The ability to precisely control residence time, primarily by adjusting channel length and flow rate, is fundamental for ensuring targeted conversion and selectivity. Enhanced thermal management is another strongpoint, with the reactor's small channels enabling rapid heat dissipation—vital for handling exothermic reactions. Furthermore, scalability is commonly achieved through numbering-up, where multiple microreactors operate in parallel, allowing increased production without sacrificing reaction control or efficiency (El zanati et al., 2018; Hessel, 2015).

Overall, thoughtful application of these principles results in microreactor systems that provide uniform product quality, facilitate efficient energy and resource use, and support the robust scaling needed in modern chemical engineering.

2.2.2 How Microreactors Work

Microreactors operate through a network of tiny channels, typically ranging from 10 to 1000 micrometers, which are etched or molded into diverse solid materials such as metal, glass, polymers, or ceramics. Reactants are introduced into these channels via micro- or syringe pumps. The small size of the channels results in predominantly laminar flow, characterized by smooth, layered movement of fluids rather than turbulent mixing. Due to the absence of turbulence, mixing relies heavily on diffusion, which is naturally efficient at small scales, and on carefully designed channel geometries like serpentine, split-and-recombine, or herringbone structures to enhance mixing efficiency (Brás, 2021).

Within the microchannels, reactants meet and chemically interact in well-defined reaction zones. The high surface-to-volume ratio present in these microstructures allows for rapid heat and mass transfer, which helps prevent hot spots, ensures uniform temperature distribution, and improves product selectivity. Rather than generating energy internally, microreactors typically depend on external sources tailored to the reaction requirements, including heating and cooling systems, light sources for photochemical reactions, electrodes for electrochemical processes, or more advanced energy fields like microwaves and ultrasound (Yang et al., 2023).

Finally, products exit the microreactor continuously through the outlet of these channels. The output stream can be used directly or further processed with separation and purification methods, either integrated on-chip or handled off-chip. The precision in flow control, mixing, and efficient energy management makes microreactors highly effective for controlled and selective chemical synthesis (Yang et al., 2024).

2.2.3 Applications of Microreactors

Microreactors are revolutionizing a diverse set of application domains due to their ability to offer precise control, enhanced safety, and scalable integration of chemical and biological processes. In pharmaceutical and fine chemical manufacturing, these microscale devices are utilized for the rapid synthesis of active pharmaceutical ingredients (APIs), often enabling multistep reactions to occur seamlessly on a single chip. Their controlled environment not only accelerates reaction times but also significantly minimizes risks when dealing with sensitive or hazardous intermediates, such as diazonium salts. This risk reduction means safer handling and improved product purity without compromising on efficiency (Schmidt et al., 2019; Bastan and Kazemeini, 2022).

Energy and fuel industries also benefit from microreactor technologies. For example, microreactors are used to generate hydrogen via steam reforming or partial oxidation, providing a compact and efficient platform crucial for decentralized fuel production. Biodiesel synthesis through continuous transesterification processes is another area where microreactors excel, delivering consistent output and clear advantages in scalability. Furthermore, they are instrumental in preparing fuel for portable fuel cells, due to their precise control over the reaction environment and product quality (Schmidt et al., 2019).

Environmental protection is yet another key field where microreactors deliver significant impact. Their advanced designs enable effective degradation of volatile organic compounds (VOCs), helping to address air and water pollution at the source. Microreactors facilitate water purification technologies by supporting rapid oxidation and photocatalytic processes, which can be implemented either individually or as part of an integrated water treatment chip (Schmidt et al., 2019).

Polymer manufacturing benefits from microreactor continuous flow production, which allows for precise adjustment of reaction parameters and efficient heat management. This leads to polymers with tightly controlled molecular weight distribution, an essential property for specialty materials used in many industrial and consumer products.

The biotechnology and diagnostics sector leverages microreactors for a variety of key processes. Enzyme-catalyzed reactions proceed efficiently within the microchannels, and advanced setups incorporate DNA amplification technologies, such as polymerase chain reaction (PCR), directly on a chip. Additionally, microreactor platforms support the integration of biosensors, which enable real-time monitoring and analysis of biological markers, providing fast and accurate diagnostics in both research and clinical settings. Microreactors play a transformative role across several fields by combining efficiency, safety, and adaptability in miniature platforms designed for continuous processing. In the pharmaceutical and fine chemicals industry, these devices enable rapid synthesis of active pharmaceutical ingredients and facilitate multistep reactions on a single chip, dramatically improving throughput and safety. For compounds that are volatile or hazardous, such as diazonium salts, microreactors allow for much safer handling and minimized risk, thanks to their controlled microenvironments. In the area of energy and fuels, microreactors are used for the production of hydrogen through steam reforming or partial oxidation methods, offering compact solutions suitable for decentralized energy generation. Biodiesel synthesis using transesterification benefits from the controlled flow and reaction conditions provided by microreactors, while fuel processing for portable fuel cells is optimized through their precise operation (Pandey and Khanduri, 2023).

Environmental engineering applications can also be benefitted, as microreactors are used to break down VOCs and purify water via oxidation and photocatalytic reactions, contributing to pollution control and water treatment advancements. Within polymer chemistry, continuous flow synthesis ensures polymers have consistent and accurately tailored molecular weight distributions, which is essential for advanced material properties. In biotechnology and diagnostics, microreactors support enzymatic reactions and DNA amplification processes such as PCR on chips, providing fast, miniaturized solutions for research and clinical laboratory workflows. Integration with biosensors allows real-time monitoring and detection of specific biological or chemical markers, further extending their utility in analytical and diagnostic applications (Li et al., 2023).

2.2.4 Dimensions and Production Capacity

Table 2.1 presents some microreactor characteristics such as dimensions, channel lengths, applications, and production scales.

2.2.5 Key Design Parameters

Microreactor design incorporates several critical parameters that directly impact flow dynamics, reaction efficiency, and safety. Channel dimensions—specifically the width, depth, and aspect ratio—play a fundamental role in determining the resistance to flow, the nature of RTD, and the length available for diffusion of reactants. Typically, smaller channels favor improved mass transfer due to increased surface area but result in higher pressure drops, which must be managed for optimal operation. To facilitate thorough mixing of reactants, microreactors employ both passive

TABLE 2.1
Microreactor Dimensions, Channel Lengths, Applications, and Production Scales

Design Type	Chip/Device Size	Channel Length (Effective)	Typical Application	Production Capacity
Straight channel microreactor	~2–5 cm × 2–5 cm × 1–2 mm	1–10 cm	Fast, single-step reactions (acid–base, neutralization, quenching)	μL–mL/min (mg–g/hr)
Serpentine/meandering channel	~3–10 cm × 3–10 cm × 1–3 mm	10–100 cm (folded in chip)	Pharmaceutical synthesis, polymerization, mixing-limited reactions	mL/min (g/day)
Split-and-recombine (SAR) microreactor	~5–10 cm × 5–10 cm × 2–5 mm	20–200 cm	Multiphase reactions (gas–liquid hydrogenations, oxidations)	mL/min to L/day (10–100 g/day)
Spiral microreactor	~5–15 cm diameter (disk-like)	0.5–2 m	Photocatalysis, nanoparticle synthesis	Up to 0.5–2 L/day
Multichannel (parallelized) microreactor	Each module: 5–20 cm × 5–20 cm × few cm	10–50 cm per channel × 10–100 channels	Scale-up for fine chemicals and APIs	Tens of L/day (100 g–10 kg/day)
Micro-packed-bed reactor	~5–20 cm length, tubular (1–2 cm dia.)	10–50 cm (with catalyst bed)	Heterogeneous catalysis (hydrogenations, oxidations, reforming)	g/hr to kg/day
Modular industrial microreactor (numbering-up)	Each module: ~20–50 cm × 10–30 cm × 5–20 cm	0.5–2 m effective per module	Large-scale continuous manufacturing (e.g., ibuprofen synthesis, fuel processing)	10–1000 tons/year

and active strategies. Passive mixing leverages unique channel geometries, such as zig-zag or herringbone patterns, to disrupt laminar flow and enhance mixing without external energy input. In contrast, active mixing uses external fields, including ultrasound, electric, or magnetic forces, to stimulate more efficient blending of reactants within the channel (Hessel et al., 2005; Liu et al., 2022; Kazemi et al., 2020).

The RTD is another essential parameter in microreactor design, as a narrow RTD is preferable for achieving highly selective chemical reactions. The design and configuration of channels directly influence RTD, making it critical to optimize geometric factors based on required selectivity. Heat management is also a priority in microreactors. Their thin walls and the use of materials with high thermal conductivity such as metals and ceramics allow for rapid heat exchange, enabling fast heating and cooling of reactants. This capability is essential for preventing thermal runaway and ensuring a uniform temperature profile throughout the reaction, both of which contribute to safer and more efficient reactor operation.

2.3 MATERIALS FOR MICROREACTORS

The choice of material is crucial as it determines chemical compatibility, thermal performance, cost, and transparency.

2.3.1 Selection Criteria

The choice of materials for microreactors is a multifaceted decision, deeply influencing chemical, thermal, and mechanical performance, as well as fabrication practicability and cost. A core selection criterion is thermal conductivity, particularly vital for exothermic reactions or processes with significant heat generation. High-conductivity materials such as certain metals or ceramics ensure that heat is rapidly and evenly dissipated, preventing local hotspots and fostering process safety and efficiency. Another critical aspect is mechanical strength; materials must withstand potentially high operating pressures, especially in systems dealing with gases or employing supercritical conditions, which could cause structural failure if inadequately robust materials are used (Oak Ridge National Laboratory, n.d.).

Chemical resistance is equally essential. Microreactors routinely encounter aggressive reagents—acids, bases, oxidizers, or organic solvents. Materials such as specialized polymers, borosilicate glass, silicon carbide, or noble metal alloys are selected to resist corrosion and prevent leaching or degradation that might compromise both device integrity and product purity. Optical transparency is valued when processes require real-time visual or spectroscopic monitoring; glass, transparent polymers, and even some ceramics offer both chemical resilience and a window for observation or sensor integration. Finally, fabrication feasibility is a practical limitation for selected materials that must be compatible with lithography, etching, bonding, or molding technologies typically used in microfabrication. This criterion affects not only the choice but also eventual device scalability and production cost. In practice, designers often balance these priorities, choosing metals for heat and strength, ceramics for thermal and chemical resistance, polymers for easy processing and transparency, and glass for a combination of chemical stability and optical clarity, thus optimizing device functionality for the intended application.

2.3.2 Commonly Used Materials

Microreactors utilize various materials tailored to specific operational requirements, balancing mechanical strength, chemical resistance, thermal conductivity, and transparency. Metals like stainless steel, titanium, and nickel alloys are favored for their robustness and excellent heat transfer properties, enabling their use in high-pressure and high-temperature environments. These attributes, however, come with drawbacks, including susceptibility to corrosion and lack of transparency, limiting their suitability in optical monitoring applications. Glass, on the other hand, is chemically inert and transparent, making it ideal for reactions that require visual or spectroscopic observation. Its brittleness and limited pressure resistance restrict its application under more demanding mechanical stresses. Polymers such as PDMS, polymethyl methacrylate (PMMA), and polyether ether ketone (PEEK) provide lightweight and cost-effective options, particularly valuable in disposable or low-cost microreactors, but they typically exhibit poorer thermal stability and limited solvent resistance. Ceramics like silicon carbide and alumina stand out for their outstanding corrosion resistance and capability to withstand harsh chemical and thermal conditions, although they involve more complex manufacturing processes and higher material costs. Each material choice impacts reactor longevity, chemical compatibility, thermal management, and feasibility in scaling production, compelling careful selection aligned to the specific chemical processes involved (Azizian & Mohammadi, 2021; Smith & Lee, 2025; Ahmed & Zhang, 2023).

2.4 WORKED CASE STUDY FOR DESIGNING MICROREACTOR

Design of a Liquid-Phase Microreactor for a First-order Reaction

Problem statement (example):

2.4.1 Case A

Design a single microchannel microreactor to convert reactant A → B following first-order kinetics $r_A = -kC_A$ with:

- Rate constant $k = 0.1\,s^{-1}$.
- Feed concentration $C_{A0} = 1.0\,mol/L = 1000\,mol/m^3$.
  Desired conversion $X = 0.90\left(90\%\right)$.
- Available pump throughput target (case A) $Q = 5.0\,mL/min$.
- Channel cross section chosen as square: $w = h = 500\,\mu m = 5.00\times10^{-4}\,m$.
- Fluid ≈ water: $\rho = 1000\,kg/m^3$, $\mu = 1.0\times10^{-3}$ Pas, $cp = 4180\,Jkg^{-1}K^{-1}$.
- Reaction enthalpy (exotherm) $\Delta H = -50\,kJ/mol = -5.0\times10^4\,J/mol$.
- Molar mass of product (for production capacity example) $M = 150\,gmol^{-1}$.

Step 1: Channel geometric properties

1. Channel width $w = 5.00\times10^{-4}\,m$.
2. Channel height $h = 5.00\times10^{-4}$ m.

3. Cross-sectional area:

$$A_c = w \cdot h = (5.00\times10^{-4})(5.00\times10^{-4}) = 2.50\times10^{-7}\,m^2.$$

1. Hydraulic diameter (for rectangular with w = h, D_h = w):

$$D_h = \frac{2wh}{w+h} = \frac{2(5\times10^{-4})(5\times10^{-4})}{(5\times10^{-4})+(5\times10^{-4})} = 5\times10^{-4}\,m.$$

Step 2: Convert volumetric flow to SI and compute average velocity

2. Flow

$$Q = 5.00\,mL\,min^{-1} = 5.00\times10^{-6}\,m^3\cdot min^{-1}.$$

To get m³/s: $Q = \frac{5\times10^{-6}}{60} = 8.33\times10^{-8}\,m^3/s.$

3. Average velocity:

$$u = \frac{Q}{A_c} = \frac{8.33\times10^{-8}}{2.5\times10^{-7}} = 0.333\,m/s.$$

Step 3: Required channel length L for target conversion (first-order, isothermal plug flow reactor (PFR))

For a first-order reaction in plug-flow:

$$L = \frac{u}{k}\ln\left(\frac{1}{1-X}\right).$$

- $u = 0.33333\,m/s.$
- $k = 0.1\,s^{-1}.$

  $X = 0.90 \Rightarrow \ln\frac{1}{1-0.9} = \ln(10) = 2.302585$
- So

$$L = \frac{0.33333}{0.1}\times 2.302585 = 7.675\,m.$$

Interpretation: To reach 90% conversion at 5 mL/min with these kinetics and a 500 μm square channel requires an *effective* channel length ≈ **7.68 m**. In practice, a long straight channel is folded into a serpentine/spiral (or you reduce flow, increase residence time, or use many channels in parallel).

Step 4: Hydrodynamics: Reynolds number and pressure drop for Case A

4. Reynolds number:

$$\mathrm{Re} = \frac{\rho u D_h}{\mu} = \frac{1000 \times 0.33333 \times 5 \times 10^{-4}}{1.0 \times 10^{-3}} = 166.67$$

(so laminar flow is consistent with microreactor operation).

5. Friction factor for laminar flow

$$f = \frac{64}{\mathrm{Re}} = \frac{64}{166.67} = 0.384$$

6. Pressure drop (Darcy–Weisbach form used earlier):

$$\Delta p = f \frac{L}{D_h} \frac{\rho u^2}{2} = 0.384 \times 15350.567 \times 55.556 = 327478.7678\, Pa$$

$$\Delta p \approx 3.27 \times 10^5\, Pa \approx 3.27\, bar$$

Interpretation: At 5 mL/min, the pressure drop across the ~7.68 m effective channel is ~3.3 bar—notable but manageable with appropriate pumps and fittings; however, such a high effective length and pressure might be impractical in a single thin chip and drives alternative design choices.

Step 5: Thermal check (adiabatic temperature rise)

Maximum possible adiabatic temperature rise for conversion X:

$$\Delta T_{adiabatic} = \frac{|\Delta H| C_{AO} X}{\rho C_P} = \frac{50000 \times 1000 \times 0.9}{1000 \times 4180} = 10.7655\, K$$

Interpretation: If the reaction is exothermic with −50 kJ/mol and 90% conversion, the adiabatic temperature rise is only ≈ **10.8 °C** for water-like fluid—modest and easily controlled by chip heat exchange capabilities (jacketed plate, cooling fluid, or high-thermal-conductivity substrate).

Heat generation rate (power) at $\mathrm{Q} = 5\,\mathrm{mL} / \mathrm{min}$:

$$\dot{Q}_{rxn} = Q \cdot |\Delta H| \cdot C_{AO} \cdot X = 8.33 \times 10^{-8} \times 1000 \times 0.9 \times 50000 \approx 3.75W$$

So, the reactor generates only **~3.75 W** thermal power at this flow—trivial to remove for a microreactor.

2.4.2 Case B: Design Alternative: Fit an Effective Channel Length Lchip = 0.50 m

A practical chip might contain a folded/serpentine channel with effective length $\mathrm{L_{chip}} = 0.50$ m. To get the same conversion with the same kinetics, find the allowable flow rate $\mathrm{Q_{req}}$ (and velocity $\mathrm{u_{req}}$) (Table 2.2).

TABLE 2.2
Result Summary of Design Parameters of a Single Microchannel Microreactor for Case A

Parameter	Value
Rate constant (k)	0.1 s^{-1}
Feed (C_{AO})	1.0 mol/L
Conversion (X)	0.9
Flow rate	5 mL/min
Channel size	500 µm
Channel length (L)	7.68 m
Velocity (u)	0.333 m/s
Reynolds number	166.5
Pressure drop	3.27 kPa
Reaction heat	3.75 W

From

$$L = \frac{u}{k}\ln\left(\frac{1}{1-X}\right)$$

$$0.5 = \frac{u_{req}}{0.1}\, ln\left(\frac{1}{1-0.9}\right)$$

$$\Rightarrow u_{req} = 0.0217147\, ms^{-1}$$

Corresponding volumetric flow:

$$Q_{req} = u_{req}\cdot \mathrm{Ac} = 0.0217147\times 2.50\times 10^{-7} = 5.4274310237906475 \times 10^{-9}\,\mathrm{m}^3\cdot\mathrm{s}^{-1} = 0.32572\,\mathrm{mLmin}-1.$$

Interpretation: To achieve 90% conversion in a 0.5-m effective channel with the same kinetics and the chosen channel cross section, the flow must be reduced to **~0.326 mL·min**$^{-1}$ (≈ 16× lower than 5 mL/min) (Table 2.3).

Hydrodynamic checks at this flow:

- New average velocity $u_{req} = 0.0217147\,\mathrm{ms}^{-1}$.
- Reynolds:

$$\mathrm{Re} = \frac{1000\times 0.021714724095162\,59\times 5.00\times 10-4}{1.0\times 10^{-3}} = 10.85736.$$

- Friction factor f = 64 / Re = 5.892
- Pressure drop:

$$\Delta p = f\,\frac{L}{D_h}\,\frac{\rho u^2}{2}.$$

$$\Delta \mathrm{p} = 5.892\times 1000\times 0.23587 \approx 1389.7\,\mathrm{Pa} \approx 0.0137\,\mathrm{bar}$$

TABLE 2.3
Result Summary of Design Parameters of a Single Microchannel Microreactor for Case B

Parameter	Value
Rate constant (k)	0.1 s^{-1}
Feed (C_{AO})	1.0 mol/L
Conversion (X)	0.9
Flow rate	5 mL/min
Channel size	500 µm
Channel length (L)	0.5 m
Velocity (u)	0.0217 m/s
Reynolds number	10.857
Pressure drop	0.0137 kPa
Reaction heat	0.24 W

Heat generation at this lower flow:

$$\dot{Q}_{rxn} = Q \cdot |\Delta H| \cdot C_{AO} \cdot X \approx 0.2443W.$$

Interpretation: For a compact 0.5-m chip, the flow/specific throughput is lower, pressure drop is tiny (~0.014 bar), and heat removal is trivial (0.24 W).

2.4.3 Production/Scaling Example (Parallelization)

Take the Case B single-channel throughput and compute **product mass per day**:

- Moles produced per second (per channel):

$$n^{\bullet} = Q_{req} \cdot C_{A0} \cdot X = 5.4274310 \times 10^{-9} \times 1000 \times 0.9 = 4.884687 \times 10^{-6}\,\text{mols}^{-1}.$$

Mass produced per day:

$$m_{day} = \dot{n} \times M \times 86400.$$

$$m_{day} = 4.884687 \times 10^{-6} \times 150 \times 86400 \approx 63.32\,\text{gday}^{-1}.$$

So, **one 0.5-m folded channel** produces **≈63.3 g/day** of product (for M = 150 g/mol).

To reach **1 kg/day**:

$$N = \frac{1000}{63.32} \approx 15.79.$$

$$N \Rightarrow 16\,\text{parallel channels / modules}$$

To reach **10 kg/day**, ~160 channels, and so on. This is the standard **numbering-up** approach.

2.4.4 Design Takeaways and Recommendations

2.4.4.1 Tradeoffs Are Explicit

- High volumetric throughput (5 mL/min) + modest k → very long channels (≈7.7 m) and high Δp (≈3.3 bar).
- Compact chip (0.5-m channel) requires lower flow (≈0.326 mL/min) or faster kinetics (higher k) to obtain the same conversion.

2.4.4.2 Ways to Meet Higher Throughput without Excessive Length

Increase reaction rate (via higher temperature, better catalyst)—increases k.

- Reduce channel cross-section to increase residence time at the same Q (but watch Δp).
- Accept lower conversion per channel and add downstream recycle or series modules.
- Numbering-up (parallel channels or parallel modules)—practical for scale-up.

2.4.4.3 Thermal and Safety

- For the example exotherm (−50 kJ/mol), the adiabatic rise is ~11 °C at high conversion—manageable with standard cooling.
- Heat generation power was low (3.75 W at 5 mL/min, 0.24 W at 0.326 mL/min)—easy to remove using the chip's conductive walls.

2.4.4.4 Hydrodynamics

- Re in both cases remains laminar (Re ≈ 10–170). Use appropriate Nu and Sh correlations for rectangular channels to size heat/mass transfer surfaces accurately if detailed design is needed.

2.5 FABRICATION METHODOLOGIES OF MICROREACTORS

2.5.1 Fabrication Methodologies

Fabrication methodologies ranging from conventional micromachining and lithography to advanced manufacturing are presented, highlighting their advantages, limitations, and suitability for diverse applications (Bojang et al., 2020).

2.5.1.1 Conventional Micromachining

Conventional micromachining applies miniaturized forms of traditional machining techniques such as milling, drilling, and EDM to create microchannels on metal or polymer substrates. This process typically involves precision CNC (computer numerical control) milling or microdrilling to form the channels, which are then sealed using diffusion bonding, welding, or adhesives to complete the microreactor structure. An advantage of this method is its suitability for metals like stainless steel and titanium, providing microreactors with high mechanical strength and excellent thermal conductivity. These metals enable the microreactors to operate under high-pressure and high-temperature conditions, making them ideal for demanding applications such as petrochemical microreactors, reformers, and high-pressure catalytic processes. The conventional micromachining approach is also scalable for moderate production volumes, allowing practical deployment beyond laboratory scales (Jähnisch et al., 2004; Wikipedia contributors, 2004).

However, there are limitations to this technique. Achieving resolutions below 50 micrometers is challenging, hindering the fabrication of very fine features. Additionally, surface roughness generated during machining can cause increased pressure drop and fouling within microchannels, potentially reducing efficiency. Tool wear and burr formation are common issues that require additional post-processing to ensure smooth channel surfaces and maintain device performance. Despite these challenges, conventional micromachining remains a widely used and valuable fabrication method for producing robust and high-performance microreactors, especially when metal substrates are required (Jain, n.d.).

2.5.1.2 Photolithography and Etching

Photolithography is a precise patterning technique adapted from the semiconductor industry to fabricate microchannels on silicon or glass wafers with high accuracy. The process begins by coating the wafer with a light-sensitive material called photoresist, which uniformly covers the surface after cleaning and preparation. A photomask containing the desired channel pattern is then aligned over the wafer, and ultraviolet (UV) light is projected through this mask, transferring the pattern onto the photoresist by inducing a chemical change in exposed areas. Following exposure, the wafer undergoes development, where the reacted photoresist parts are selectively removed, revealing the underlying material (Azhar & Banerjee, 2025).

Next, either wet or dry etching techniques remove the exposed substrate regions to create the actual microchannels. Wet etching uses chemical solutions to dissolve the material, though it can be isotropic, sometimes eroding beneath the photoresist layer. Dry etching, including plasma etching or reactive ion etching, offers more directional and controlled material removal, essential for finely detailed and vertically defined channels. After etching, the wafer is bonded to a cover plate through methods such as anodic or fusion bonding to seal the channels within the microreactor structure. Advantages of photolithography include the ability to achieve very fine features with channel widths under 10 micrometers and consistent reproducibility with excellent surface finish. Additionally, this process allows integration of functional elements like sensors, heaters, or microelectromechanical systems (MEMS) components within the reactor. However, photolithography requires expensive cleanroom environments, and the brittle nature of silicon or glass substrates may limit mechanical robustness, while processing large-area devices can be time-consuming. Its precision and integration capability make photolithography ideal for applications such as lab-on-chip systems, analytical instruments, and high-precision pharmaceutical synthesis (Fattahi & Kim, 2024; Finlayson, n.d.; Wikipedia contributors, 2001).

2.5.1.3 Soft Lithography (PDMS Casting)

Soft lithography is a cost-effective and rapid prototyping technique extensively used in research laboratories to fabricate microfluidic devices, particularly through PDMS casting. The process begins with the fabrication of a master mold, which can be created using photolithography or 3D printing methods. Once the mold is ready, liquid PDMS is poured over it and cured by heating, solidifying into the desired shape. The cured PDMS layer, which contains the negative pattern of the mold, is then carefully peeled off and bonded to a flat substrate such as glass or another PDMS layer to enclose the microchannels. This approach is favored for its low cost, ease of use, and rapid turnaround from design to device, as well as the flexibility and biocompatibility of PDMS. These properties make it suitable for biological assays and applications requiring low-pressure conditions (Xia & Whitesides, 1998).

However, PDMS also has limitations; it is permeable to gases and some organic solvents, which can lead to swelling and deformation of microchannels during certain chemical exposures. Additionally, it possesses limited resistance to high temperatures and pressures, restricting its use in more rigorous chemical processes. Despite these constraints, soft lithography remains popular for fabricating cell culture devices, microfluidic chips used in diagnostic applications, and educational research tools given its versatility, affordability, and relatively simple procedure (Nickels, 2022; Elveflow, 2020).

2.5.1.4 Laser Micromachining

Laser micromachining employs high-energy, focused laser beams to precisely ablate material and create microchannels on a substrate surface. The process uses pulsed lasers such as CO_2, excimer, or femtosecond lasers, which scan over the material, vaporizing it layer by layer to form the desired features. This direct-write method requires no photomasks, offering flexibility and rapid fabrication of custom designs across diverse materials, including polymers, glass, ceramics, and metals. Key advantages include high precision, adaptability to various substrates, and suitability for rapid prototyping or small-batch production. However, the resulting surface can have roughness, necessitating post-processing, and there is a risk of heat-affected zones that may alter the material's properties. Laser micromachining is widely used for producing polymer and glass microreactors and offers an etching-free fabrication route, adding to its appeal in the microfabrication industry (Georgieva & Utkin, 2014; GFMS, 2015; Zhao & Geng, 2019; Amada Welding Technologies, n.d.).

2.5.1.5 Advanced Manufacturing (3D Printing/Additive Manufacturing)

Recent advancements in additive manufacturing, commonly known as 3D printing, have significantly enhanced the fabrication of microreactors by enabling the creation of complex three-dimensional architectures that were previously difficult or impossible to achieve with traditional methods. The fabrication process starts with designing computer-aided design (CAD) models, which are sliced into thin layers that a 3D printer builds sequentially using technologies such as stereolithography, 2PP, or selective laser melting. These techniques allow unparalleled freedom in design, including the integration of intricate features like mixers, sensors, and fluid manifolds directly within a single print, streamlining the production process (Table 2.4).

TABLE 2.4
Comparative Summary Results

Method	Resolution	Material Compatibility	Cost	Strength	Applications
Micromachining	50–500 μm	Metals, polymers	Medium	High	High-pressure catalytic reactors
Photolithography	<10 μm	Silicon, glass	High	Medium	Lab-on-chip, sensors
Soft lithography	20–100 μm	PDMS	Low	Low	Bioassays, low-pressure chips
Laser micromachining	20–100 μm	Polymers, glass, ceramics	Medium	Medium	Rapid prototyping
3D printing	50–500 μm	Polymers, metals	Medium	Medium	Complex 3D reactors

The rapid iteration capabilities of 3D printing facilitate quick prototyping and small-batch production, enabling researchers and manufacturers to efficiently test and optimize reactor designs. However, limitations persist, such as constrained resolution for very small channels typically below 50 micrometers and the need for post-processing steps like cleaning and curing to finalize the devices. Additionally, the range of compatible materials remains narrower compared to conventional machining techniques. Despite these challenges, 3D-printed microreactors have found applications across various fields, including custom microreactor chips for process development, bioreactors for biological studies, energy conversion devices, and multiphase reactors that handle complex chemical reactions within integrated systems (Sydoryk & Hsu, 2022; Abdelrahman & Ahmed, 2020).

2.6 CONCLUSIONS

This chapter has highlighted the fundamental principles of microreactor design and fabrication, emphasizing the key designs of microreactor at the microscale. Various fabrication methodologies, from traditional micromachining and lithography to advanced manufacturing techniques, were discussed in the context of their benefits, limitations, and practical applications. Case studies further demonstrated the versatility of microreactors in fields such as design of a liquid-phase reactor.

Looking ahead, emerging trends, including digital design tools, sustainable materials, and hybrid fabrication approaches, are expected to expand the scope and impact of microreactor technology. By integrating these advancements, microreactors will continue to play a pivotal role in enabling more efficient, flexible, and environmentally responsible chemical manufacturing.

REFERENCES

Abdelrahman, S., & Ahmed, H. (2020). 3D printing and continuous flow chemistry technology to advance pharmaceutical manufacturing in developing countries. *Arabian Journal of Chemistry*, *13*(9), 6665–6674. https://doi.org/10.1016/j.arabjc.2020.03.023

Ahmed, F., & Zhang, T. (2023). Microreactor-based micro/nanomaterials: Fabrication, characterization, and applications. *Materials Horizons*, *10*(7), 1825–1840. https://doi.org/10.1039/D3MH00329A

Amada Welding Technologies. (n.d.). *Laser micromachining: 5 things to know about this precision manufacturing technology*. Retrieved October 6, 2025, from https://amadaweldtech.com/blog/laser-micromachining-5-things-to-know-about-this-precision-manufacturing-technology/

Azhar, R., & Banerjee, S. (2025, September 28). *Photolithography: A step-by-step guide*. AZoNano. Retrieved October 6, 2025, from https://www.azonano.com/article.aspx?ArticleID=6376

Azizian, S., & Mohammadi, P. (2021). Current scenario and prospects in manufacture strategies of microreactors: A comprehensive review. *Chemical Engineering Journal*, *414*, 128810. https://doi.org/10.1016/j.cej.2021.128810

Bastan, F., & Kazemeini, M. (2022). Activated pharmaceutical ingredients produced by microreactors versus batch processes: A review. *Journal of Particle Science and Technology*, *8*(2), 63–78.

Bojang, A. A., Zhang, Q., & Sun, Z. (2020). Design, fundamental principles of fabrication, and applications of microreactors: A review. *Processes*, *8*(8), 891.

Brás, E. J. S. (2021). Recent developments in microreactor technology for biocatalysis applications. *Reaction Chemistry & Engineering*, *6*(4), 675–693. https://doi.org/10.1039/D1RE00024A

El Zanati, E., et al. (2018). Design and scale-up of continuous flow micro-reactor. *Journal of Engineering Science and Technology*, *13*(8), 2481–2494.

Elveflow. (2020, December 13). *Microfluidic chip fabrication techniques: Wet and dry etching*. Retrieved October 6, 2025, from https://elveflow.com/microfluidic-reviews/soft-lithography-fabrication-technics/

Engelbrecht, N., Everson, R. C., Bessarabov, D., & Kolb, G. (2020). Microchannel reactor heat-exchangers: A review of design strategies for the effective thermal coupling of gas phase reactions. *Chemical Engineering and Processing – Process Intensification*, *157*, 108164. https://doi.org/10.1016/j.cep.2020.108164

Fattahi, S., & Kim, J. (2024). Photolithography – An overview. In *Encyclopedia of Microsystems* (pp. 145–158). ScienceDirect. Retrieved October 6, 2025, from https://www.sciencedirect.com/topics/physics-and-astronomy/photolithography

Finlayson, W. (n.d.). *How do you make a microreactor?* University of Washington. Retrieved October 6, 2025, from https://faculty.washington.edu/finlayso/che475/microreactors/Group_C/how%20do%20you%20make%20a%20microreactor.htm

Gateway for Accelerated Innovation in Nuclear. (2025). *DOE Microreactor Program Review 2024*. https://gain.inl.gov

Georgieva, D., & Utkin, A. (2014). Laser micro-milling of microchannel on copper sheet as a microreactor. *International Journal of Heat and Mass Transfer*, *78*, 442–450. https://doi.org/10.1016/j.ijheatmasstransfer.2014.06.046

GFMS. (2015, December 31). *Laser micromachining – Achieve small, high-precision features*. Retrieved October 6, 2025, from https://www.gfms.com/com/en/machines/laser/laser-micromachining.html

Hessel, V. (Ed.). (2015). *Design and engineering of microreactor and smart-scaled flow processes*. Multidisciplinary Digital Publishing Institute (MDPI). *Processes* 2015, 3(1), 19–22. https://doi.org/10.3390/pr3010019

Hessel, V., Löwe, H., & Schönfeld, F. (2005). Micromixers – A review on passive and active mixing principles. *Chemical Engineering Science*, *60*(8–9), 2479–2501.

Hessel, V., Mukherjee, S., Mitra, S., Goswami, A., Tran, N. N., Ferlin, F., Vaccaro, L., Galogahi, F. M., Nguyen, N., & Escribà-Gelonch, M. (2024). Sustainability of flow chemistry and microreaction technology. *Green Chemistry*, *26*(18), 9503–9528. https://doi.org/10.1039/D4GC01882F

Oak Ridge National Laboratory. (n.d.). *Assessment of microreactor safety analysis challenges and recommendations for utilization of the comprehensive reactor analysis bundle*. September 2023, from https://info.ornl.gov/sites/publications/Files/Pub204154.pdf

Jähnisch, K. F., Hessel, V., Löwe, H., & Baerns, M. (2004). Microreactors: New technology for modern chemistry. *Angewandte Chemie International Edition*, *43*(46), 6554–6573.

Jain, V. K. (n.d.). *Micro machining processes*. IIT Kanpur Lecture Notes. Retrieved October 6, 2025, from https://home.iitk.ac.in/~vkjain/MMPs_L1_Intro._010113.pdf

Kazemi, M., et al. (2020). Microchannel reactor heat-exchangers: A review of design, applications, and performance. *Materials Chemistry and Physics*, *250*, 122984.

Li, B., Liu, S., & Zhang, Q. (2023). Microreactor-based synthesis of fine chemicals and pharmaceuticals. *Journal of Industrial and Engineering Chemistry*, *108*, 83–99.

Liu, W., et al. (2022). Shape design of channels and manifolds in a multichannel microreactor: From basic rules to advanced technologies. *Frontiers in Chemical Engineering*, *2*, Article 838336.

Nickels, P. C. (2022). *Polycarbonate masters for soft lithography*. University of Illinois Chicago, UIC Indigo Repository. Retrieved October 6, 2025, from https://indigo.uic.edu/articles/thesis/Polycarbonate_Masters_for_Soft_Lithography/20253258

Pandey, P. N., & Khanduri, S. K. (2023). Applications of microreactors in energy, environment, and biotechnology. *Environmental Science & Technology*, *57*(14), 7788–7802.

Reid, B., Sweetland, K., & Ward, M. (2023). *Advances in microreactor fabrication (LA-UR-23-23350)*. Los Alamos National Laboratory, U.S. DOE/NNSA.

Rodríguez-Guerra, J., Gerling, M., López-Guajardo, E., Lozano-García, J., Nigam, K. D. P., & Montesinos-Castellanos, A. (2016). Design of micro- and milli-channel heat exchanger reactors for homogeneous exothermic reactions in the laminar regime. *Industrial & Engineering Chemistry Research*, *55*(22), 6435–6442.

Schmidt, T. W., Baghery, J. J., & Rao, K. S. S. R. K. (2019). Microreactor technology in chemical synthesis: Principles and applications. *Chemical Engineering & Technology*, *42*(3), 623–638.

Smith, J., & Lee, K. (2025). Reactor fabrication, design, and cutting-edge applications. *Journal of Chemical Engineering Research and Design*, *171*, 13–35. https://doi.org/10.1016/j.cherd.2024.09.001

Suryawanshi, P. L., Ghosh, P. K., & Rathod, V. K. (2018). A review on microreactors: Reactor fabrication, design, and cutting-edge applications. *Chemical Engineering Journal*, *355*, 492–514. https://doi.org/10.1016/j.cej.2018.08.139

Sydoryk, F., & Hsu, W. (2022). Mathematical modeling of 3D printing for microreactors: Optimizing layer resolution and flow parameters. *International Journal of Research in Pharmaceutical Sciences*, *6*(5), 357–368.

Wikipedia contributors. (2001, August 13). *Photolithography*. Wikipedia. Retrieved October 6, 2025, from https://en.wikipedia.org/wiki/Photolithography

Wikipedia contributors. (2004, June 15). *Microreactor*. In Wikipedia, The Free Encyclopedia. Retrieved October 6, 2025, from https://en.wikipedia.org/wiki/Microreactor

Xia, Y., & Whitesides, G. M. (1998). Soft lithography. *Annual Review of Materials Science*, *28*(1), 153–184. https://doi.org/10.1146/annurev.matsci.28.1.153

Yadav, D., Lu, X., Vishwakarma, C. B., & Jing, D. (2023). Advancements in microreactor technology for hydrogen production via steam reforming: A comprehensive review of experimental studies. *Journal of Power Sources*, *585*, 233621. https://doi.org/10.1016/j.jpowsour.2023.233621

Yang, T., Tian, Z., Fang, H., et al. (2023). Fundamentals and development of microreaction technology for fine chemicals synthesis. *Journal of Chemical Engineering of Japan*, *56*(1), 1–15. https://doi.org/10.1080/00219592.2023.2197011

Yang, T., et al. (2024). Apparent kinetics and thermodynamics study of methyl ethyl ketone peroxide in an efficient continuous-flow microreaction system. *Chemical Engineering Science*, *285*, 119472. https://doi.org/10.1016/j.ces.2023.119472

Zhang, J., Chen, X., & Li, Z. (2021). Process intensification approach using microreactors for synthesizing nanomaterials: A critical review. *Nanomaterials*, *11*(1), 121.

Zhao, Y., & Geng, X. (2019). The fabrication of micro/nano structures by laser machining. *Micromachines*, *10*(12), 857. https://doi.org/10.3390/mi10120857

3 Organic Chemical Synthesis Using Microreactor

Thangamani Suppan, Kiruthika Paramasivam, Moorthi Pichumani, M. Habibur Rahman, and Rama Ranjan Bhattacharjee

3.1 INTRODUCTION AND HISTORICAL DEVELOPMENT

Microreactor technology, where chemical reactions are carried out in channels or compartments with dimensions ranging from tens to hundreds of micrometers, has emerged in recent decades as a transformative approach to organic synthesis (Hessel and Lowe, 2005). The idea of miniaturization dates back to Richard Feynman's famous 1959 lecture, while the first practical steps were taken with microfluidic analytical devices developed in the 1980s. However, true chemical microreactors began to appear in the 1990s, with the maturation of microfabrication techniques, enabling the first demonstrations that conventional reactions could be moved from round-bottom flasks to continuous flow in microchannel devices (Ehrfeld et al., 2000).

In the early 2000s, several important milestones were reached. Henke et al. proved that nitrations could be performed more safely and selectively in microreactors, avoiding the risk of runaway reactions (Henke and Winterbauer, 2005). Burange and collaborators achieved multistep synthesis of pharmacologically active compounds using an integrated microfluidic platform (Burange et al., 2022). At MIT, Jensen's group introduced reconfigurable flow systems that functioned as "mini chemical plants," demonstrating how such devices could accelerate process development (Adamo et al., 2016). These successes cemented microreactors as a viable and promising tool for organic synthesis.

Over the following years, the field grew rapidly. Microreactor-based flow chemistry evolved from a laboratory curiosity to a powerful tool in both academic and industrial research (Noel et al., 2019; Porta et al., 2016). Reviews during this period emphasized its key advantages: excellent heat and mass transfer, tight residence time control, increased safety, and straightforward scalability, summarized under the concept of "process intensification" (Plutschack et al., 2017; Watts and Wiles, 2013). Importantly, microreactors enable reactions under conditions unattainable in batches, such as superheating solvents or stabilizing highly reactive intermediates. Yoshida introduced the concept of "flash chemistry," where microreactors allow the

 DOI: 10.1201/9781003633761-3

immediate generation and consumption of short-lived cationic or anionic intermediates before they can decompose (Yoshida, 2010). This marked a paradigm shift by opening up access to reactive species that are too unstable or dangerous in conventional settings.

Another breakthrough was the demonstration of continuous-flow multistage synthesis of complex molecules. In 2016, Porta's team presented a microreactor system capable of performing five consecutive reactions to produce an active pharmaceutical ingredient (API) without isolating intermediates (Porta et al., 2016; Vijaya Bhaskar et al., 2024). Subsequently, Tsubogo, Oyamada, and Kobayashi achieved an eight-step synthesis of the drug candidate rolipram using sequential flow reactors with heterogeneous catalysts (Bogdan et al., 2009; Tsubogo et al., 2015). In 2007, a refrigerator-sized continuous manufacturing platform was developed that not only performed the multistage synthesis but also integrated separation and dose formulation, approaching true pharmaceutical manufacturing (Ahmed-Omer et al., 2007). These advances attracted the attention of regulatory agencies, and in 2017, the FDA approved the first drug produced entirely by continuous-flow methods. By the late 2010s, continuous-flow microreactor processes were being tested for the production of pharmaceuticals and fine chemicals such as ibuprofen, artemisinin, and ciprofloxacin (Bogdan et al., 2009; Fanelli et al., 2017; Richardson and Kolmel, 2021). Today, microreactor technology is not only an emerging field but an established strategy in both research and industry, offering chemists new tools to rethink organic synthesis.

3.1.1 Advantages of Microreactor Technology in Organic Synthesis

Microreactors offer significant advantages, primarily due to their extremely high surface-area-to-volume ratios and the micron-sized flow channels (Ekinci et al., 2024; Fletcher et al., 2002; Hessel et al., 2005). These features translate into higher heat and mass transfer efficiencies compared to traditional batch reactors. Rapid heat dissipation allows highly exothermic transformations, such as nitrations and peroxidations, to be carried out safely under near-isothermal conditions, thus minimizing the risk of thermal runaways (Ahmed-Omer et al., 2007; Balakrishnan et al., 2023). Furthermore, reactions can be carried out at temperatures above the normal boiling points of solvents, operating at elevated pressures with precise thermal regulation (Fanelli et al., 2017; Geyer et al., 2006).

Mass transfer improvements derived from segmented flow regimes and micromixing designs facilitate rapid mixing, making microreactors ideal for multiphase interactions such as gas–liquid and liquid–liquid reactions (Feng et al., 2022). Diffusion-limited gas–liquid hydrogenations and oxidations can reach near-plug flow conditions within microreactors, resulting in improved conversions and selectivities (Fanelli et al., 2017; Zong and Yue, 2022). Tunable reaction residence time through flow rate control allows chemists to efficiently intercept transient reactive intermediates or kinetically competitive pathways. For example, organolithium compounds, whose reactive half-lives are expressed in milliseconds at low temperatures, can be reliably generated and consumed in flow (Fanelli et al., 2017; Willms et al., 2020).

The very small internal volumes of microreactors (microliter to milliliter scale) limit the amount of hazardous materials exposed at any one time, significantly improving safety (Ahmed-Omer et al., 2007; Fletcher et al., 2002). This allows for on-demand generation and immediate use of reactive species such as diazomethane, avoiding the need for hazardous storage (Geyer et al., 2006). Furthermore, microreactors provide safer and more reproducible conditions for nitrations previously considered risky in batches (Willms et al., 2020). In conclusion, superior heat/mass transfer, tunable short residence times, and increased safety contribute to the indispensability of microreactor technology for both fundamental and applied organic synthesis efforts.

3.1.2 Classification of Microreactors

Microreactors can be classified based on their geometry, fluid type, flow pattern, and reaction type, as shown in Figure 3.1. In terms of their geometry, they can be designed as channel or manifold systems, and channel microreactors are divided into single-channel, multichannel, and serpentine configurations (Hessel et al., 2005). In terms of fluid type, microreactors can handle liquids or gases. Liquid systems can utilize channels with circular, elliptical, triangular, rectangular, trapezoidal, or hexagonal cross-sections, while gaseous systems can be classified as compressible or incompressible (Ehrfeld et al., 2000). In terms of flow patterns, microreactors can

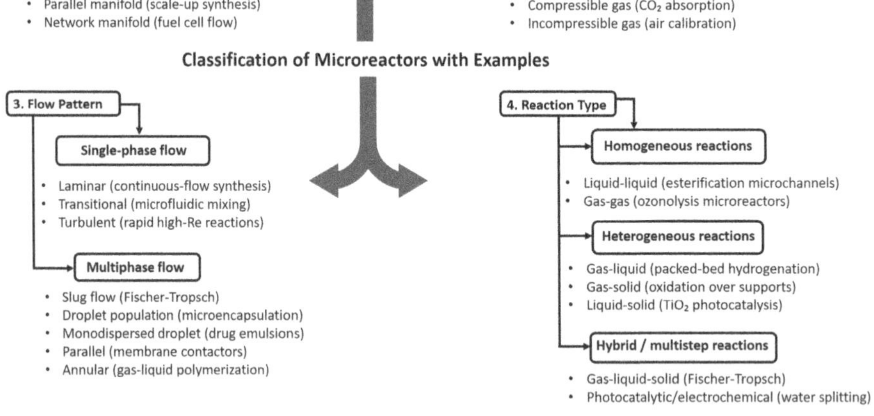

FIGURE 3.1 Classification of microreactors according to geometry, fluid type, flow pattern, and reaction type, highlighting representative examples of their applications.

operate in laminar, transitional, or turbulent regimes or involve multiphase flows such as slug flow, monodisperse flow, droplet population flow, parallel flow, and annular flow (Tanimu et al., 2017). Finally, with respect to the type of reaction, microreactors allow for various chemistries, including homogeneous liquid–liquid catalysis and heterogeneous gas–liquid catalysis, among others (Fang et al., 2010; Nieuwelink et al., 2021).

3.1.2.1 Geometry

Microreactors are classified according to their physical architecture, including channel-based and manifold-based designs. Channel-based designs include single-channel systems used in capillary hydrogenation, multichannel configurations for parallel photochemistry, and serpentine channels for rapid mixing. Manifold-based designs include parallel manifolds for synthesis scale-up and network manifolds present in fuel cell flows. These designs allow for precise control of reaction environments and scalability (Hessel et al., 2005).

3.1.2.2 Fluid Type

Microreactors handle different fluid phases, primarily liquid and gas, which influences their design and performance. Liquid-phase systems feature cross-sections such as circular droplets (emulsions), rectangular channels (DNA amplification), and complex shapes, such as triangular and hexagonal, for improved mixing in lab-on-chip. Gas-phase reactors handle both compressible (e.g., CO_2 absorption) and incompressible (e.g., air calibration) gases, which require distinct designs due to their different physical properties (Abidi et al., 2021).

3.1.2.3 Flow Pattern

Flow in microreactors can be single-phase or multiphase. Single-phase flow includes laminar flow observed in continuous synthesis, transitional flow favoring microfluidic mixing, and turbulent flow for fast reactions at high Reynolds numbers. Multiphase flow includes slug flow used in Fischer–Tropsch synthesis, droplet populations for microencapsulation, monodisperse droplets for drug emulsions, parallel flows observed in membrane contactors, and annular flow relevant to gas–liquid polymerizations; all of which improve mixing efficiency and mass transfer (Zhao, 2013).

3.1.2.4 Reaction Type

Microreactors support a variety of reaction types: homogeneous reactions including liquid–liquid esterification and gas–gas ozonolysis; heterogeneous reactions such as gas–liquid packed bed hydrogenation, supported gas–solid oxidation, and liquid–solid TiO_2 photocatalysis; and hybrid or multistep reactions integrating gas–liquid–solid phases, e.g., Fischer–Tropsch synthesis and photocatalytic/electrochemical water splitting, demonstrating the adaptability of microreactors to diverse chemical processes. This classification provides a framework for selecting or designing microreactors tailored to specific chemical processes, emphasizing the importance of geometry, fluid dynamics, and reaction mechanisms, and aligning with broader challenges in flow control and reactor configuration.

3.2 MICROREACTOR IN ORGANIC SYNTHESIS

The addition of a catalyst in microreactor organic synthesis typically involves the immobilization of catalytic materials, such as metals, metal oxides, or organometallic complexes, on solid supports within the microreactor channels using techniques such as sol-gel coating, chemical deposition, or layer-by-layer assembly. These supported catalysts continuously convert reactants as they flow through the microchannels, offering high selectivity and reducing byproduct formation. The choice of support, such as alumina, silica, zeolite, or carbon, is crucial, as it affects catalytic activity, stability, and mass transfer dynamics. This approach accelerates reaction rates by lowering the activation energy, simplifies product isolation, ensures reproducible conversions, and improves safety and scalability compared to traditional batch operations (Feng et al., 2022).

3.2.1 Catalyst-Induced Reactions in Microreactors

Catalyst-induced reactions in microreactors refer to organic and inorganic transformations initiated or accelerated by catalysts immobilized or suspended within the microreactor system, as shown in Figure 3.2. These reactions benefit from the unique characteristics of microreactors, such as improved mass transfer, superior thermal management, and precise control of reaction selectivity compared to conventional batch methods. In these systems, heterogeneous catalysts such as Pd/C, Pt nanoparticles, or organocatalysts immobilized on solid supports are frequently employed. For example, catalytic hydrogenations with Pd/C in microreactors typically achieve faster conversion rates and higher yields under milder conditions than in batch reactors.

The reported cases demonstrate complete conversion within minutes in flow configurations, while batch processes can require hours or even days at similar temperatures and pressures. Similarly, organocatalytic condensations and other base- or

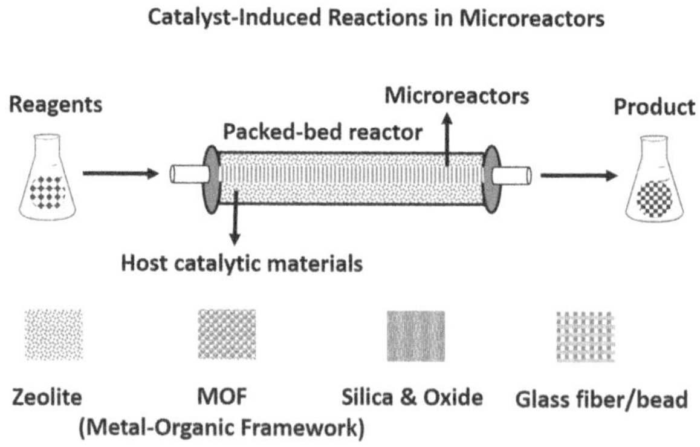

FIGURE 3.2 Schematic representation of catalyst-induced reactions in microreactors using a packed-bed reactor. Reactants flow through catalytic supports such as zeolites, metal-organic frameworks (MOFs), silica and oxides, or glass fibers/beads, where the immobilized catalysts drive continuous and selective transformations to obtain the desired product.

acid-catalyzed reactions exhibit accelerated kinetics and greater selectivity in microchannels, thanks to higher mixing efficiency and tightly controlled residence times.

Another advantage of microreactor systems is the continuous removal of unreacted products and reactants, which shifts equilibrium positions and improves overall performance. Furthermore, the platform allows for precise tuning of critical parameters such as temperature gradients, residence time, and catalyst exposure, which directly impact catalytic performance (Feng et al., 2022; Tanimu et al., 2017).

Catalysts immobilized on silica, polymer supports, or carbon structures within microchannels provide not only increased stability and recyclability but also precise control of catalyst loading and surface properties. This immobilization minimizes catalyst leaching, extends catalyst lifetime, and significantly reduces secondary purification steps, thereby lowering operating costs. Furthermore, supported catalysts often allow for specific tuning of activity and selectivity, which is particularly advantageous for stereoselective or multistage organic transformations. The confined microchannel environment further amplifies mass and heat transfer rates, ensuring uniform exposure of reactants to the catalyst surface. This eliminates the diffusion limitations common in batch reactors, allowing reactions to proceed with higher conversion efficiencies and cleaner product profiles. Furthermore, the ability to rapidly adjust flow rates, residence times, and reaction parameters provides unprecedented flexibility for fine-tuning reaction kinetics.

From a sustainability perspective, these features directly contribute to resource efficiency by reducing excessive reagent use, minimizing solvent consumption, and lowering energy consumption through improved thermal control. Waste generation is significantly reduced, and when photocatalysts or electrocatalysts are used, the integration of renewable energy sources, such as visible light or electrochemical potential, further improves the process's environmental profile.

At an industrial scale, catalyst-induced reactions in microreactors support continuous manufacturing paradigms, enabling safer handling of hazardous intermediates, reproducible scale-up, and real-time process monitoring. This makes them highly attractive for pharmaceutical synthesis, fine chemical production, and green catalytic applications. Overall, microreactor-based catalysis exemplifies process intensification by combining stability, efficiency, scalability, and sustainability, offering a greener and more resource-efficient alternative to traditional batch catalytic systems.

3.2.2 Catalyst-Controlled Reactions in Microreactors

Catalyst-controlled reactions, as shown in Figure 3.3, in microreactors highlight how the nature, structure, and position of the catalyst directly influence reaction outcomes, such as selectivity, stereochemistry, and product distribution. Unlike simple catalytic acceleration, the catalyst in these systems actively guides the reaction pathway and regulates the formation of specific products. Microreactor platforms allow precise control of catalyst exposure, temperature, and mixing, enabling fine regulation of reaction kinetics and selectivity compared to batch reactors. For example, supported chiral catalysts in packed or monolithic microreactors have been shown to produce enantioselective products in high yields and enantiomeric excess while maintaining controlled residence times and thermal stability (Fanelli et al., 2017).

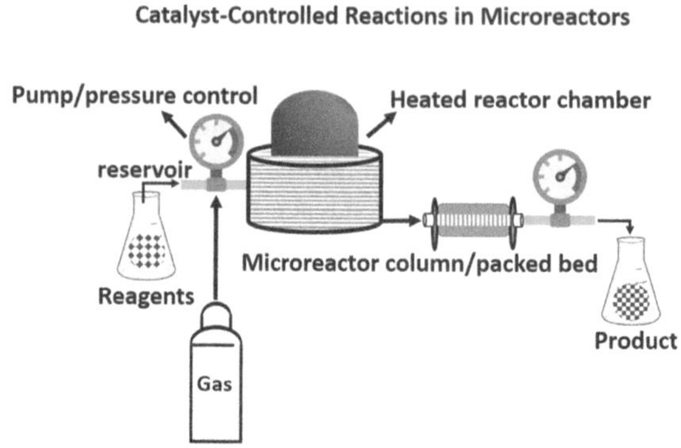

FIGURE 3.3 Schematic diagram of catalyst-controlled reactions in microreactors, showing the flow configuration where reactants and gas are introduced from reservoirs, under pump and pressure control, into a heated reactor chamber. The reaction mixture passes through a column or packed bed of the microreactor containing the catalyst, and the resulting product is collected at the outlet. This configuration allows precise control of reaction conditions, such as temperature, pressure, and flow rate, for improved catalytic performance and continuous chemical synthesis.

Typical examples include Pd-catalyzed cross-coupling reactions, such as Suzuki–Miyaura and Mizoroki–Heck reactions, where catalyst immobilization and microchannel design regulate catalyst accessibility and turnover rates, thereby improving selectivity and yield. Similarly, organocatalytic transformations performed in flow benefit from these conditions, enabling rapid condensations and asymmetric transformations with increased efficiency (Fletcher et al., 2002). Furthermore, the fixed catalyst environment minimizes leaching and enables catalyst recycling, providing both economic and environmental advantages. This precise regulation helps mitigate challenges such as heterogeneous catalyst poisoning and ensures sustainable and reproducible synthesis. In summary, catalyst-controlled reactions leverage the microreactor environment to fine-tune reaction pathways, optimize selectivity, and improve overall synthetic efficiency in continuous-flow chemistry.

Catalyst-controlled reactions in microreactors rely on an integrated, continuous-flow configuration, where liquid or gaseous reactants are supplied from reservoirs using precision pumps and pressure regulators, ensuring constant and precise flow rates. Within the system, the microreactor chamber, typically consisting of microchannels, coils, or packed beds coated with catalytic materials such as metals, oxides, or organometallic complexes, is heated to maintain an optimal temperature that enhances catalytic activity. This confined environment promotes efficient mass and heat transfer, minimizing gradients and ensuring the efficient use of catalytic sites. As reactants pass through the catalyst bed, rapid chemical transformations occur, resulting in high conversion rates, minimizing byproduct formation, and increasing selectivity compared to conventional batch reactors.

The system then includes cooling, filtration, or quenching modules to stabilize the products before collection, while pressure monitoring ensures reproducibility and safety even under elevated conditions. These microreactor-based catalytic configurations offer key advantages: continuous operation, reduced reaction times, compatibility with hazardous or sensitive chemicals, and seamless scalability from laboratory research to industrial production. Furthermore, the modularity of microreactors enables the integration of multistage catalytic processes, facilitating the optimized synthesis of fine chemicals, pharmaceuticals, and functional materials with superior efficiency and environmental sustainability.

Catalyst-controlled reactions in microreactors rely on an integrated, continuous-flow configuration, where liquid or gaseous reactants are supplied from reservoirs using precision pumps and pressure regulators, ensuring constant and precise flow rates. Within the microreactor chamber, typically designed as microchannels, helical capillaries, or packed beds immobilized with catalytic materials such as metals (Pd, Pt, Au), metal oxides (TiO_2, Fe_2O_3, MnO_2), or organometallic complexes, the system maintains a controlled temperature and pressure that enhances catalytic activity. The confined microscale environment provides superior surface-area-to-volume ratios, enabling efficient mass and heat transfer while maximizing catalyst utilization.

As the reactants pass through the catalyst bed, transformations such as hydrogenation (Pd beds for alkenes/alkynes), oxidation (MnO_2 or Pt-catalyzed alcohol oxidation), C–C coupling reactions (Suzuki–Miyaura with Pd catalysts), esterification, and photocatalysis (TiO_2-coated channels under UV/visible light) are carried out with high conversion rates and excellent selectivity. Post-reaction modules, including cooling, quenching, or online separation, stabilize the products and ensure continuous-flow safety. This microreactor-based catalytic approach offers numerous advantages: reduced reaction times, minimal byproduct formation, safer handling of hazardous intermediates, and scalability from laboratory-scale synthesis to industrial production. Its modularity also enables multistep catalytic sequences (e.g., oxidation followed by hydrogenation), streamlining the synthesis of pharmaceuticals, fine chemicals, and advanced materials. Notably, the controlled configuration improves sustainability by reducing energy demand and solvent use and enabling greener chemistry.

3.2.3 Catalyst-Specified Reactions in Microreactors

Figure 3.4 shows catalytically specified reactions in microreactors, focusing on transformations tailored to catalyst choice and design, including catalyst type, particle size, support material, and formulation. In these cases, the specific catalyst not only accelerates the reaction but also defines the mechanism, selectivity, and product distribution. For example, microreactors have been applied to hydrogenations, oxidations, and C–C coupling reactions using catalysts such as Pt, Pd, and Rh nanoparticles and zeolite-based catalysts, where the inherent chemical properties of the catalyst directly influence conversion efficiency and selectivity (Yue, 2022). Nanoparticle catalysts suspended as colloids in the microreactor flow typically exhibit improved reaction rates compared to bulk catalysts due to improved mass transfer and greater surface accessibility. Furthermore, the ability to recycle colloidal catalysts, such as Pt nanoparticles, and to be reused multiple times without significant loss of activity illustrates the feasibility of catalyst specification in continuous-flow configurations (Fang et al., 2010).

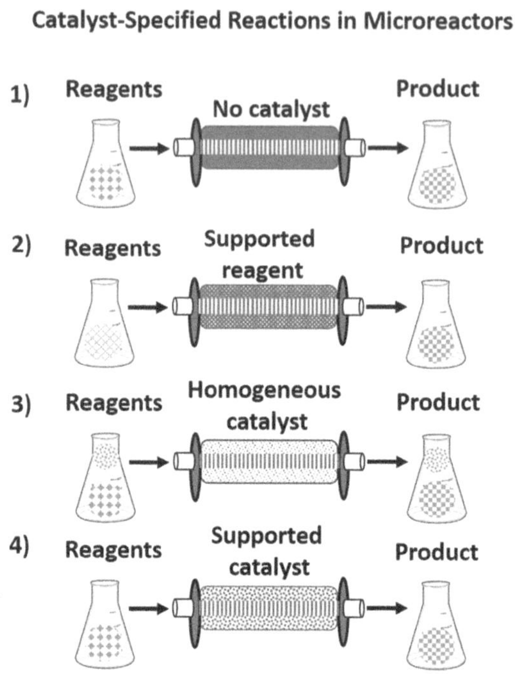

FIGURE 3.4 Schematic representation of catalyst-specified reaction modes in microreactors. From top to bottom: 1) catalyst-free reactants flow through the reactor column without catalytic assistance; 2) reactive species with supported reactants are immobilized on a solid support within the reactor; 3) homogeneous catalyst is dissolved in the reaction mixture for uniform catalytic activity; 4) solid catalyst with supported catalyst is immobilized in the reactor column, enabling heterogeneous catalysis. Each configuration influences reaction pathways and product formation, illustrating distinct strategies for integrating catalysts into flow chemistry.

This schematic illustrates four representative reaction configurations specific to catalysts in microreactors, highlighting how the presence, form, and mode of catalyst immobilization directly influence reaction pathways and efficiency. In the absence of a catalyst, reactants pass through the microreactor column without catalytic mediation, and product formation depends solely on thermal, photochemical, or pressure activation, which generally reduces selectivity and efficiency. In the supported-reactant configuration, a reactive species is immobilized on a solid matrix within the microchannel or packed bed. As they flow, the reactants undergo selective transformations mediated by the anchored reactant, enabling reuse and minimizing contamination of the product stream.

The homogeneous catalyst mode involves the direct dissolution of the catalysts into the reaction fluid, ensuring close molecular interactions between the catalyst and reactants. This configuration provides uniform catalytic activity and allows for high turnover frequencies, although separation of the catalyst from downstream products may be necessary. In contrast, the supported catalyst configuration immobilizes metallic, oxide, or organometallic catalysts on solid supports such as silica, alumina, zeolite, or carbon within the microreactor column. In this case, heterogeneous catalysis occurs when the reactants come into contact with the catalyst surface during flow, promoting robust activity, greater thermal stability, and easy catalyst recovery.

By selecting among these catalyst modes or integrating them into hybrid systems, microreactors offer tunable platforms to optimize efficiency, product selectivity, and scalability across diverse chemical transformations, from hydrogenation and oxidation to C-C bond formation and photocatalytic reactions. This versatility underpins their growing role in sustainable continuous-flow chemical manufacturing. Specialized strategies also include catalyst coatings on channel walls or packed-bed designs, designed to favor specific mechanisms or optimize activity for specific substrate classes. These catalyst-specific systems are particularly valuable for reactions requiring specific acidity, basicity, or redox behavior, such as selective oxidations or hydrocarbon functionalizations (Nieuwelink ct al., 2021). In general, catalyst-specified reactions highlight how deliberate catalyst design and selection enable efficient, selective, and reusable catalytic processes in microreactors, tailored to diverse synthetic and industrial objectives (Fang et al., 2010; Nieuwelink et al., 2021).

3.3 MICROREACTORS AND GREEN CHEMISTRY

Microreactor technology plays a pivotal role in advancing green chemistry by enabling the use of sustainable, environmentally friendly solvents that reduce environmental impact and improve reaction efficiency. Performing chemical transformations in miniaturized, continuous-flow environments minimizes solvent waste, improves safety, and enables the systematic exploration of new solvent systems for more environmentally friendly processes (Hessel et al., 2024; Porta et al., 2016).

3.3.1 Ionic Liquids in Microreactor Systems

Ionic liquids (ILs) are organic salts that remain liquid at relatively low temperatures and exhibit unique characteristics such as negligible vapor pressure, high thermal stability, and tunable polarity. Their non-volatile nature reduces the emission of hazardous volatile organic compounds (VOCs), contributing to safer and more environmentally friendly processes. In microreactors, ILs form stable droplet microenvironments that allow precise control of reaction parameters, thereby improving mixing, mass transfer, and catalytic activity (Zhang et al., 2025).

ILs accelerate reaction rates and improve selectivity under mild conditions compared to conventional solvents. Their ability to dissolve both inorganic salts and organometallic catalysts expands the synthetic scope achievable in microreactors. Furthermore, the integration of IL-based systems with digital microfluidics enables the manipulation of IL droplets for automated, high-throughput synthesis, ideal for pharmaceutical and fine chemical applications Recent developments include the coupling of ILs with nanocatalysts, which improves dispersion and reusability while reducing catalyst leaching, in line with the goals of sustainable chemistry. Ongoing research is focused on the design of biodegradable ILs with reduced toxicity, while maintaining their beneficial properties for microreactor applications (Hessel et al., 2024).

3.3.2 Water as a Reaction Medium

Water is the most abundant and environmentally benign solvent. While the solubility of hydrophobic organic compounds has historically limited their use, microreactor technology overcomes these challenges through segmented flow, micellar catalysis, and two-phase reaction systems (Porta et al., 2016). In aqueous microreactors, water's high heat capacity and polarity improve thermal management and accelerate reaction kinetics. Micellar catalysis enables hydrophobic substrates to react efficiently by encapsulating them in nanoscale micelles, reducing the need for organic solvents. This approach has been applied to enzymatic catalysis, selective oxidations, and multicomponent cascade reactions, achieving high atom economy with reduced byproduct formation (Chen et al., 2021). Fluid phase separation and integrated product extraction further reduce purification requirements and minimize solvent waste.

3.3.3 Supercritical CO_2 and Alternative Green Media

Supercritical CO_2 ($scCO_2$) combines gas-like diffusivity with liquid-like solvating power, making it a safe and clean alternative to solvents due to its non-toxicity, non-flammability, and easy removal after depressurization (Badgujar et al., 2021). Microreactors exploit the tunable properties of $scCO_2$ to improve mass transfer, selectivity, and reaction safety, particularly in oxidation, hydrogenation, and extraction processes (Hessel et al., 2005). In addition to $scCO_2$, deep eutectic solvents and bio-based solvents from renewable feedstocks are gaining importance as biodegradable and low-toxicity alternatives for microreactor applications. These solvent systems expand the range of green media compatible with continuous-flow reactors and offer viable avenues for sustainable chemical manufacturing (da Silva Jr et al., 2025).

3.3.4 Microreactors for Thermochemical Gas–Solid Reactions

Microreactors have become increasingly important tools for the measurement and analysis of gas–solid thermochemical reactions thanks to their precise control of reaction conditions and their improved heat and mass transfer characteristics. These small-scale reactors allow for detailed investigation of the kinetics, thermodynamics, and transport phenomena in gas–solid catalytic reactions, offering significantly greater safety and efficiency than conventional large-scale systems. Recent advances highlight that microreactors designed for gas–solid studies can incorporate advanced heating systems to achieve rapid temperature changes, precise regulation of gas flow and composition, and minimization of thermal gradients. Their compact geometry ensures uniform heating and efficient contact between the gas and solid phases, which is especially critical for highly exothermic or endothermic processes. Measurement techniques employed in these systems include infrared (IR) thermography for spatial temperature mapping, spectroscopic methods for reactant and product identification, and time-resolved gas chromatography or mass spectrometry for quantitative concentration profiling.

The advantages of microreactors for thermochemical gas–solid analysis include:

1. High surface-area-to-volume ratios, which improve heat transfer and minimize hot spots.
2. Transient operation capability, which enables investigation of reaction kinetics and catalyst deactivation.
3. Increased safety, especially when handling hazardous or highly reactive gases.
4. Catalyst testing under realistic thermal and flow conditions with reduced material requirements.
5. Integration with in-situ analytical tools, which enables real-time monitoring of reaction progress and catalyst status.

Innovative reactor architectures include packed-bed microreactors with integrated heating elements, microfluidic channels coated with catalytic films, and hybrid membrane-assisted designs that enable selective reaction control. These advanced platforms facilitate rapid selection of catalytic materials and precise tuning of operating parameters, thereby accelerating catalyst discovery and process intensification.

3.4 APPLICATIONS TO A BROAD RANGE OF REACTIONS

Microreactor applications cover a wide range of chemical processes, enabling improved reaction rates, selectivity, and sustainability compared to conventional batch methods. Their precise control of reaction environments makes them well suited for a variety of transformations, from oxidation and multicomponent condensations to photochemical and catalytic reactions.

3.4.1 Oxidation Reactions in Microreactors

Oxidation reactions greatly benefit from microreactor technology thanks to increased control over heat and mass transfer, which improves selectivity and safety. The high surface-area-to-volume ratio of microreactors enables rapid and uniform heat dissipation, avoiding hot spots that often lead to unwanted side reactions or decomposition. For example, the electrochemical oxidation of sulfides and thiols has been efficiently achieved in continuous-flow microreactors using water as the oxygen source, enabling environmentally friendly processes with a wide range of substrates and excellent functional group tolerance (Fanelli et al., 2017; Zhao, 2013). These reactions are scalable and allow optimization of parameters such as residence time and applied voltage to maximize yield and selectivity. Microreactors also enable safer handling of hazardous oxidants and improved mixing, which increases reaction rates and reduces waste generation. The ability to precisely control reaction conditions facilitates the selective formation of sulfoxides, sulfones, and disulfides, which are valuable intermediates in pharmaceuticals and fine chemicals (Hessel et al., 2005; da Silva et al., 2025).

Recent advances in catalyst design tailored for microreaction systems further reinforce their green potential. For example, Jeffamine®-modified phosphotungstic acid (PTA-Jeffamine®) has been explored as a liquid catalyst compatible with aqueous and continuous-flow conditions. These catalysts integrate well into microreactors, where confined channels provide improved contact between the oxidant, substrate, and catalyst, ensuring efficient turnover and easy product separation. Studies have shown that PTA-Jeffamine® not only improves the oxidation efficiency of benzyl alcohol but also facilitates catalyst recovery and reuse, making it well suited for microflow applications where sustainability and stability are crucial (Bhattacharjee et al., 2017).

Similarly, PTA-Jeffamine® derivatives have demonstrated catalytic activity in porphyrin synthesis under aqueous conditions. When applied in microreaction setups, the simultaneous promotion of condensation and oxidation steps benefits from controlled residence time and uniform mixing, overcoming equilibrium limitations that typically hinder porphyrin formation in bulk aqueous media. Importantly, the confined microreactor environment minimizes side reactions, while maintaining high singlet oxygen efficiency and product yield (Suppan et al., 2018). Thus, the integration of microreactor platforms with advanced catalysts such as Jeffamine®-modified PTA exemplifies how microreaction engineering and catalyst innovation together enable greener, safer, and more selective oxidation pathways for pharmaceutical and fine chemical synthesis.

3.4.2 Multicomponent Reactions: Biginelli Condensation and Porphyrin Synthesis

Microreactors have been widely applied in multicomponent reactions (MCRs), such as Biginelli condensation and porphyrin synthesis, where the advantages of continuous flow, precise control, and optimized mixing translate into higher yields, improved selectivity, and significantly shorter reaction times (Yue, 2022). Biginelli condensation, which combines aldehydes, urea, and β-ketoesters, particularly benefits from the microreactor environment, where reaction times are shortened from hours or even days in batches to just minutes, while maintaining high efficiency and promoting greener synthesis pathways through solvent and reagent optimization (Fanelli et al., 2017).

In addition to conventional approaches, the catalytic activity of PTA-Jeffamine® hybrids has been demonstrated in one-pot Biginelli reactions, starting directly from benzyl alcohol in aqueous media. The system, which uses hydrogen peroxide as a benign oxidant instead of nitrate salts, achieved up to 87% yields of dihydropyrimidinones (DHPMs) in 5.5 h at neutral pH, without the need for added mineral acids. The PTA-Jeffamine® catalyst not only simplified pH control but also provided a scalable and green platform for the synthesis of DHPMs with potential pharmaceutical relevance, including promising anticancer activity in selected products (Thangamani et al., 2020). When applied in microreactors, these hybrid catalysts can further exploit the advantages of continuous flow, accelerating the condensation and oxidation stages while ensuring uniform mixing and precise residence time control.

Similarly, porphyrin synthesis, often hampered in bulk aqueous systems by low yields and long reaction times, is enhanced in microreactors through efficient heat and mass transfer. PTA-Jeffamine® derivatives have proven particularly effective in catalyzing the condensation and oxidation steps required for porphyrin macrocycle assembly. Incorporating these catalysts into microreaction setups alleviates equilibrium constraints, suppresses side reactions, and provides porphyrins with high singlet oxygen generation efficiencies under sustainable aqueous conditions (Suppan et al., 2018). Taken together, the integration of advanced hybrid catalysts such as PTA-Jeffamine® with microreactor technology demonstrates how catalyst innovation and flow engineering converge to enable greener, safer, and more selective routes for pharmaceutically important scaffolds such as DHPMs and porphyrins.

Table 3.1 illustrates the broad applicability of microreactor technology in diverse organic transformations, such as oxidation, multicomponent, photochemical, and organometallic reactions. By integrating precisely designed catalysts, controlled residence times, and optimized solvent systems, microreactors offer significant improvements in reaction efficiency, selectivity, and product yield compared to conventional batch methods. Notably, reactions such as the Biginelli condensation and porphyrin synthesis benefit from improved mixing, heat transfer, and the minimization of side reactions, highlighting the importance of microreaction environments in enabling sustainable and scalable processes. Overall, these examples underscore the potential of microreactor platforms to optimize complex syntheses while reducing waste, energy consumption, and environmental impact.

3.4.3 Photocatalysis in Microreactors

Microreactor photocatalysis offers significant advantages over conventional batch systems thanks to improved light penetration, efficient heat dissipation, and rapid mixing in microchannels. Small reaction volumes ensure uniform photon distribution, which improves reaction selectivity and minimizes byproduct formation. Microreactor platforms also allow for safer handling of reactants or photosensitive materials and enable continuous-flow operation for scalable synthesis (Fanelli et al., 2017; Yue, 2022).

Ultrasound-assisted microreactors have further improved photocatalytic performance by promoting mass transfer, reducing phase boundary limitations, and preventing channel clogging. The combination of sonochemistry and microreaction technology improves catalyst dispersion, reaction kinetics, and overall process efficiency (Trinh et al., 2025). Microreactors are widely employed for photochemical oxidation, selective transformations, and environmental remediation reactions, often using immobilized photocatalysts such as TiO_2 or ZnO under LED or UV irradiation. These systems promote greener and more sustainable synthetic pathways by reducing solvent usage, energy consumption, and reaction times (Das et al., 2016).

3.4.4 Catalytic Transformations and Continuous Catalysis

Continuous-flow microreactors offer significant advantages for catalytic transformations thanks to their high surface-area-to-volume ratio, improved heat and mass transfer, and precise control of reaction parameters such as temperature, pressure, and

TABLE 3.1
Comparative Overview of Microreactor-Mediated Reactions: Catalysts, Reaction Conditions, Solvents, and Observed Outcomes

S. No	Reaction Type	Catalyst/ Catalyst System	Microreactor Setup/ Conditions	Solvent/ Medium	Key Outcomes/ Observations	Ref.
	Oxidation of sulfides and thiols	Electrochemical cell	Continuous-flow microreactor, controlled voltage	Water	High selectivity, broad substrate scope, scalable	Fanelli et al. (2017)
	Oxidation of benzyl alcohol	PTA-Jeffamine®	Microreaction, H_2O_2 oxidation, room temperature	Water	Enhanced conversion to benzaldehyde, easy catalyst separation	Bhattacharjee et al. (2017)
	Biginelli condensation	PTA-Jeffamine®	One-pot, one-step microreactor, 5.5 h, pH 7.5	Water	Yield 87%, no additional acid, scalable	Suppan et al. (2020)
	Porphyrin (H_2TPP) synthesis	PTA-Jeffamine®	Continuous-flow microreactor	Water	Simultaneous condensation and oxidation, high singlet oxygen efficiency	Suppan et al. (2018)
	Electrochemical sulfoxidation	Electrochemical cell	Microreactor with segmented flow	Water	Improved selectivity, safer handling	Hessel et al. (2024)
	Oxidation of sulfides	Pd/C nanoparticles	Flow microreactor	Water/ ethanol	Faster conversion rates, high yield	Yue (2022)
	Aldol Reaction	Organocatalyst	Continuous-flow microreactor, room temperature	Organic solvent	Increased rate vs batch, improved selectivity	Wiles et al. (2001)

residence time. These characteristics enable faster reactions, greater selectivity, and improved safety compared to traditional batch processes, promoting more efficient catalysis and process intensification (Fanelli et al., 2017).

Heterogeneous catalysis in microreactors has advanced with various catalyst immobilization strategies, including packed-bed, monolithic, and coated-wall systems. Immobilized catalysts improve contact between reactants and active sites, enable continuous operation, and facilitate catalyst recovery and reuse. Continuous-flow operation also minimizes catalyst degradation and waste generation, which favors more sustainable chemical synthesis for pharmaceuticals, fine chemicals, and materials (Porta et al., 2016).

Homogeneous catalysis has been demonstrated under various conditions, including elevated temperatures and pressures, particularly for organometallic and acid-/base-catalyzed reactions. Microreactors enable rapid screening, rapid optimization of catalyst systems and reaction conditions, and sequential multistep synthesis without intermediate isolation. Continuous catalytic processes, such as oxidation, amination, and hydrogenation in flow, show remarkable intensification, increased productivity, and reproducibility (Fang et al., 2010).

Key examples and enabling technologies include:

1. Flow screening systems: Integrating injection into autosamplers with capillary reactors allows multiple catalytic conditions to be analyzed sequentially with minimal reagent consumption and high reproducibility.
2. Supported catalysts: Immobilized catalysts in packed beds or monolithic reactors efficiently catalyze gas–liquid and liquid–liquid reactions, exhibiting improved conversion rates and selectivities.
3. Acoustic or ultrasonic irradiation: This method reduces solids formation and clogging, extending reactor lifetime for heterogeneous catalysis.
4. Advanced catalyst design: Stable immobilization of nanoparticles and functional coatings improves catalytic activity and expands the diversity of achievable transformations.

3.4.5 Role of Microreactors in Green Chemistry and Catalytic Applications

Figure 3.5 illustrates the integration of continuous-flow microreactor technology with the principles of green chemistry, highlighting its versatility in diverse fields of science and chemical engineering. The diagram shows how reactants are introduced into a continuous-flow microreactor, where they undergo controlled transformations under precisely controlled conditions of temperature, pressure, and catalyst presence. The result is an optimized, efficient, and scalable process that delivers high-quality products with less waste and energy consumption compared to traditional batch synthesis. Essentially, microreactors serve as crucial platforms for catalysis studies and catalyst development, offering superior control over reaction kinetics and mass transfer. Their microstructured environments allow for the immobilization of catalysts on supports such as zeolites, silica, polymers, or carbon frameworks, minimizing leaching and enabling recyclability. This makes them excellent tools for studying heterogeneous, homogeneous, and photocatalytic reactions.

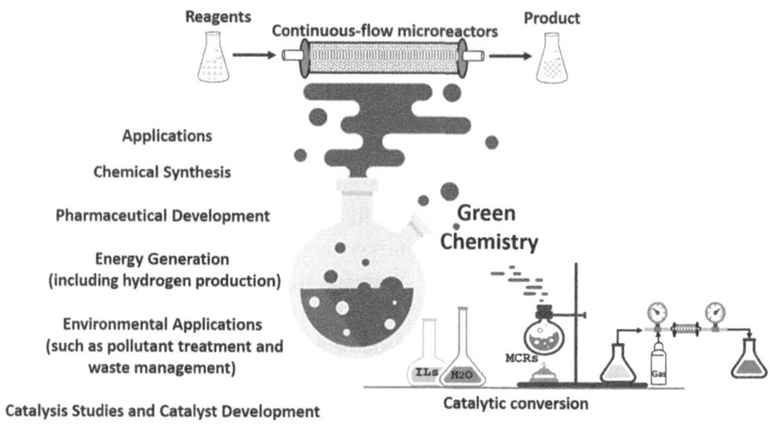

FIGURE 3.5 Schematic illustration of the integration of microreactors with green chemistry principles and their wide range of applications. The pyramid highlights application areas, from fundamental catalyst studies and environmental remediation to pharmaceutical development, energy generation, and advanced chemical synthesis. The core concept emphasizes the use of green solvents (e.g., ionic liquids, water) and continuous-flow microreactor systems, which improve reaction efficiency, selectivity, and sustainability. Together, these characteristics position microreactors as versatile platforms for advancing green chemistry across multiple sectors.

Using the bottom-up approach, the figure links the use of microreactors to environmental applications, such as pollutant degradation, water treatment, and sustainable waste management. By employing environmentally friendly solvents such as ILs, water, or $scCO_2$, microreactors promote green chemistry while reducing dependence on toxic organic solvents. Similarly, in energy generation, they have shown promise in processes such as hydrogen production, CO_2 reduction, and biofuel enhancement, where precise control of reaction pathways is crucial.

In the pharmaceutical sector, microreactors facilitate rapid drug discovery and development by enabling safer handling of hazardous intermediates, miniaturized scale-up strategies, and the reproducible synthesis of APIs. Their compatibility with MCRs, such as the Biginelli condensation or porphyrin synthesis, further strengthens their role in synthetic innovation. At the top of the pyramid are applications in large-scale chemical synthesis, where microreactors ensure safer, cleaner, and more energy-efficient synthesis routes. In short, this figure underscores how microreactors connect green chemistry with catalytic conversion, driving innovation across multiple sectors through sustainable and scalable approaches.

Overall, continuous catalysis in microreactors provides a robust platform for rapid, safe, and scalable chemical synthesis, accelerating innovations in catalyst development and green chemistry. Microreactors constitute an ideal platform for photochemical transformations thanks to their small reaction volumes, which ensure efficient light penetration and uniform photon distribution. Continuous-flow photoreactors, often equipped with LED light sources, allow precise control of irradiation intensity and residence time, resulting in higher selectivity, faster reactions, and minimal byproduct formation. Reactions such as the photocatalytic oxidation of thiols to disulfides and the selective oxidation of benzyl alcohol benefit significantly from microreactor platforms, especially in green solvents such as water or ethanol (Yue, 2022).

3.5 OUTLOOK AND CONCLUSION

3.5.1 Outlook

The integration of catalysis, photochemistry, and multicomponent reactions in microreactors has opened new horizons for process intensification and green chemistry. Microreactor systems continue to evolve thanks to advances in materials, reactor design, and technologies such as ultrasound and light sources. The trend is toward modular, automated, and scalable microreactor platforms that connect laboratory-scale discoveries to industrial applications.

Future research is expected to focus on:

1. Hybrid catalytic systems, combining homogeneous and heterogeneous strategies to leverage complementary advantages.
2. Immobilized nanocatalysts and photocatalysts, ensuring high activity, reusability, and environmentally friendly operation.
3. Advanced photoreactor designs, including LED arrays, fiber optics, and microstructured materials, for uniform photon distribution.
4. Process–reaction integration, enabling multistage synthesis in a single flow system, such as tandem Biginelli condensation–oxidation cascades or porphyrin synthesis.
5. AI-based digitalization and optimization, where machine learning algorithms predict optimal parameters for catalytic and photochemical reactions in real time.
6. Sustainable chemistry objectives, with an emphasis on solvent minimization, the use of environmentally friendly oxidants such as hydrogen peroxide, and the design of recyclable catalysts.

These directions will not only improve the efficiency and sustainability of microreactor processes, but also accelerate their adoption in pharmaceutical, fine chemical, and material manufacturing.

3.5.2 Conclusion

Microreactors offer a transformative platform for chemical synthesis, integrating the principles of catalysis, photochemistry, and MCRs under continuous-flow conditions. Their inherent advantages, such as enhanced heat and mass transfer, precise parameter control, high surface-area-to-volume ratios, and safe operation, enable faster, cleaner, and more selective reactions compared to traditional batch systems. Heterogeneous and homogeneous catalysis in microreactors have demonstrated remarkable potential for sustainable synthesis, while photochemical processes benefit from uniform light distribution and faster reaction rates. Ultrasound-assisted methods further extend reactor lifetimes and expand the scope of accessible transformations. MCRs such as the Biginelli condensation and porphyrin synthesis exemplify how microreactors can shorten reaction times, improve yields, and align with the principles of green chemistry.

Overall, the integration of microreactor technology into catalysis and photochemistry represents not only a scientific breakthrough, but also a step toward industrial sustainability and scalable innovation. By combining continuous-flow methodologies with emerging catalyst designs and digital optimization, microreactors are poised to play a pivotal role in the future of chemical manufacturing.

REFERENCES

Abidi, A., Ahmadi, A., Enayati, M., et al. 2021. A Review of the Methods of Modeling Multi-Phase Flows within Different Microchannels Shapes and Their Applications. *Micromachines* 12(9), 1113. https://doi.org/10.3390/mi12091113

Adamo, A., Beingessner, R.L., Behnam, M., Chen, J. et al. 2016. On-Demand Continuous-Flow Production of Pharmaceuticals in a Compact Reconfigurable System. *Science* 352 (6281): 61–67.

Ahmed-Omer, B., Brandt, J. C., and Wirth, T. 2007. Advanced Organic Synthesis Using Microreactor Technology. *Organic & Biomolecular Chemistry* 5: 733–740. https://doi.org/10.1039/B615072A

Badgujar, K. C., Dange, R., and Bhanage, B. M. 2021. Recent Advances of Use of Supercritical Carbon Dioxide for the Biomass Pre-Treatment and Extraction: A Mini-Review. *Journal of the Indian Chemical Society* 98 (1): 100018. https://doi.org/10.1016/j.jics.2021.100018

Balakrishnan, D., El Maiss, J., Olthuis, W., and Garcia, C. P. 2023. Miniaturized Control of Acidity in Multiplexed Microreactors. *ACS Omega* 8: 7587–7594. https://doi.org/10.1021/acsomega.2c06897

Bhattacharjee, R. R., Suppan, T., and Mal, S. S. 2017. A Liquid Derivative of Phosphotungstic Acid as Catalyst for Benzyl Alcohol Oxidation in Water: Facile Separation and Stability of Benzaldehyde at Room Temperature. *ChemistrySelect* 2 (16): 4368–4375. https://doi.org/10.1002/slct.201700443

Bogdan, A. R., Poe, S. L., Kubis, D. C., Broadwater, S. J., and McQuade, D. T. 2009. The Continuous-Flow Synthesis of Ibuprofen. *Angewandte Chemie International Edition* 48 (45): 8547–8550. https://doi.org/10.1002/anie.200903055

Burange, A., Osman, S., Luque, R. 2022. Understanding Flow Chemistry for the Production of Active Pharmaceutical Ingredients. *iScience* 25: 103892. https://doi.org/10.1016/j.isci.2022.103892.

Chen, J., Zhu, M., Xiang, F. et al. 2021. Research Progress on Microreactor Technology in Oxidation Reactions. *Current Organic Chemistry* 25 (10): 1235–1245. https://doi.org/10.2174/1385272825666210319092545

Das, S., and V. Srivastava. 2016. Microfluidic-Based Photocatalytic Microreactor for Environmental Application: A Review of Fabrication Substrates and Techniques, and Operating Parameters. *Photochemical & Photobiological Sciences* 15. https://doi.org/10.1039/C5PP00469A

Ehrfeld, W. Hessel, V. and Lowe, H. 2000. *Microreactors: New Technology for Modern Chemistry*. Weinheim: Wiley-VCH.

Ekinci, F., Guzel, M. S., Acici, K., and Asuroglu, T. 2024. The Future of Microreactors: Technological Advantages, Economic Challenges, and Innovative Licensing Solutions with Blockchain. *Applied Sciences*, 14(15), 6673. https://doi.org/10.3390/app14156673

Fanelli, F., Parisi, G., Degennaro, L., and Luisi, R. 2017. Contribution of Microreactor Technology and Flow Chemistry to the Development of Green and Sustainable Synthesis. *Beilstein Journal of Organic Chemistry* 13: 520–542. https://doi.org/10.3762/bjoc.13.51

Fang, H., Xiao, Q., Wu, F., Floreancig, P. E., and Weber, S. G. 2010. Rapid Catalyst Screening by a Continuous-Flow Microreactor Interfaced with Ultra-High-Pressure Liquid Chromatography. *Journal of Organic Chemistry* 75 (16): 5619–5626. https://doi.org/10.1021/jo100981e

Feng, H., Y. Zhang., J. Liu., and D. Liu. 2022. Towards Heterogeneous Catalysis: A Review on Recent Advances of Depositing Nanocatalysts in Continuous-Flow Microreactors. *Molecules* 27 (22): 8052. https://doi.org/10.3390/molecules27228052

Fletcher, P. D., Haswell, I., Pombo-Villar, S. J., et al. 2002. Micro Reactors: Principles and Applications in Organic Synthesis. *Tetrahedron* 58 (24): 4735–4757. https://doi.org/10.1016/S0040-4020(02)00432-5

Geyer, K., J. D. C. Codee, and P. H. Seeberger. 2006. Microreactors as Tools for Synthetic Chemists-the Chemists' Round-Bottomed Flask of the 21st Century?. *Chemistry – A European Journal* 12 (33): 8434–8442. https://doi.org/10.1002/chem.200600596

Henke, L., and Winterbauer, H. 2005. A Modular Micro Reactor for Mixed Acid Nitration. *Chemical Engineering & Technology* 28: 749–752. https://doi.org/10.1002/ceat.200500096

Hessel, V., Angeli, P., Gavriilidis, A., and Lowe, H. 2005. Gas-Liquid and Gas–Liquid–Solid Microstructured Reactors: Contacting Principles and Applications. *Industrial & Engineering Chemistry Research* 44 (25): 9750–9769. https://doi.org/10.1021/ie0503139

Hessel, V., and Lowe, H. 2005. Organic Synthesis with Microstructured Reactors. *Chemical Engineering & Technology* 28 (3): 267–284. https://doi.org/10.1002/ceat.200407124

Hessel, V., Mukherjee, S., Mitra, S., Goswami, A., Tran, N. N., Ferlin, F., Vaccaro, L., Galogahi, F. M., Nguyen, N.-T., & Escribà-Gelonch, M. 2024. Sustainability of Flow Chemistry and Microreaction Technology. *Green Chemistry* 26: 9503–9528. https://doi.org/10.1039/D4GC01882F

Da Silva Jr., J. L., Wang, H., and Oliveira, F. A. 2025. Microreactor Technology Applied to Catalytic Processing of Hydrogen-Based Reactions. *Journal of Industrial and Engineering Chemistry* 143: 65–84. https://doi.org/10.1016/j.cej.2022.138270

Nieuwelink, A.E., Vollenbroek, J., Ferreira de Abreu, C., et al. 2021. Single Catalyst Particle Diagnostics in a Microreactor for Performing Multiphase Hydrogenation Reactions. *Faraday Discussions* 229: 267–280. https://doi.org/10.1039/D0FD00006J

Noel, T., Cao, Y., and Laudadio, G. 2019. Flow Reactors in Electrochemistry: Fundamentals and Applications. *Accounts of Chemical Research* 52 (11): 2858–2869. https://doi.org/10.1021/acs.accounts.9b00248

Plutschack, M. B., Pieber, B., Gilmore, K. and Seeberger, P. H. 2017. The Hitchhiker's Guide to Flow Chemistry. *Chemical Reviews* 117 (18): 11796–11893. https://doi.org/10.1021/acs.chemrev.7b00183

Porta, R., Benaglia, M., and Puglisi, A. 2016. Flow Chemistry: Recent Developments in the Synthesis of Pharmaceutical Products. *Organic Process Research & Development* 20: 2–25. https://doi.org/10.1021/acs.oprd.5b00325

Richardson, P., and Kolmel, D. K. 2021. Safer Handling of Diazomethane through Continuous Generation and Consumption. *Synfacts* 17 (4): 472. https://doi.org/10.1055/s-0040-1720336

Suppan, T., Kunjunni, M. K., Barik, A., and Bhattacharjee, R. R. 2018. Effect of Jeffamine®-Modified Phosphotungstic Acid on Porphyrin Synthesis in Water. *ChemistrySelect* 3 (4): 1275–1281. https://doi.org/10.1002/slct.201702845

Suppan, T., Mahendran, H. P., Jeyaraj, S., Mohanta, K., and Bhattacharjee, R. R. 2020. Phosphotungstic acid–Jeffamine® hybrid catalyst for one-pot Biginelli reaction starting from benzyl alcohol. *Applied Catalysis A: General* 603: 117734. https://doi.org/10.1016/j.apcata.2020.117734

Tanimu, A., Jaenicke, S., and Alhooshani, K. 2017. Heterogeneous Catalysis in Continuous Flow Microreactors: A Review of Methods and Applications. *Chemical Engineering Journal* 327: 792–821. https://doi.org/10.1016/j.cej.2017.06.161

Thangamani, S., Mahendran, H. P., Jeyaraj, S., Mohanta, K., and Bhattacharjee, R. 2020. Phosphotungstic Acid–Jeffamine® Hybrid Catalyst for One-Pot Biginelli Reaction Starting from Benzyl Alcohol. *Applied Catalysis A: General* 603: 117734. https://doi.org/10.1016/j.apcata.2020.117734

Trinh, Q. T., Golio, N., Cheng, Y. et al. 2025. Sonochemistry and Sonocatalysis: Current Progress, Existing Limitations, and Future Opportunities in Green and Sustainable Chemistry. *Green Chemistry* 27: 4926–4958. https://doi.org/10.1039/D5GC01098E

Tsubogo, T., Oyamada, H., and Kobayashi, S. 2015. Multistep Continuous-Flow Synthesis of (R)- and (S)-Rolipram Using Heterogeneous Catalysts. *Nature* 520: 329–332. https://doi.org/10.1038/nature14343

Vijaya Bhaskar, K., Rashed, M., Bhat, K. S., et al. 2024. A Continuous Flow Microreactor as a Practical Tool for the Synthesis of Active Pharmaceutical Ingredients. *Process Safety and Environmental Protection* 192: 437–449. https://doi.org/10.1016/j.psep.2024.09.122

Watts, P., and Wiles, C. 2013. Application of Microreactor Methodology for Organic Synthesis: General Methods and Strategies. *Wiley-VCH* https://doi.org/10.1002/9781118596784.ssd006

Wiles, C., Watts, P., Haswell, S. J., and Pombo-Villar, E. 2001. The Aldol Reaction of Silyl Enol Ethers within a Microreactor. *Lab on a Chip* 1: 100–101. https://doi.org/10.1039/B107861E

Willms, T., Kryk, H., and Hampel. U. 2020. Microreactor Studies for Efficient Organic Oxidation Processes. *Catalysis Today* 346: 3–9. https://doi.org/10.1016/j.cattod.2019.05.008

Yoshida, J. 2010. Flash Chemistry: Flow Microreactor Synthesis Based on High-Resolution Reaction Time Control. *Chemical Records* 10 (5): 332–341.

Yue, J. 2022. Green Process Intensification Using Microreactor Technology for the Synthesis of Biobased Chemicals and Fuels. *Chemical Engineering and Processing -Process Intensification* 177: 109002. https://doi.org/10.1016/j.cep.2022.109002

Zhang, Y., He, N., Jiang, Y., Shi, A., Faria, J. A., Jiao, B., and Wang, Q. 2025. Anisotropy in Pickering Emulsions Enables Catalysis at the Interface. *ChemRxiv.* https://doi.org/10.26434/chemrxiv-2025-vt8rq

Zhao, C. X. 2013. Multiphase Flow Microfluidics for the Production of Single or Multiple Emulsions for Drug Delivery. *Advanced Drug Delivery Reviews* 65 (11–12): 1420–1446. https://doi.org/10.1016/j.addr.2013.05.009

Zong, J., and Yue, J. 2022. Continuous Solid Particle Flow in Microreactors for Efficient Chemical Conversion. *Industrial & Engineering Chemistry Research* 61: 6269–6291. https://doi.org/10.1021/acs.iecr.2c00473

4 Polymer-Based Material Synthesis Using Microreactors

Avik Mukherjee, Papita Das, and Saswata Bose

4.1 INTRODUCTION

Advancements in science and technology have prompted large-scale developments in the design of microreactors (<1 mm). A microreactor, also sometimes referred to as a microchannel reactor, is generally considered to be a device designed meticulously to execute chemical reactions within a confined zone, which is typically less than 1 mm in length (Yoshida et al., 2013b; Baraldi & Hessel, 2012). The most widespread structure of a microreactor is the microchannel model. In 1959, microreactors were developed by integrating the concepts of micromachining and microsized relays. Generally, microreactors are operated in continuous-flow mode. Unlike batch-mode conventional reactors, microreactors offer better control over the reaction mechanism, rate, and quality of products. Other advantages include energy-efficient heat transfer, high yield of products, safety measures, scalability, and many more. Owing to their versatility and numerous advantages, microreactors have been exploited in a range of applications. Microprocess engineering is a novel arena where microreactors have also been implemented. Over the past years, the technology concerning microreactors has undergone various state-of-the-art standardizations, in particular for batch-mode operations. The microreactor design consisting of microdimensional capillaries is considered to be the best operational one. It consists of a network of microchannels delivering to the substrate. The technology of microreactors has been scaled up beyond the domain of micromachining (Baraldi & Hessel, 2012; Becker & Gärtner, 2001; Haswell et al., 2001). Various chemical reactions involving complexities have been carried out within microreactors, which encompass critical side reactions and reaction mechanisms. Such reactions include esterification, chemical hydrogenation, and free radical reactions. The microreactor technology has been explored extensively in analytical chemistry and produced better results in terms of yield, visibility of biological cells and amino acids, along with enhanced transport properties. This fact is attributed to the inception of the microreactor technology in pharmaceutical applications (Fletcher et al., 2002a; Pennemann & Kolb, 2016a; Plouffe et al., 2014; Rizkin et al., 2019). The prevailing types of microreactors are difficult to be categorized in terms of their selectivity and overall performance (Gavriilidis et al., 2002; Haswell & Watts, 2003). Based on the parameters such

DOI: 10.1201/9781003633761-4

as dimensional size, length, and resolutions, microreactors could be categorically designed. The popularity and demand for microreactors have been ever increasing, which has led to various fabrication-oriented requirements and adoption of more and more micromachining techniques (Kundu et al., 2008). To bridge the gap between academia and industries from a research viewpoint, microreactor technology and design articulations require to integrate different operational strategies in order to meet challenges (Chierichetti et al., 2014; He et al., 2020; Plouffe et al., 2016; Watts, 2004; Wohlgemuth et al., 2015). The research has dealt with various attributes of microreactors and kinetic study of different chemical reactions. Various factors have been reported to influence the functioning of microreactors, including functioning mode, construction specifications, etc. An article was published detailing the principal mathematical concepts associated with microreactors (X. Wang et al., 2001).

4.2 MICROREACTOR: CONSTRUCTIONAL DESIGN

In general, most of the well-known microdimensional fluidic fuel cells have been constructed by conventional methods used in physical science. Anatomically, such structures are featured with microchannels, a pair of electrodes, and a set-up meant for handling liquid-compact systems (Kjeang et al., 2006, 2009; Zebda et al., 2011). The conventional techniques associated with the designing of microchannels have been lithography (disbursal and soft) and rapid prototyping (Duffy et al., 1998; Kjeang et al., 2009; Love et al., 2001; Xia & Whitesides, 1998). Polydimethylsiloxane (PDMS) has been used to fabricate microfluidic electric cells featuring parallel flowing systems (Kjeang et al., 2006, 2009; Zebda et al., 2011). In the literature, various methods have been mentioned involving channel designing (Lee et al., 2003). An industrial lithography-based technique of channel construction involves a pretreatment over a substrate, followed by the application of a coating. Spin coating has also been employed to construct microchannel layers. Photoresistive materials could be selected for spin-coating operations. In order to result in a stable photoresistive mechanism, the substrate is subjected to mild heating following the coating process. Precisional accuracy is maintained in the design of microchannels by employing CAD software. The resolution of the construction is also controlled minutely. Some conventional fuel cells and enzyme-based biofuel cell processes/reactions have been found to be significantly well conducted in microreactors containing microfluidic systems. The advancements in the microreactor design and fabrication have been considered to be very critical in terms of the global economy. Analysis of different chemical processes and the related simulation are considered as two key indices related to fabrication of microreactors. Microreactors have contributed to industrial growth all across by facilitating the generation of fascinating materials possessing enhanced properties at cheaper rates. Normally, microreactors are made of materials such as ceramics (including silicon, glass), metals, and alloys (such as steel) in conjunction with various polymers and copolymers. There could be various designs applicable for microreactors. Figure 4.1 describes the different categories of microreactors based on applications.

Such designs include capillary forms and chip forms. Capillary form microreactors are specifically designed depending on the properties of the chemical (fluid)

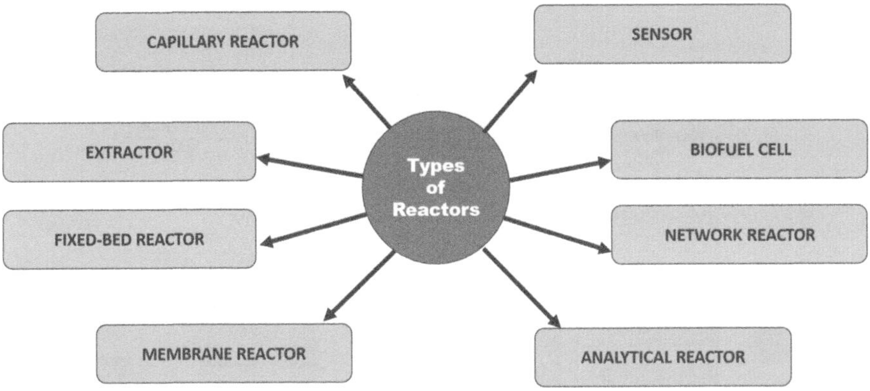

FIGURE 4.1 Different types of microreactors applied in various arenas.

involved in the reaction. On the other hand, microreactors with chip form structures are usually crafted via operations including microetching, micromachining, and various lithographic methods. Plastics, ceramic, or metals are generally used for such fabrications (Krishna et al., 2013; Lee et al., 2003; Pennemann & Kolb, 2016b; Phillips et al., 2014). Microreactors featuring specifically designed microcapillaries have been employed to chemically manufacture 1,3-butadiene from ethanol, characterized with yields at higher conversion rates (Jolliffe & Gerogiorgis, 2015; Rossetti & Compagnoni, 2016). Microreactors have been reported to offer better mass transfer and enhanced momentum transfer between molecules with better mass balance and are perfect for elevating reaction rates. Microreactors have also been found to be highly effective in methanol synthesis. According to a report, a copper-based catalytic component had been used within a fixed-bed microreactor (Lo & Wu, 2019).

4.3 MICROREACTOR FABRICATION TECHNIQUES

In order to cater to the demands to manufacture chemical products at high yielding levels, numerous methods to fabricate and design microreactors have been developed, which increase the chemical industry's economic impact. Microreactors are generally structured by means of numerous techniques, including lithography, etching (wet) molding, and micromachining (Acikgoz et al., 2011; Jena & Yue, 2012; Konstantinou et al., 2016). Micromachining, lithography, electroplating, molding (such as lithographic, galvanoforming, and abforming), and etching have been the most extensively exploited ones for the design of microreactors. Figure 4.2 portrays different materials used in the construction of microreactors along with the various methods of fabrication.

4.3.1 Etching Techniques

The process of etching typically entails corroding materials from a workpiece known as a substrate. Wet etching and dry etching are its two primary categories. Glasswork creations and the crafting of microchannels intended for laminar flow systems are

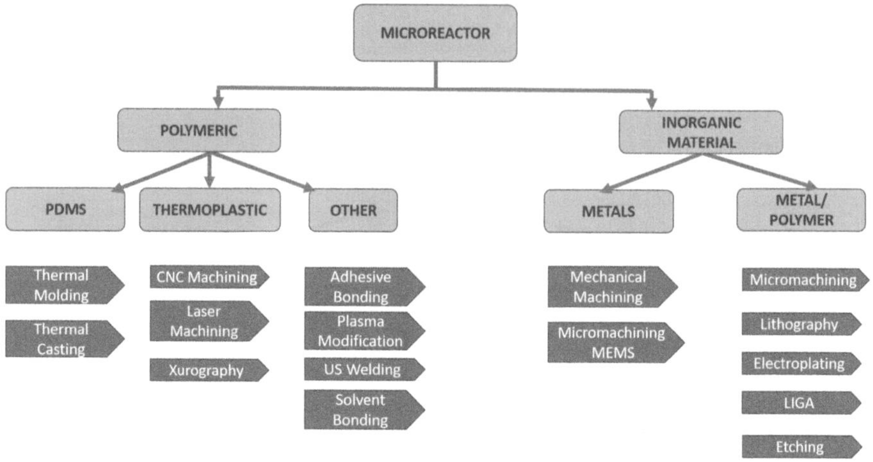

FIGURE 4.2 Various materials used in structuring of microreactors.

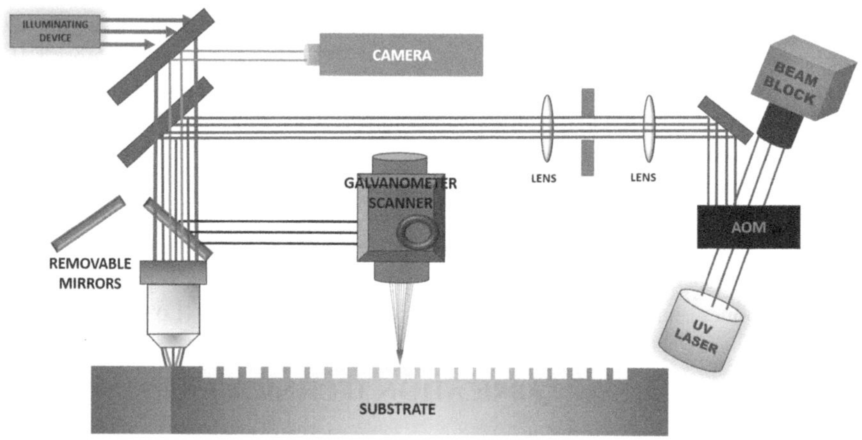

FIGURE 4.3 Schematic diagram of preparation of substrate surface by laser-based micro-etching process.

two main applications for wet etching, and the latter is often carried out experimentally in a lab. Since dry etching necessitates a voluminous amount of capital and is usually used in large-scale industrial production, high-performance processes are required to be developed (Pattekar & Kothare, 2004). Etching of metal substrates is carried out by using a photosensor. In order to facilitate the etching process, the photosensor is utilized in conjunction with a mask, which has a molecular layer that guides the photosensor to the light source. Etching applications employ an extensive use of photosensors. Figure 4.3 shows the schematic diagram of the construction of the surface of a substrate by laser-based microetching technique. A material that is to be exposed to light is segmented into two portions. The part exposed to the light is subjected to polymerization reactions, while the unexposed part is immediately dissolved in a chemical solution. This distinguishes the material into two segments, the one polymerized being insoluble, and the other being soluble (Becker & Gärtner, 2001). Since the abundance ratio is an important parameter in wet etching processes, the highest value is considered to be in the range of 0.5–0.6.

Dry etching processes, on the other hand, may be utilized to produce a wide variety of geometric structures in various sizes and ratios. The liquid medium, which is generally employed in a majority of etching applications, remains immersed inside a large etchant tub that needs to be properly managed. Most often, buffered acid is used as the etchant. Materials that are semiconducting can undergo silicon etching (Plouffe et al., 2016). For solid materials subjected to an extreme chemical reaction, wet etching is employed as an etching technique. Either the substrate is submerged in the reactant or the latter is sprayed onto the former during these procedures. It is worthwhile to mention that wet etching procedures are essentially similar, regardless of the orientation of the crystal (Hessel, 2005; Patwardhan & Joshi, 1998). The strong dependence of the print rate on the incised material is implied by high selectivity attributes. The crystalline structure is imprinted by maintaining an attacked semiconducting material, such as silicon, inside the surface's locality ("Contents List," 2019; Rizkin et al., 2019; Yoon et al., 2008; Yoshida et al., 2013a). Furthermore, an isotropic mode etching occurs when plasma irradiation etches the substrate in all directions, while when the etching occurs perpendicularly in a single direction, an anisotropic printing or etching is said to occur. Thierry's low-pressure plasma systems have been achieved by means of isotropic and anisotropic etchings (Cross, 2006; Urban et al., 2006; X. Wang et al., 2001; Waterkamp et al., 2007). The etching rate of a substrate is governed by the selectivity of the etching process. When applied to wet etching, in most occasions selectivity has been found to differ between the photosensor and the substance or materials (Cheemalapati et al., 2014; Miyawaki et al., 2012; S. Yang et al., 2014). Etching rates have also been observed to vary directionally, i.e., different rates along horizontal and vertical etching directions respectively. On the other hand, dry etching, also known as plasma etching, involves etching with plasma at a particular location, which may essentially be via anisotropic etching.

4.3.2 Micromachining Techniques

From microfluidic devices to different kinds of microreactors, micromachining has numerous units of operation across various industrial sectors that are involved in the development, use, and manufacturing of small-scale microelectronic systems, including microreactors.

Such technologies are generically referred to as micro-electro-mechanical systems (MEMSs). The concerned technology is capable of producing a wide range of microreactors designed for carrying out synthetic processes related to the manufacture of different organic and inorganic compounds ("10.9761/Jasss.129," 2000; Guckenberger et al., 2015; Roberge, 2012). The process of micromachining deals with physical drilling of a material, usually referred to as the substrate, by means of an electronic beam (Pileni, 1993; Plouffe et al., 2016). Micromachining technique is successfully employed in the industrial manufacturing of microreactors. This is due to the fact that the formation of small channels and structural layers in the microchannel is crucial as it makes the microreactor efficient and suitable to conduct multifunctional analyses in all relevant fields. The semiconductor industry is the finest application field to implement micromachining since it can aid in the mass manufacture of microreactors (He et al., 2020; Zeng et al., 2012; B. Zhang & Chen, 2019). Furthermore, micromachining, which involves numerous grinding processes to design particular shapes and forms that require multiple edges and particular geometries, is a helpful strategy for the manufacturing of small-scale reactors. The fabrication of a reactor can occasionally be facilitated by the use of less elastic or flexible materials. When compared to alternative strategies for producing microreactors, micromachining technology is more affordable to operate. It is a well-known fact that exposure to high temperatures alters a material's integrity. Such operations aid melting, evaporating, or decomposing material to transform it into a different phase for safe processing. These may be achieved via either mechanical or optical device machining (Chierichetti et al., 2014; Y. Li et al., 2015; Muhler et al., 1987).

Figure 4.4 describes the schematic view of a traditional micromachining technique and subsequent channel construction in a microreactor. As per the literature, copper materials can be mechanically machined to produce beneficial surface structure layers for good surface structure (Fletcher et al., 2002b; Friend & Yeo, 2010). Parallelly, radiations are also excellent stimuli to design materials for the construction of microreactors. One such effective mode has been lasers. Apart from radiations, there have been other sources to create microreactor designs. Certain instruments are used for cutting hard metallic pieces, depending on the ductility of the metals or materials (such as Cu and Ni). Micromachining processes also utilize drills to cut and craft materials or compounds with high hardness (Kolb, 2013; B. Zhang & Chen, 2019).

4.3.3 The (LIGA) Techniques: Lithography, Electroplating, and Molding or Lithography, Galvanoforming, and Abforming

The processes lithography, electroplating, and molding are together abbreviated as LIGA. These technologies were first developed in Germany. The reason behind the popularity of these methods is the applicability for a wide variety of materials, including metals, polymer derivatives, and even plastics. This technology makes use of electronic X-ray radiation to help mold or craft materials into various shapes, as the word lithography suggests (Mills & Nicole, 2005; Miyawaki et al., 2012). The technique comprises three essential steps: (i) pattern transfer to a photosensor (usually a

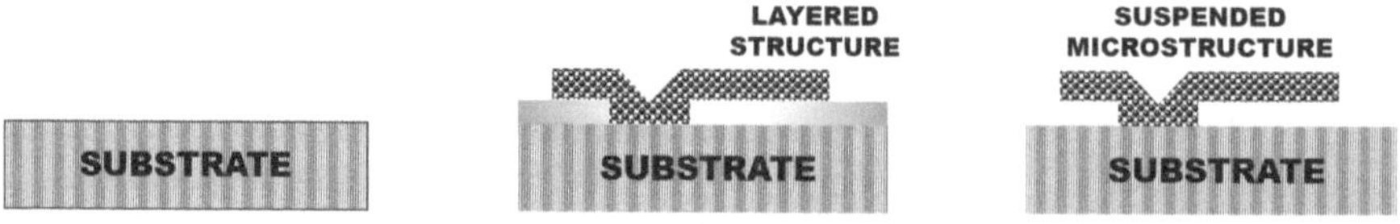

FIGURE 4.4 Schematic diagram of preparation of substrate surface by different micromachining methods.

resinous material); (ii) electroplating of the material's surface area by removing some structural layers; and (iii) withdrawal of the photoirradiation, which is supplied by the photoresistor or photosensor (Rizkin et al., 2019). Microreactors have been playing a vital role in accomplishing a spectrum of reactions in terms of their mechanism and nature, starting from chemical synthesis to biologically oriented ones. The extraction of DNA from organisms is one biological application process that can be carried out in microreactors. A PDMS–silicon-based microreactor was fabricated by a group of researchers for the chemical extraction of DNA (Effenhauser et al., 1997), featuring silicon-semiconducting materials forming the microchannels. For the operational specifications of the fabrication process, they had employed PDMS techniques, wherein lithography, electroplating, and molding were the basic processing methods.

4.4 CONTINUOUS-FLOW MICROREACTOR-BASED POLYMERIZATION REACTIONS

Polymerization reactions of various techniques have been well explored within continuous-flowing microreactors. The improved mixing, effective heat transmission, and fine-tuned control over reaction parameters such as temperature and residence time make continuous-flow microreactors an excellent choice for polymerization operations. Modern industrial developments have created a great demand for polymeric materials offering versatility and tunable properties. This demand has led to advancements in novel polymerization techniques. Such unique strategies necessitate meticulous control over operational and reaction conditions. The advent and employment of microreactor technology has definitely been a boon for the implementation of novel polymerization methods. Let us now dive deeper into various literature works concerning polymerization reactions conducted in continuous-flow microreactors.

4.4.1 Homogeneous Polymerization

Microreactors have the capability of carrying out homogeneous polymerization to produce a polymer without the process of phase separation of solutions as performed during post-treatment (Bally et al., 2010). The monomers, along with the initiators, are normally solubilized in suitable solvents in which they are soluble and are pumped into a micromixer. The reaction mix is then forced into the microchannel, and polymerization is accomplished subsequently. Following gradual technological advancements, the applicability of microreactors was extended to various categories of homogeneous polymerizations encompassing free radical polymerization, living polymerization, enzyme-catalyzed polymerization, and condensation polymerization that have been performed in microreactors.

4.4.1.1 Free Radical Polymerization (FRP)

The most prevalent and one of the simpler techniques of polymerization has been FRP, which is widely used in various laboratory and industrial applications. It is known for its simple operation specifications (X. Li et al., 2016; Iwasaki & Yoshida, 2005; Tonhauser et al., 2012). FRP requires precise temperature regulation since the majority of the FRP processes are extremely exothermic (Schimanski et al., 2006; Szwarc et al., 1956). The main limitations of conventional batch reactors are uncontrollable molecular weight distribution originating from inefficient removal of heat and temperature profile variation (X. Li et al., 2016). Contrastingly, microreactors have an excellent index of heat and mass transfer and are considered to be a superior alternative to the conventional batch reactors for the FRP processes. Iwasaki et al. carried out a comparative study concerning FRP reactions separately for five monomers (butyl acrylate, benzyl methacrylate (BzMA), methyl methacrylate, vinyl benzoate, and styrene) in microreactors with 250–1000 μm inner diameter (Iwasaki & Yoshida, 2005). The report suggested a significant control over molecular weight distribution related to a very low polydispersity index (PDI) values as obtained for the FRP reactions of all five monomers in microreactors. In addition to this, the group of authors had extended their work by scaling up the polymerization reaction using eight parallel microchannels that generated poly(methyl methacrylate) (PMMA) with controllable molecular weight and PDI and higher yield (Iwasaki et al., 2006). The relevant microreactor system was able to operate up to 6 days hassle freely, yielding a high amount of PMMA operating under continuous mode. The findings indicated that microreactors had immense prospects for relatively large-scale industrial productions. Figure 4.5a and b depicts the schematic view of a traditional microchannel construction in a microreactor.

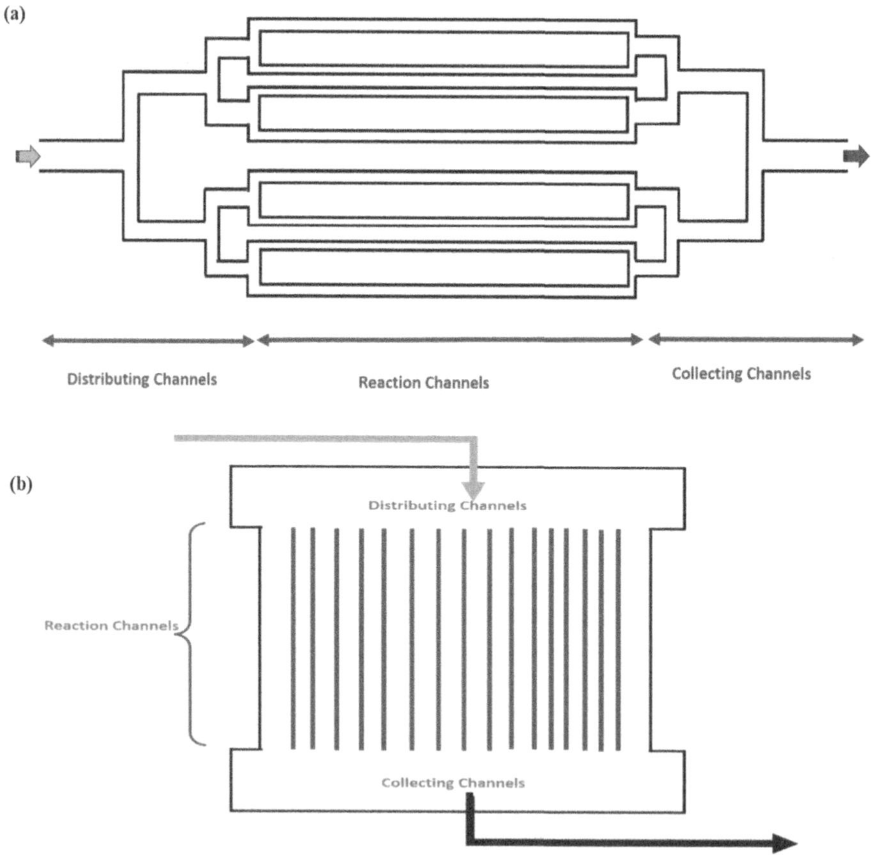

FIGURE 4.5 (a), (b) Schematic diagram of microchannels in a microreactor.

4.4.1.2 Living Chain Polymerization

Living chain polymerization is an advanced technology which had been developed in 1956 to produce polymers with controlled molecular weight distribution and PDI. It is broadly classified as anionic and cationic living polymerization based upon the nature of the initiator used (Szwarc et al., 1956). The basic principle of "living" polymerization concerns with a limited-time initiation step with simultaneous control over propagation rate of growing chains and restriction of various termination reactions, including chain transfer and others (Goto & Fukuda, 2004; Qiu et al., 2001). The name of the polymerization technique "living" had been adopted to justify its potentially ever-growing chains. Living polymerized polymers tend to possess a narrow range of molecular weight distributions along with well-defined structures (Grubbs & Grubbs, 2017). The utility of microreactors in implementing living polymerization reactions lies in the fact that such reactor designs provide intimate and uniform level of mixing of reacting monomers and initiators along with very high rates, attributes which are highly necessitated for handling very fast initiation reactions. Microreactors also provide a sealed environment which immensely facilitates the longevity of "living" propagating chains (Tonhauser et al., 2012). In short, living polymerization reactions become easy and effective when carried out in microreactor systems (Tonhauser et al., 2012).

4.4.1.2.1 Cationic Polymerization

One of the types of living polymerization is cationic polymerization. Cationic species or Lewis acids are used to initiate such reactions. A study by Yoshida et al. introduced an efficient and easy method to synthesize dendritic polymers having linear structures via living cationic polymerization in a continuous-flow mode (Nagaki et al., 2015). In their study, vinyl ethers monomers were polymerized by using dendrite-structured diarylcarbenium ions functioning as dendrite templates cum initiators, while 2, 6-di-tert-butylpyridine played the role of a proton scavenger. The synthesized polymers possessing well-defined linear dendritic structures were characterized by narrow molecular weight distribution and low PDI. The exciting results are attributed to an adequate level of mixing which got materialized within a microreactor system.

4.4.1.2.2 Anionic Polymerization

Microreactors have also been very functional in effecting living anionic polymerization of various monomers such as methacrylates. For instance, a specially designed microreactor consisting of a couple of microtubes and micromixers respectively was chosen to polymerize methacrylates initiated by 1,1-diphenylhexyllithium and using tetrahydrofuran (THF) as the solvent (Nagaki et al., 2009). The use of microreactors allowed better control over residence time and facilitated rapid mixing, and both factors resulted in maintaining narrow molecular weight distribution and low PDI of the prepared block copolymer. The microreactor was also believed to have aided the maintenance of the sequence of monomer introduction, which is a governing parameter in the fabrication of block copolymers. Living chain anionic polymerization of styrene monomers in THF solvent was successfully carried out by Nagaki et al. The microdimensional reactor set-up had been designed with two T-shaped micromixers with two microchannel reactors. Anionic initiators such as n-BuLi or s-BuLi would be efficient for such polymerization in microchannel reactors (Nagaki et al., 2016) and were reported to produce high molecular weight products having low PDI values. The obtained results were considered highly encouraging, as continuous 3 hours of operation of the microreactor had yielded 1 kg of polymer, an outcome which can be further scaled up to meet industrial standards.

4.4.2 Controlled Radical Polymerization (CRP)

Although the FRP technique is the most widely implemented and explored method adopted industrially to manufacture a lion's share of products, precise regulation of macromolecular structure and molecular weight distribution through FRP is inherently challenging due to the unpredictable nature of the reaction mechanism. On the other hand, the utilization of living polymerization is also restricted owing to its rigorous reaction conditions and the limited range of appropriate monomers (Weiss, 1981). In order to address the issue, CRP has been conceptualized, which integrates the favorable attributes of both FRP and living polymerization.

Within the CRP framework, a delicate balance between active and dormant species can be maintained. CRP operates on the principle of an optimization between these active and dormant species, which mitigates the propagation rate and facilitates consistent chain elongation while constraining the termination reactions, thereby inherently enhancing the control over polymer characteristics. By the CRP technique, a diverse range of monomers can be used, yielding polymers possessing highly defined morphologies and narrow molecular weight distributions through the CRP methodology. CRP strategies may encompass different types of polymerization techniques such as atom transfer radical polymerization (ATRP), reversible addition-fragmentation chain transfer radical polymerization (RAFT), and nitroxide-mediated polymerization (NMP) (Matyjaszewski, 2012; Qiu et al., 2001). The advancements in microreactor technology have successfully allowed CRP techniques to be implemented through continuous-flow routes.

4.4.3 Atom Transfer Radical Polymerization (ATRP)

ATRP is a novel technique of polymerization which is associated with the controlling of generated radicals and thereby produces polymers possessing specific architectures and functionalities. The process is featured with reversible deactivation of active species. Polymerization is initiated by the action of a transition metal catalyst which functions by activating a dormant species such as alkyl halides and generates active radicals. The radicals subsequently combine with the monomers and become activated to form the growing polymer chain. The process allows to have better control over the molecular weight characterized by narrow molecular weight distribution due to the mechanistic reversible activation/deactivation cycles. Microreactors are found to be extremely suitable for carrying out such precise and delicate reactions, maintaining all experimental conditions. In recent years, CRP has been executed via a continuous-flow methodology in conjunction with advancements in microreactor

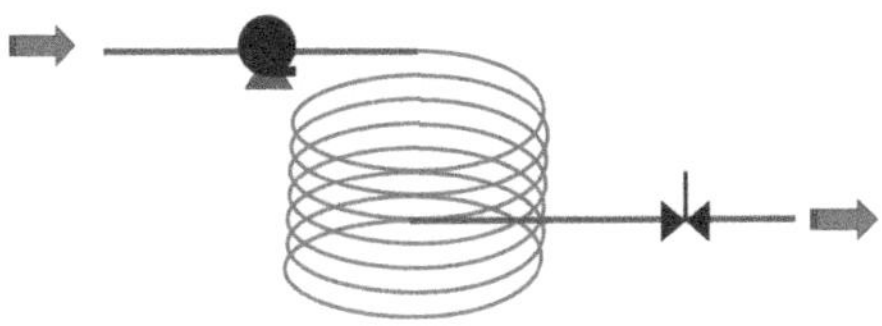

FIGURE 4.6 Schematic diagram of tubular designed continuous-flow microreactor used for carrying out complex polymerization reactions.

technology. A research group executed the synthesis of 2-(dimethylamino) ethyl methacrylate employing an ATRP technique within tubular microreactors featured with microdimensional internal diameters (Parida et al., 2014). The resulting polymers were reported to possess a molecular weight approximately equal to 2×10^4 g/mol, characterized by a low value of PDI (less than 1.7), purged at varying throughput rates. The microreactor architecture, as depicted in Figure 4.6, was optimized to enhance polymerization efficacy. The study involved optimization in the designing of microreactors, and it was noted that the coil flow inverter microreactor demonstrated superior mixing efficiency than its traditional coil counterpart, thereby facilitating the production of polymers with elevated monomer conversion rates and narrower molecular weight distributions. The study also focused on the influence of micromixers on polymer forming efficiency utilizing the ATRP approach to generate copolymers from 2-dimethyl amino ethyl methacrylate (DMAEMA) and BzMA as a representative process (Parida et al., 2014). A copolymerization reaction time was found to be significantly lowered with microreactors with reference to the same in batch reactors. Drawing inference, it could be expected that the rate of ATRP reactions would be accelerated as well if carried out within microreactors, adding to the superior product properties.

4.4.4 Reversible Addition–Fragmentation Chain-Transfer (RAFT) Polymerization

A study carried out by Hornung and coworkers involved the RAFT polymerization of a diverse array of monomers, inclusive of acrylamides, acrylates, and vinyl acetate (Vac), achieving conversion rates that varied from 80% to 100% at thermal conditions ranging from 70 to 100 °C, employing a variety of initiators, solvents, and RAFT agents using microreactors (Hornung et al., 2011). According to the report, the resulting polymer products exhibited narrowly defined molecular weight distributions, characterized by a PDI fluctuating between 1.15 and 1.20. The use of microreactors operated in continuous-flow mode, facilitated with a closed reaction condition set-up conducive to polymerization, thereby efficiently mitigating radical deactivation, which is a consequence of the oxygen-sensitive characteristics inherent to the RAFT mechanism. The radical bearing end groups of the growing polymer chains were separated from the reaction medium by means of a steel tube flow reactor system (Hornung, Postma, et al., 2012). The resulting outcome exhibited excellent attributes of number average molecular weight and PDI values, which spurred immense potentialities of the use of microreactors for RAFT polymerization reactions.

In another study, a succession of unit operations was integrated within a microreactor system tailored for RAFT processes (Hornung, Nguyen, et al., 2012). The polymerization of N,N-dimethylacrylamide (DMA) was accomplished utilizing trithiocarbonate (TTC) as a RAFT agent. Polymers obtained were characterized by narrow molecular weight distributions, and a yield of 90% was attainable with reaction durations of 30 to 90 minutes at temperature domains maintained at 70 to 80 °C. The desired level of polymer chain lengths could be achieved via strategic sequentialization of reactions. By amalgamating continuous operation synthesis with some extended processes, the synthesis of polymers via the RAFT method could be efficiently scaled up. A report had claimed the synthesis of poly(N-isopropylacrylamide) (PNIPAM) with molecular weights approximating 20 kg/mol and possessing high architectural integrity, by employing TTC as a chain transfer agent in continuous-flow tubular reactors through the homogeneous RAFT polymerization technique (Diehl et al., 2010). It has been proved experimentally that RAFT polymerization reactions could be sped up with the use of microreactors. Another group of researchers explored the RAFT polymerization of n-butylacrylate within a microreactor and concluded that owing to the rapid heat transfer capabilities of the microreactor, side reactions could be significantly minimized (Derboven et al., 2015). The obtained results were also corroborated by means of mathematical simulations which predicted the successful formation of well-designed products, viz block copolymers, within microreactors. Additionally, RAFT polymerization of acrylates was accomplished, in which polymeric products were employed as macro-RAFT agents and were subsequently reacted with butyl acrylate to form a block. This outcome thereby portrayed that the entire polymerization process would be best carried out in microreactors as the same provides with highly regulated control of reaction conditions. Some further design-based modifications could be adopted to extend the duration of RAFT processes. Some commonly used methodologies toward attaining such a goal include continuous and looping processing. An initiative by a group of researchers dealt with the RAFT polymerization of butyl acrylate within a tube reactor operated in three distinct modes ("Masthead: Macromol. React. Eng. 2/2017," 2017). The experimental data analysis had pointed out the effectiveness of microreactors in yielding polymer chains with well-distributed chain lengths and offered better control over other attributes. Furthermore, it was proved that the looping tube reactor, in conjunction with living polymerization strategy, represented a formidable instrument for tailoring polymers possessing multimodal molecular weight distributions, merely by altering the residence duration and recycle proportions.

4.4.5 Nitroxide-mediated Polymerization (NMP)

Microreactors have also been implemented in the past to carry out polymerization reactions by means of NMP strategies. One such study synthesized styrene, butyl acrylate, or methyl methacrylate (Ryu et al., 2012). Microreactor-polymerized products have been reported to possess higher molecular weight along with low PDI, in contrast to those obtained from conventional batch reactors. Another research activity involved the polymerization of styrene or n-butyl acrylate via the NMP route in a microreactor made of stainless steel having an inner diameter of 900 μm (Rosenfeld et al., 2007). The concerned microreactor set-up was fitted with a regulating device developing back pressure in order to maintain the system pressure in a specific range. On comparing the efficacies of microreactors and the conventional ones, it was concluded that although for styrene they performed almost at par, for the polymerization of n-butyl acrylate, microreactors produced lower-PDI polymers.

4.4.5.1 Polycondensation

Polycondensation reactions are considered to be a typical category of step-growth polymerization wherein monomeric entities, which typically possess two reactive terminal functionalities, integrate, concurrently yielding small molecular byproducts such as water, ammonia, methanol, etc. (Weiss, 1981). Step-growth linear polymers are an outcome of monomers with bifunctionalities (Yokozawa & Yokoyama, 2009). By virtue of reactivity traits, polycondensation reactions are predominantly associated with high reaction exotherms. Consequently, the reactor must be capable of effective heat dissipation. Microreactors have also made their inroads into polycondensation reaction domains considering their efficient heat transfer characteristics. Various research works have involved the exploration of microreactors for carrying out polycondensation reactions. Poly(p-phenylene terephthalamide) (PPTA) was reported to be synthesized following a two-step polycondensation reaction in a specially designed microreactor. The reaction was associated with a rapid radical generation, while the majority of the propagation phase was carried out in delay loop, followed by mechanical stirring (P. Wang et al., 2015). The stirring action had been reported to have played a vital part in almost doubling the value of the weight-average molecular weight. The researchers further investigated the reaction kinetics in relation to the microreactor design. A study conducted by Kessler et al. involved the polycondensation of trialkoxysilane to yield poly(silsesquioxanes)s characterizing tunable properties, in a uniquely designed continuous-flow microreactor set-up (Kessler et al., 2009). The results suggested that the obtained molecular weight distribution was far more efficiently controlled by carrying out the reaction in a microreactor, in comparison to that in a conventional batch reactor. This enhancement in property was attributed to the better mixing features offered by a microreactor. The use of microreactors has also been useful to generate high yield of polymer products.

4.4.6 Heterogeneous Polymerization

Heterogeneous polymerization concerns with a polymerization technique in which the reaction is carried out in different phases. Such environments are usually encountered in heterogeneous mixtures, namely emulsions, suspensions, or dispersions. In contrast to homogeneous polymerization, heterogeneous polymerization comprises distinct phases, such as monomer phases dispersed in a solvent medium. Inherently, different phases coexist distinctly owing to their immiscibility. One of the major advantages that heterogeneous polymerization possesses over homogeneous is the facility to use high monomer concentrations without the requirement of cumbersome purification methods. One of the key requisites in heterogeneous polymerization is the attainment of high dispersibility of the different immiscible phases (mostly liquids). In order to facilitate polymer growth and maintain steady reaction rates, excess/unreacted monomers and produced byproducts are immediately removed from the reaction medium by suitable means. Moreover, polymerization reactions may be confined within segmented liquid slugs during liquid–liquid heterogeneous polymerization carried out in microreactors. In heterogeneous polymerization reactions, convectional heat transfer is caused due to internal recirculation in liquid slugs, thereby leading to enhanced mixing efficiencies overpowering molecular diffusion processes which are found in homogeneous polymerization. Further effectiveness in mixing characteristics occurring within liquid slugs may be achieved by adopting the use of microchannels specifically designed in winding pattern, thereby disrupting recirculation flow symmetry (J. Zhang et al., 2017). As per studies, segmented slugs may function as miniature level batch reactors featuring excellent mixing capabilities, efficient heat and mass transfer attributes, and uniformly distributed residence time. The execution of polymerization in microreactors possesses risks of channel clogging, particularly when the viscosity of monomers or polymers is high. The heterogeneous polymerization strategies offer better preventive measures to avoid such risks. Considering the fact that channel walls possess different wettabilities when applied to various immiscible liquids, a liquid interfacial layer may be developed between the produced polymer and the inward lining of the microreactor. This phenomenon functions as a blessing in disguise and prevents fouling and clogging of the reactor (K. Wang et al., 2017).

Polymer granules with coarser particle size are highly desirable due to improved qualities of the manufactured products. There has been a serious industrial demand to look for specific synthetic strategies and/or routes in order to reach the goal. Key polymer characteristics associated with achieving such targets involve size distributions and molecular weights of the products. Hence, precise control over these properties is required to be maintained. A study from the past reported the polymerization of styrene or methyl methacrylate (MMA) designed inside microemulsion templated droplets. A microreactor system equipped with a K-M micromixer and delay-type tubing within slugs was used for the polymerization. The researchers concluded that heat removal in free-radical polymerization processes could be effectively accomplished via the implementation of internal recirculation within the slugs, resulting in enhanced controllable properties (Okubo et al., 2010). In another study, coaxial capillary microreactors were tested, which led to the formation of controlled polymerization of butyl acrylate (BA) within suspension droplets. Uniform droplet size was reportedly obtained by the researchers. To qualitatively estimate the temperature profile inside each monomer droplet, a simplified simulation method, the computational fluid dynamics (CFD) model, was recommended by the research group. By employing simulation models, a satisfactory level of performance was concluded in terms of polymerization conditions and polymer properties. Such attributes included narrow molecular weight distributions, etc. It was further inferred that almost an isothermal environment could be maintained within the emulsion droplets, which accounted for effective heat dissipation and ultimately resulted in controlling molecular weight of the polymer chains (Liu et al., 2011). A microdimensional environment was adopted in order to carry out such specialized reactions. Phosphine-doped polystyrene microcapsules could be synthesized by means of suspension polymerization in a continuous mode of operation maintained by a microfluidic system. Such operational mode and microenvironmental reaction assisted the maintenance of particle size of the produced microcapsules at a reasonable residence time. Microcapsules are highly utilitarian substances which offer various advantages. They are known to immobilize $Pd(PPh_3)_4$ catalysts, and thereby enhance catalytic activities in comparison to homogeneous bare palladium-based catalysts, highly desirable for fixed-bed reactors (K. Wang et al., 2017).

Over the years, polymers possessing microdimensional particles have become highly desirable. Synthesis of such products requires precision control over the reaction. A study carried out by Chu et al. reported the synthesis of poly (N-isopropylacrylamide) microparticles featured with hole-shell morphology in coaxial capillary microreactors. Another research reported the successful synthesis of polymer capsules of spherical geometry having unimodal molecular weight distribution in a continuous-mode microreactor. The researchers have claimed to adopt a single-step continuous method. The reaction was based on heterogeneous polymerization where one phase was silicon oil and the other, the aqueous phase, in which the monomeric tripropyleneglycol diacrylate (TPGDA) or ethyleneglycol dimethacrylate (EGDMA) was dissolved. The two different phases were introduced into the specially designed microreactor by means of a microfluidic flow-focusing system equipped with a coaxial jet. The polymer chain dimension and droplet morphologies were maintained by means of the specific microreactor design. The polymer was ultimately photopolymerized within the microfluidic reactor (Nie et al., 2005). The synthesis of polyacrylamide (PAM) hydrogel possessing microspherical structures was reported by Luo et al. in microchannels. The study claimed the attainment of monodispersed morphologies for the synthesized polymers. A research group studied the effects of influencing factors such as the composition of monomers and the reaction temperatures on the structural morphology and capacity of absorption of polyacrylamide (B. Yang et al., 2012). A single-step polymerization was carried out to produce poly(acrylamide-co-sodium acrylate) hydrogel beads adopting an inversion suspension mechanism, by using a microfluidic reactor. The formulations were found to possess core–shell morphological structures and the authors expressed high expectations for the same in various applications including optical devices and in different types of sensors (S. Yang et al., 2014).

4.4.7 Photopolymerization in Microreactors

Photopolymerization reaction rates are found to be immensely dependent on the light intensity, which is used for the polymerization reaction. On the other hand, very high light intensity is detrimental to molecular weight distribution, particularly at higher conversion rates (Chen et al., 2016). Synthesis of controlled molecular weight distribution (MWD) polymers requires that the reaction duration must be high and the light used must be of low intensity. Moreover, it has been proved by various studies that the use of microreactors facilitates the maintenance of irradiation uniformities during photopolymerization processes (A. Braun et al., 2012; A.

M. Braun et al., 1993). In this context, the design of reactor plays more governing role in case of photo-driven polymerization reactions compared to those in thermal polymerizations. However, optimization in reactor designs is required to simultaneously address both scalability and product property issues. Consequently, specialized reactors need to be designed for carrying out photopolymerization reactions, adequately equipped to maintain irradiation intensity levels and produce products with desirable molecular weight attributes at high yielding rates. The use of continuous-flow reactors has been considered to be ideally suited for maintaining irradiation intensity uniformly in literature (Elvira et al., 2013; Knowles et al., 2012; Kockmann et al., 2011; Lobry et al., 2014; Razzaq & Kappe, 2010; Su et al., 2014; Tonhauser et al., 2012). Photochemical reactions are found to exhibit low residence time when executed in microreactors. Unlike those obtained from conventional batch reactors, the polymers procured from microphotoreactors possess better selectivity and lower dosage requirements of photocatalyst. The significance of microreactors has been justified by various studies reported in the literature. The flux density of photons has been determined to be largely increased by the use of microreactors operated in continuous mode as compared to conventional batch mode reactors. The higher level of photon flux density has been considered to account for the accelerated reactions in microphotoreactors (Aillet et al., 2014; Jamali et al., 2013; Su et al., 2015). Microreactors have been exploited for the production of varied macromolecular and small molecular substances by means of photochemical processes (Cambié et al., 2016; W. Xu et al., 2018).

4.4.7.1 Photo-induced Radical Polymerization

A light-stimulated copper-infested radical polymerization involving methyl acrylate (MA) housed by microreactors operated in flow, such as a glass-chip photomicroreactor and a tubular photoreactor, was described by a group of researchers (Wenn et al., 2014). Me6TREN was employed as the catalyst/ligand combination, and the reaction solvent was composed of a combination of dimethylsulfoxide and $CuBr_2$. The resulting polymer had been reported to possess a low PDI value, and the average MW had been verified to be proportional during the course of polymerization. The management of the polymerization kinetics was also found to be highly desirable in flow. Furthermore, compared to a traditional tank reactor, the glass-chip photomicroreactor was reported to be successful in speeding up the photo-induced polymerization of MA, achieving nearly 90% monomer conversion in a short span of just 20 minutes. Poly(methyl acrylate) (PMA) possessing a desirable average molecular weight of 9800 $gmol^{-1}$ was synthesized using microreactor technology. The research group went on further to successfully synthesize a block copolymer poly(MA)-block-poly(nBA) in flow reactor, along with the homopolymer PMMA. The resulting block copolymer exhibited comparable low dispersibility (Wenn et al., 2014). Another study reported the successful fabrication of a mixture of acrylate/methacrylate (MA and methyl methacrylate (MMA)) block copolymers by free-radical photosensitive reaction in a microreactor (Chuang et al., 2015). To achieve ideal performance, the right ligand and polymerization sequence had been chosen. The tubular photomicroreactor was found to significantly accelerate the polymerization reaction, reduce product dispersity, and simplify the conditions for block copolymer production when compared to the batch photoreactor. With the help of an initiation-transfer-terminator (iniferter) under visible irradiations, a group produced the diblock and triblock polymethacrylate copolymers in a microreactor via photopolymerization. The synthesized polymers' PDI value was reported to exhibit a narrow molecular weight distribution (PDI = 1.2–1.3) (Rubens et al., 2017). In a microreactor irradiated by ultraviolet light, Kermagoret and coworkers had carried out Co-based free-radical polymerization of VAc or vinyl acetate/1-octene (VAc/1-Oct) under visible light (Kermagoret et al., 2015). The concerned polymerization reaction was initiated by the alkyl cobalt(III) adduct ($[Co(acac)_2(CH(OAc)\text{–}CH_2)_{<4}R_0]$, which produced the radicals when exposed to light. Their findings had shown that the microreactor technique greatly accelerated the polymerization process in addition to helping to prepare very pure poly(vinyl acetate) (PVA). The traditional batch reactor was found to generate a significant amount of byproducts during the UV photopolymerization process used to prepare PVA. It is interesting to note that PVA had shown a distinct monomodal mode when the reaction was conducted in a microreactor. With the aid of an iniferter, this photomicroreactor set-up was further used to copolymerize VAc with reactive olefins (1-octene). To fabricate telechelic PNIPAM and related block copolymers, Johnson and Chen had developed a straightforward and effective continuous-mode microreactor set-up for executing photo-driven living radical polymerization using TTC as an inducer (Chen & Johnson, 2015). The study had inferred that metal-free and switchable photopolymerization may be done under flow circumstances. High molecular weight polymers could be prepared more quickly in the microreactor in comparison to that in a batch reactor using photo-controlled live radical polymerization. Externally stimulated light was believed to have directly influenced the control over living polymerization, as demonstrated by intermittent photo-cycles living radical polymerization in a continuous-flow microreactor. It is to be mentioned that the molecular weights and molecular weight distributions could be easily controlled by switching out this photomicroreactor arrangement for reactions with varying monomer/TTC ratios. It was argued that such a continuous-flow technology would be a beneficial approach for both industry and academia due to its operational simplicity and capacity to generate high-quality polymers in a scalable manner. A study documented a controlled-radical polymerization of iridium-catalyzed photo-redox reaction in a microreactor to produce poly(methyl methacrylate) (Melker et al., 2015). The polymerization process was initiated via ethyl α-bromophenyl acetate, which was believed to have been attributed to visible light stimulation of the MMA conversion. Specifically, for this photo-stimulated polymerization, several capillary-based microreactors made of four commonly used tube components, including perfluoroalkoxy polymer, fluorinated ethylene propylene, tetrafluoroethylene (Tefzel), and ethylene chlorotrifluoroethylene (Halar), had been utilized. The accomplishment of this photopolymerization was found to depend on the employment of the microcapillary-based reactor made of Halar, which has a reduced oxygen permeability. It is worthwhile to mention over here that Halar is referred to as a fluoropolymer coating which is designed to offer excellent chemical resistance, including acids, caustic alkalies, and a variety of solvents. Traditional batch reactors as well as microreactors had been reported to have demonstrated first-order kinetics, and the continuous-flow mode of operation had boosted the polymerization rate by at least 50% when compared to traditional batch modes. A branched poly(butyl acrylate) was reported to be synthesized via photo-stimulated free radical-based polymerization of normal/tertiary-butyl acrylate in a continuous microreactor in the presence of tri(propylene glycol) diacrylate (TPGDA) as a crosslinking agent and dodecanethiol (DDT) in varying concentrations as a chain transfer agent to avoid gel formation (Eckardt et al., 2017). UV radiations along with solventless environment was maintained to synthesize branch-structured poly(normal-butyl acrylate)s with naturally reduced bulk viscosity, achieving maximum polymerization in a shorter time duration, which was highly unusual for radical polymerization operated in continuous mode. The resulting polymer materials' average molecular weight showed high values, with the monomer–thiol ratio having a significant influence. As stated by the authors, a batch weight of 300 g of branched-architectural poly(normal-butyl acrylate) was synthesized continuously for 24 hours using a scalable continuous-flow photoreactor. This was attributed to the enhancement of monomer conversion, thereby offering improved control over reaction parameters by preventing local hot spots and by allowing the polymerization to proceed in the presence of uniform irradiation.

4.4.7.2 PhotoRAFT Processes

RAFT is referred to as a CRP reaction technique employed to synthesize polymers having controlled molecular weights and narrow molecular weight distributions. The strategy is associated with a versatile method facilitating the generation of various polymer architectures, which include block copolymers and star polymers. The mechanism involving the RAFT process is based on the use of a chain transfer agent, also commonly referred to as a RAFT agent, to facilitate the polymerization via a reversible chain transfer. Such polymerization reactions have frequently been executed via thermal initiators. As per reports, it has been inferred that traditional methods of polymerization do not result in proper outcomes. The inefficiency of conventionally induced RAFT polymerization has eventually led to the adoption of new strategies like photo-electron transfer RAFT (PET-RAFT), as reported to be an effective technique for a range of intricate synthetic targets (J. Xu et al., 2015). A study carried out had described RAFT polymerization of methacrylates or acrylamides utilizing standard photoinitiators in a microreactor while being exposed to a range of LED arrays or a 150-W mercury lamp (Gardiner et al., 2016). To generate a high yield of RAFT polymers, the researchers had employed millimeter-scale fluoropolymer capillaries with superior light transmission capabilities. Investigations were conducted into how various photoinitiators and wavelength variations affected the quality of the final product and the overall performance of polymerization. By employing wavelengths between 310 and 380 nm, it was discovered that the photoinitiated polymerization yielded polymers characterized by low polydispersity (< 1.3) and good conversion. The results pointed out to the fact that wider wavelength ranges proportionately led to faster conversion but increased polydispersity by virtue of scission and simultaneous termination reactions brought on by greater activation of the RAFT end groups, and a careful examination of the relationship between the range of wavelengths employed and residence duration is necessary. By means of using high initiator and RAFT agent loadings, the study revealed exceptionally high polymerization rates with virtually 100% conversion at 30 °C. This finding was found to differ with those obtained by traditional thermal RAFT, which typically requires high temperatures or extended reaction periods. Photoinitiated RAFT polymerization carried out in the flow operational mode appears to have a wide spectrum of possible applications in polymer research and industry. In contrast to the conventional thermal RAFT method, which usually necessitates high temperatures or prolonged reaction times, the process followed by the researchers demonstrated extraordinarily rapid polymerization reaching up to 100% completion almost instantaneously at about room temperature range in the presence of high dosage of initiator and RAFT agent. The authors therefore had fancied high hopes for range of potential uses for photoinitiated RAFT polymerization in flow in both polymer research and industry. A microreactor was chosen by a group of researchers to carry out the polymerization reaction involving N,N'-dimethylacrylamide (DMAm) along with N,N'-diethylacrylamide (DEAm) when exposed to green light from an LED stripe (Corrigan et al., 2016). The product analysis was done. A poly(vinyl chloride) (PVC) pipe (inner diameter-15 cm) with polytetrafluoroethylene (PTFE) tubing coiled around the exterior of a 10-cm outer diameter PVC pipe had an energy-efficient LED panel attached to the interior. As seen in Figure 4.6, this made it possible for the reaction mixture to flow within the PTFE tube. Zinc tetraphenylporphyrin was believed to have served as the PET-RAFT process's photoredox catalyst. The specific reaction system featuring a large illuminated area had provided a congenial environment to carry out the polymerization. The same photomicroreactor set-up had been utilized to successfully synthesize the PDEAm-co-PDMAm copolymer, which exhibited a high MW and a low PDI without the requirement for prior deoxidation. Another investigation involved the use of a glass chip microreactor to accomplish the RAFT radical chain transfer polymerization of n-butyl acrylate, methyl acrylate, or their combination in the presence of traditional photoinitiators (Wenn & Junkers, 2016). The polymerization was assisted by 2-(dodecylsulfanyl)thiosulfanyl]sulfanyl)propionic acid (DoPAT), which functioned as the RAFT agent, and the UV light source was used. The influencing parameters concerned with the reaction were found to be the ratio of concentration of RAFT agent to that of the initiator, reaction temperature, light intensity, and initiator type. The best results were reported at specified reaction conditions including reaction temperature (of 60°C) and a light intensity of 30 mW/cm^2 (365 nm) when the concentration proportions of initiator benzoin were 0.25 to the initial concentration of the RAFT agent. The photoinitiators underwent rapid decomposition as the reaction occurred in the continuous-flow operational mode. Regulation of the polymer's chain length was achieved following the depletion of the initiator by a photoiniferter mechanism for this photoRAFT method. Under visible light irradiation, Corrigan and group were able to produce controlled MWD in a continuous microreactor featuring the photo-stimulated electron transfer-reversible addition–fragmentation chain-transfer (PET–RAFT) polymerization (Corrigan et al., 2017). The reaction comprised DMAm as the monomer; 5-((dodecylthio)-carbonothioyl) thio)propanoic acid (DTPA), the RAFT agent; 5,10,15,20-tetraphenyl-21H,23H-porphine zinc (ZnTPP), the effective catalyst; and dimethyl sulfoxide (DMSO), the reaction solvent. Through alterations in the chemical concentrations, the light source's wavelength and intensity, and the residence times of the reactant streams, the researchers could significantly control the molecular weight distributions.

Ramsey et al. (2017) employed microreactors consisting of continuous-flow capillaries in order to perform photo-stimulated catalyst-driven ATRP of several methacrylate-derived monomers (Ramsey et al., 2017). The uniquely designed photomicroreactor system demonstrated a variety of methacrylate monomers, the adaptation of several visible-light absorbing photoredox catalysts, the synthesis of controlled polymers with relatively reduced photocatalyst dosage, and the benefit on understanding the catalytic mechanism. Microreactors' consistent radiation and excellent mass transfer properties produced better polymerization results with measurable initiator efficiencies and a reduced MWD.

4.4.7.3 Various Photo-stimulated Polymerizations

Using microreactor technology, certain additional photopolymerization processes that did not fit into the above-described categories have also been documented. According to a study by Lobry and coworkers, photopolymerization in a microreactor was used to synthesize polymeric nanoparticles (Lobry et al., 2014). A syringe pump was used in the concerned experiment to facilitate the injection of the reaction mix, which included acrylate and thiol-ene miniemulsions, into the micro-sized reactor. A short-wavelength UV source capable of emitting photons served as the light source for the experiment. It was asserted by the authors that the droplet size, temperature, and residence span inside the microreactor could directly regulate the molecular weight of polymer nanoparticles.

In another paper, two assistive methods were described concerning the synthesis of carbohydrate-activated sequential oligo(amidoamine)s using the thiol-ene photo-driven coupling reaction in a micro-sized reactor (Wojcik et al., 2013). The technique involved conjugating different olefins to unprotected thiocarbohydrates after synthesizing polymer or oligomeric amidoamines from those supported on a solid substrate. In order to fabricate a unique sequence-defined carbohydrate-functionalized poly/oligo(amidoamine), the thiol-ene photo-driven coupling reaction was executed in a Teflon-lined capillary-based microreactor with short-wavelength UV irradiation. As per the findings, the approach had been successful in producing a specified homologous sequence in a single step by fully functionalizing every reaction site on the poly/oligo (amidoamine) backbone. Another alternative method involved employing a thiol-ene coupling to synthesize a fluorenylmethyloxycarbonyl (Fmoc)-protected glycosylated building block in a capillary microreactor with UV illumination. The solid product yielded from the first step was subsequently converted to a carbohydrate-functionalized sequence-defined oligomeric amidoamine. The second fabrication approach had been reported to be suitable enough, enabling structural control of various carbohydrates and large-scale synthesis. The two concerned fabrication methods had exploited the use of the microreactors' shortened photon transport path length, devoid of the use of any radical initiators.

4.5 CONCLUSION

Microreactors have been found to perform quite brilliantly in housing the production of organic compounds, biopolymers, nanoparticles, and several other materials. The overall properties of the synthesized products generally depend on microreactors.

The microreactor's microdimensionality allows it to carry out even very complex chemical reactions with critical paths and reaction processes. The procedures, principles, and structural materials utilized in the construction of small-scale reactors were covered in detail in this chapter. By efficiently synthesizing additives and offering embedded devices, microreactors can assist the pharmaceutical sector in building medication delivery systems. Additionally listed are the general materials or compounds utilized in the construction of microreactors.

Small-scale reactors' performance, efficiency, and high and rapid mass-transfer mechanism have recently attracted a remarkable amount of attention from the research community. Microreactors are believed to possess high potentialities to completely replace traditional batch reactors in the future. Microreactors have a faster rate of nanoparticle synthesis than batch reactors. Another special characteristic of microreactors is their ability to integrate multiple reaction sequences into a single reactor. Continued efforts have been put into the development of microreactor technology than ever before, and its successes in a number of domains are well known. Parallel ventures of renowned academic research centers and multinational institutions have created commercially available microreactors and accessory equipment in response to market demands.

Nonetheless, certain polymer syntheses find it challenging to function in microreactors due to economic issues and particular reaction integration. In order for microreactors to be widely used in all pertinent disciplines of study, further careful research into a few essential procedures, including synthesis ability, measurement analysis, extraction, and detection, must be done. Polymerization processes intrinsically exhibit unique characteristics like high fluid viscosity, the autoacceleration of polymerization rate, as well as the involvement of solid products. These characteristics lead to reduced transport properties, thereby increasing challenges encountered in carrying out mixing and dispersion of reactive fluids, consequently causing significant channel clogging in microreactors. Thus, there still remains a major requirement for thorough research on basic engineering concepts like fluid dynamics and transport processes associated with polymerization in flow. It has been experimentally observed that microreactors based on membrane-dispersion technology produce large droplet swarms with a strikingly low diameter, useful for heterogeneous polymerization reactions. Furthermore, the choice and structural morphology of microreactors intended for polymerization are usually made without referring to the reaction kinetics statistics. Polymerization kinetics in microreactors are also rarely documented. Often, the batch process data are generally referred to for the purpose. However, the use of batch reactor kinetics' data in microreactor design is limited since the polymerization kinetics measured in batch reactors might occasionally differ significantly from those measured in microreactors. One crucial step toward obtaining workable solutions for the accurate management of polymerization processes and the research of its associated complicated reaction is the simulation of software-based characterization and automation techniques in the microreactor system. Such innovative technical approaches possess high potential in obtaining acute control over polymerization reaction mechanism kinetics, developing novel polymerization processes to create unique polymers. In addition, the technology also helps in carrying out kinetic studies concerning complex mechanisms of polymerization reactions. The majority of polymerization processes encounter a major issue relating to the lower throughput values in comparison to those required by industrial standards. Continuous-flow microreactor-based processes have been known to adopt scale-up approaches, and pertinent techniques like the numbering-up strategy have been developed. Exploration of such strategies has been very limitedly done as far as polymerization reactions are concerned. The employment of a numbered-up microreactor with eight parallel micro-sized channels for the manufacture of PMMA is one standout example; during the course of six days, the productivity of PMMA reached 4 kg. Actually, the numbering-up technique is influenced by the parallel microchannels, and microreactors achieve an excellent flow distribution. This is fairly challenging since even slight variations in the pressure drop within each microchannel lead to an uneven flow distribution, particularly in polymerization processes where fluids' physical characteristics vary drastically. The performance of capillaries or tubular reactors with millidimensional inner diameters has been found to be much better when applied to polymerization processes. They can offer a significantly higher throughput than conventional microreactors, which have a characteristic dimension of hundreds of micrometers and still maintain good polymerization performance. Therefore, the fabrication of a powerful method to boost the synthesis of polymer products utilizing microreactor technology is highly necessitated that would facilitate the implementation of the scale-up strategy to a certain degree in conjunction with the numbering-up method.

REFERENCES

Acikgoz, C., Hempenius, M. A., Huskens, J., & Vancso, G. J. (2011). Polymers in Conventional and Alternative Lithography for the Fabrication of Nanostructures. *European Polymer Journal*, *47*(11), 2033–2052. https://doi.org/10.1016/j.eurpolymj.2011.07.025

Aillet, T., Loubiere, K., Dechy-Cabaret, O., & Prat, L. (2014). Accurate Measurement of the Photon Flux Received Inside Two Continuous Flow Microphotoreactors by Actinometry. *International Journal of Chemical Reactor Engineering*, *12*(1), 257–269. https://doi.org/10.1515/ijcre-2013-0121

Bally, F., Serra, C. A., Hessel, V., & Hadziioannou, G. (2010). Homogeneous Polymerization: Benefits Brought by Microprocess Technologies to the Synthesis and Production of Polymers. *Macromolecular Reaction Engineering*, *4*(9–10), 543–561. https://doi.org/10.1002/mren.201000006

Baraldi, P. T., & Hessel, V. (2012). Micro Reactor and Flow Chemistry for Industrial Applications in Drug Discovery and Development. *Green Processing and Synthesis*, *1*(2). https://doi.org/10.1515/gps-2012-0008

Becker, H., & Gärtner, C. (2001). Polymer Based Micro-reactors. *Reviews in Molecular Biotechnology*, *82*(2), 89–99. https://doi.org/10.1016/s1389-0352(01)00032-0

Braun, A., Peschl, G., & Oliveros, E. (2012). Industrial Photochemistry. In *CRC Handbook of Organic Photochemistry and Photobiology, Third Edition - Two Volume Set* (pp. 1–19). CRC Press. https://doi.org/10.1201/b12252-2

Braun, A. M., Jakob, L., Oliveros, E., & do Nascimento, C. A. O. (1993). Up-Scaling Photochemical Reactions. In *Advances in Photochemistry* (pp. 235–313). Wiley. https://doi.org/10.1002/9780470133491.ch3

Cambié, D., Bottecchia, C., Straathof, N. J. W., Hessel, V., & Noël, T. (2016). Applications of Continuous-Flow Photochemistry in Organic Synthesis, Material Science, and Water Treatment. *Chemical Reviews*, *116*(17), 10276–10341. https://doi.org/10.1021/acs.chemrev.5b00707

Cheemalapati, S., Ladanov, M., Winskas, J., & Pyayt, A. (2014). Optimization of Dry Etching Parameters for Fabrication of Polysilicon Waveguides with Smooth Sidewall Using a Capacitively Coupled Plasma Reactor. *Applied Optics*, *53*(25), 5745. https://doi.org/10.1364/ao.53.005745

Chen, M., & Johnson, J. A. (2015). Improving Photo-controlled Living Radical Polymerization from Trithiocarbonates through the Use of Continuous-flow Techniques. *Chemical Communications*, *51*(31), 6742–6745. https://doi.org/10.1039/c5cc01562f

Chen, M., Zhong, M., & Johnson, J. A. (2016). Light-Controlled Radical Polymerization: Mechanisms, Methods, and Applications. *Chemical Reviews*, *116*(17), 10167–10211. https://doi.org/10.1021/acs.chemrev.5b00671

Chierichetti, F., Kumar, R., & Mahdian, M. (2014). The Complexity of LSH Feasibility. *Theoretical Computer Science*, *530*, 89–101. https://doi.org/10.1016/j.tcs.2014.02.030

Chuang, Y.-M., Wenn, B., Gielen, S., Ethirajan, A., & Junkers, T. (2015). Ligand Switch in Photoinduced Copper-mediated Polymerization: Synthesis of Methacrylate–Acrylate Block Copolymers. *Polymer Chemistry*, *6*(36), 6488–6497. https://doi.org/10.1039/c5py00592b

Contents List. (2019). *Catalysis Science & Technology*, *9*(23), 6537–6546. https://doi.org/10.1039/c9cy90112d

Corrigan, N., Almasri, A., Taillades, W., Xu, J., & Boyer, C. (2017). Controlling Molecular Weight Distributions through Photoinduced Flow Polymerization. *Macromolecules*, *50*(21), 8438–8448. https://doi.org/10.1021/acs.macromol.7b01890

Corrigan, N., Rosli, D., Jones, J. W. J., Xu, J., & Boyer, C. (2016). Oxygen Tolerance in Living Radical Polymerization: Investigation of Mechanism and Implementation in Continuous Flow Polymerization. *Macromolecules*, *49*(18), 6779–6789. https://doi.org/10.1021/acs.macromol.6b01306

Cross, J. (2006). MEDLINE, PubMed, PubMed Central, and the NLM. *Editors' Bulletin*, *2*(1), 1–5. https://doi.org/10.1080/17521740701702115

Derboven, P., Van Steenberge, P. H. M., Vandenbergh, J., Reyniers, M., Junkers, T., D'hooge, D. R., & Marin, G. B. (2015). Macromol. Rapid Commun. 24/2015. *Macromolecular Rapid Communications*, *36*(24), 2117. https://doi.org/10.1002/marc.201570094

Diehl, C., Laurino, P., Azzouz, N., & Seeberger, P. H. (2010). Accelerated Continuous Flow RAFT Polymerization. *Macromolecules*, *43*(24), 10311–10314. https://doi.org/10.1021/ma1025253

Duffy, D. C., McDonald, J. C., Schueller, O. J. A., & Whitesides, G. M. (1998). Rapid Prototyping of Microfluidic Systems in Poly(dimethylsiloxane). *Analytical Chemistry*, *70*(23), 4974–4984. https://doi.org/10.1021/ac980656z

Eckardt, O., Wenn, B., Biehl, P., Junkers, T., & Schacher, F. H. (2017). Facile Photo-flow Synthesis of Branched Poly(butyl acrylate)s. *Reaction Chemistry & Engineering*, *2*(4), 479–486. https://doi.org/10.1039/c7re00013h

Effenhauser, C. S., Bruin, G. J. M., Paulus, A., & Ehrat, M. (1997). Integrated Capillary Electrophoresis on Flexible Silicone Microdevices: Analysis of DNA Restriction Fragments and Detection of Single DNA Molecules on Microchips. *Analytical Chemistry*, *69*(17), 3451–3457. https://doi.org/10.1021/ac9703919

Elvira, K. S., i Solvas, X. C., Wootton, R. C. R., & deMello, A. J. (2013). The Past, Present and Potential for Microfluidic Reactor Technology in Chemical Synthesis. *Nature Chemistry*, *5*(11), 905–915. https://doi.org/10.1038/nchem.1753

Fletcher, P. D. I., Haswell, S. J., Pombo-Villar, E., Warrington, B. H., Watts, P., Wong, S. Y. F., & Zhang, X. (2002a). Micro Reactors: Principles and Applications in Organic Synthesis. *Tetrahedron*, *58*(24), 4735–4757. https://doi.org/10.1016/s0040-4020(02)00432-5

Fletcher, P. D. I., Haswell, S. J., Pombo-Villar, E., Warrington, B. H., Watts, P., Wong, S. Y. F., & Zhang, X. (2002b). Micro Reactors: Principles and Applications in Organic Synthesis. *Tetrahedron*, *58*(24), 4735–4757. https://doi.org/10.1016/s0040-4020(02)00432-5

Friend, J., & Yeo, L. (2010). Fabrication of Microfluidic Devices Using Polydimethylsiloxane. *Biomicrofluidics*, *4*(2). https://doi.org/10.1063/1.3259624

Gardiner, J., Hornung, C. H., Tsanaktsidis, J., & Guthrie, D. (2016). Continuous Flow Photo-initiated RAFT Polymerisation Using a Tubular Photochemical Reactor. *European Polymer Journal*, *80*, 200–207. https://doi.org/10.1016/j.eurpolymj.2016.01.033

Gavriilidis, A., Angeli, P., Cao, E., Yeong, K. K., & Wan, Y. S. S. (2002). Technology and Applications of Microengineered Reactors. *Chemical Engineering Research and Design*, *80*(1), 3–30. https://doi.org/10.1205/026387602753393196

Goto, A., & Fukuda, T. (2004). Kinetics of Living Radical Polymerization. *Progress in Polymer Science*, *29*(4), 329–385. https://doi.org/10.1016/j.progpolymsci.2004.01.002

Grubbs, R. B., & Grubbs, R. H. (2017). *50th Anniversary Perspective*: Living Polymerization—Emphasizing the *Molecule* in *Macromolecules*. *Macromolecules*, *50*(18), 6979–6997. https://doi.org/10.1021/acs.macromol.7b01440

Guckenberger, D. J., de Groot, T. E., Wan, A. M. D., Beebe, D. J., & Young, E. W. K. (2015). Micromilling: A Method for Ultra-rapid Prototyping of Plastic Microfluidic Devices. *Lab on a Chip*, *15*(11), 2364–2378. https://doi.org/10.1039/c5lc00234f

Haswell, S. J., Middleton, R. J., O'Sullivan, B., Skelton, V., Watts, P., & Styring, P. (2001). The Application of Micro Reactors to Synthetic Chemistry. *Chemical Communications*, *5*, 391–398. https://doi.org/10.1039/b008496o

Haswell, S. J., & Watts, P. (2003). Green Chemistry: Synthesis in Micro Reactors. *Green Chemistry*, *5*(2), 240–249. https://doi.org/10.1039/b210539j

He, L., Fan, Y., Luo, L., Bellettre, J., & Yue, J. (2020). Preparation of Pt/γ-Al2O3 Catalyst Coating in Microreactors for Catalytic Methane Combustion. *Chemical Engineering Journal*, *380*, 122424. https://doi.org/10.1016/j.cej.2019.122424

Hessel, V. (2005). From Microreactor Design to Microreactor Process Design. *Chemical Engineering & Technology*, *28*(3), 243. https://doi.org/10.1002/ceat.200590004

Hornung, C. H., Guerrero-Sanchez, C., Brasholz, M., Saubern, S., Chiefari, J., Moad, G., Rizzardo, E., & Thang, S. H. (2011). Controlled RAFT Polymerization in a Continuous Flow Microreactor. *Organic Process Research & Development*, *15*(3), 593–601. https://doi.org/10.1021/op1003314

Hornung, C. H., Nguyen, X., Dumsday, G., & Saubern, S. (2012). Integrated Continuous Processing and Flow Characterization of RAFT Polymerization in Tubular Flow Reactors. *Macromolecular Reaction Engineering*, *6*(11), 458–466. https://doi.org/10.1002/mren.201200029

Hornung, C. H., Postma, A., Saubern, S., & Chiefari, J. (2012). Macromol. React. Eng. 6–7/2012. *Macromolecular Reaction Engineering*, *6*(6–7). https://doi.org/10.1002/mren.201290015

Iwasaki, T., Kawano, N., & Yoshida, J. (2006). Radical Polymerization Using Microflow System: Numbering-up of Microreactors and Continuous Operation. *Organic Process Research & Development*, *10*(6), 1126–1131. https://doi.org/10.1021/op060127u

Iwasaki, T., & Yoshida, J. (2005). Free Radical Polymerization in Microreactors. Significant Improvement in Molecular Weight Distribution Control. *Macromolecules*, *38*(4), 1159–1163. https://doi.org/10.1021/ma048369m

Jamali, A., Vanraes, R., Hanselaer, P., & Van Gerven, T. (2013). A Batch LED Reactor for the Photocatalytic Degradation of Phenol. *Chemical Engineering and Processing: Process Intensification*, *71*, 43–50. https://doi.org/10.1016/j.cep.2013.03.010

Jena, R. K., & Yue, C. Y. (2012). Cyclic Olefin Copolymer Based Microfluidic Devices for Biochip Applications: Ultraviolet Surface Grafting Using 2-methacryloyloxyethyl Phosphorylcholine. *Biomicrofluidics*, *6*(1). https://doi.org/10.1063/1.3682098

Jolliffe, H. G., & Gerogiorgis, D. I. (2015). Process Modelling and Simulation for Continuous Pharmaceutical Manufacturing of Ibuprofen. *Chemical Engineering Research and Design*, *97*, 175–191. https://doi.org/10.1016/j.cherd.2014.12.005

Kermagoret, A., Wenn, B., Debuigne, A., Jérôme, C., Junkers, T., & Detrembleur, C. (2015). Improved Photo-induced Cobalt-mediated radical Polymerization in Continuous Flow Photoreactors. *Polymer Chemistry*, *6*(20), 3847–3857. https://doi.org/10.1039/c5py00299k

Kessler, D., Löwe, H., & Theato, P. (2009). Macromol. Chem. Phys. 10/2009. *Macromolecular Chemistry and Physics*, *210*(10). https://doi.org/10.1002/macp.200990014

Kjeang, E., Djilali, N., & Sinton, D. (2009). Microfluidic Fuel Cells: A Review. *Journal of Power Sources*, *186*(2), 353–369. https://doi.org/10.1016/j.jpowsour.2008.10.011

Kjeang, E., Sinton, D., & Harrington, D. A. (2006). Strategic Enzyme Patterning for Microfluidic Biofuel Cells. *Journal of Power Sources*, *158*(1), 1–12. https://doi.org/10.1016/j.jpowsour.2005.07.092

Knowles, J. P., Elliott, L. D., & Booker-Milburn, K. I. (2012). Flow Photochemistry: Old Light through New Windows. *Beilstein Journal of Organic Chemistry*, *8*, 2025–2052. https://doi.org/10.3762/bjoc.8.229

Kockmann, N., Gottsponer, M., & Roberge, D. M. (2011). Scale-up Concept of Single-channel Microreactors from Process Development to Industrial Production. *Chemical Engineering Journal*, *167*(2–3), 718–726. https://doi.org/10.1016/j.cej.2010.08.089

Kolb, G. (2013). Review: Microstructured Reactors for Distributed and Renewable Production of Fuels and Electrical Energy. *Chemical Engineering and Processing: Process Intensification*, *65*, 1–44. https://doi.org/10.1016/j.cep.2012.10.015

Konstantinou, D., Shirazi, A., Sadri, A., & Young, E. W. K. (2016). Combined Hot Embossing and Milling for Medium Volume Production of Thermoplastic Microfluidic Devices. *Sensors and Actuators B: Chemical*, *234*, 209–221. https://doi.org/10.1016/j.snb.2016.04.147

Krishna, K. S., Li, Y., Li, S., & Kumar, C. S. S. R. (2013). Lab-on-a-chip Synthesis of Inorganic Nanomaterials and Quantum Dots for Biomedical Applications. *Advanced Drug Delivery Reviews*, *65*(11–12), 1470–1495. https://doi.org/10.1016/j.addr.2013.05.006

Kundu, A., Ahn, J. E., Park, S.-S., Shul, Y. G., & Han, H. S. (2008). Process Intensification by Micro-channel Reactor for Steam Reforming of Methanol. *Chemical Engineering Journal*, *135*(1–2), 113–119. https://doi.org/10.1016/j.cej.2007.02.026

Lee, J. N., Park, C., & Whitesides, G. M. (2003). Solvent Compatibility of Poly(dimethylsiloxane)-Based Microfluidic Devices. *Analytical Chemistry*, *75*(23), 6544–6554. https://doi.org/10.1021/ac0346712

Li, X., Mastan, E., Wang, W.-J., Li, B.-G., & Zhu, S. (2016). Progress in Reactor Engineering of Controlled Radical Polymerization: A Comprehensive Review. *Reaction Chemistry & Engineering*, *1*(1), 23–59. https://doi.org/10.1039/c5re00044k

Li, Y., Yan, L., Liu, Y., Qian, K., Liu, B., Yang, P., & Liu, B. (2015). High-efficiency Nano/Micro-reactors for Protein Analysis. *RSC Advances*, *5*(2), 1331–1342. https://doi.org/10.1039/c4ra12333f

Liu, Z., Lu, Y., Yang, B., & Luo, G. (2011). Controllable Preparation of Poly(butyl acrylate) by Suspension Polymerization in a Coaxial Capillary Microreactor. *Industrial & Engineering Chemistry Research*, *50*(21), 11853–11862. https://doi.org/10.1021/ie201497b

Lo, I.-C., & Wu, H.-S. (2019). Methanol formation from carbon dioxide hydrogenation using Cu/ZnO/Al2O3 catalyst. *Journal of the Taiwan Institute of Chemical Engineers*, *98*, 124–131. https://doi.org/10.1016/j.jtice.2018.06.020

Lobry, E., Jasinski, F., Penconi, M., Chemtob, A., Croutxé-Barghorn, C., Oliveros, E., Braun, A. M., & Criqui, A. (2014). Continuous-flow Synthesis of Polymer Nanoparticles in a Microreactor via Miniemulsion Photopolymerization. *RSC Advances*, *4*(82), 43756–43759. https://doi.org/10.1039/c4ra06814a

Love, J. C., Anderson, J. R., & Whitesides, G. M. (2001). Fabrication of Three-Dimensional Microfluidic Systems by Soft Lithography. *MRS Bulletin*, *26*(7), 523–528. https://doi.org/10.1557/mrs2001.124

Masthead: Macromol. React. Eng. 2/2017. (2017). *Macromolecular Reaction Engineering*, *11*(2). https://doi.org/10.1002/mren.201770005

Matyjaszewski, K. (2012). Atom Transfer Radical Polymerization (ATRP): Current Status and Future Perspectives. *Macromolecules*, *45*(10), 4015–4039. https://doi.org/10.1021/ma3001719

Melker, A., Fors, B. P., Hawker, C. J., & Poelma, J. E. (2015). Continuous Flow Synthesis of Poly(methyl methacrylate) via a Light-mediated Controlled Radical Polymerization. *Journal of Polymer Science Part A: Polymer Chemistry*, *53*(23), 2693–2698. https://doi.org/10.1002/pola.27765

Mills, P. L., & Nicole, J. F. (2005). Multiple Automated Reactor Systems (MARS). 2. Effect of Microreactor Configurations on Homogeneous Gas-Phase and Wall-Catalyzed Reactions for 1,3-Butadiene Oxidation. *Industrial & Engineering Chemistry Research*, *44*(16), 6453–6465. https://doi.org/10.1021/ie048830z

Miyawaki, Y., Kondo, Y., Sekine, M., Ishikawa, K., Hayashi, T., Takeda, K., Kondo, H., Yamazaki, A., Ito, A., Matsumoto, H., & Hori, M. (2012). Highly Selective Etching of SiO2 over Si3N4 and Si in Capacitively Coupled Plasma Employing C5HF7 Gas. *Japanese Journal of Applied Physics*, *52*(1R), 16201. https://doi.org/10.7567/jjap.52.016201

Muhler, M., Schlögl, R., Eder, S., & Ertl, G. (1987). Design of a Continuous Flow Microreactor Attached to a Surface Analysis System: First Results with an Iron Oxide Based Catalyst. *Surface Science*, *189–190*, 69–79. https://doi.org/10.1016/s0039-6028(87)80417-x

Nagaki, A., Nakahara, Y., Furusawa, M., Sawaki, T., Yamamoto, T., Toukairin, H., Tadokoro, S., Shimazaki, T., Ito, T., Otake, M., Arai, H., Toda, N., Ohtsuka, K., Takahashi, Y., Moriwaki, Y., Tsuchihashi, Y., Hirose, K., & Yoshida, J. (2016). Feasibility Study on Continuous Flow Controlled/Living Anionic Polymerization Processes. *Organic Process Research & Development*, *20*(7), 1377–1382. https://doi.org/10.1021/acs.oprd.6b00158

Nagaki, A., Takumi, M., Tani, Y., & Yoshida, J. (2015). Polymerization of Vinyl Ethers Initiated by Dendritic Cations Using Flow Microreactors. *Tetrahedron*, *71*(35), 5973–5978. https://doi.org/10.1016/j.tet.2015.05.096

Nagaki, A., Tomida, Y., Miyazaki, A., & Yoshida, J. (2009). Microflow System Controlled Anionic Polymerization of Alkyl Methacrylates. *Macromolecules*, *42*(13), 4384–4387. https://doi.org/10.1021/ma900551a

Nie, Z., Xu, S., Seo, M., Lewis, P. C., & Kumacheva, E. (2005). Polymer Particles with Various Shapes and Morphologies Produced in Continuous Microfluidic Reactors. *Journal of the American Chemical Society*, *127*(22), 8058–8063. https://doi.org/10.1021/ja042494w

Okubo, Y., Maki, T., Nakanishi, F., Hayashi, T., & Mae, K. (2010). Precise Control of Polymer Particle Properties Using Droplets in the Microchannel. *Chemical Engineering Science*, *65*(1), 386–391. https://doi.org/10.1016/j.ces.2009.06.004

Parida, D., Serra, C. A., Garg, D. K., Hoarau, Y., Muller, R., & Bouquey, M. (2014). Flow Inversion: An Effective Means to Scale-Up Controlled Radical Polymerization Tubular Microreactors. *Macromolecular Reaction Engineering*, *8*(8), 597–603. https://doi.org/10.1002/mren.201400002

Pattekar, A. V., & Kothare, M. V. (2004). A Microreactor for Hydrogen Production in Micro Fuel Cell Applications. *Journal of Microelectromechanical Systems*, *13*(1), 7–18. https://doi.org/10.1109/jmems.2004.823224

Patwardhan, A. W., & Joshi, J. B. (1998). Design of Gas-Inducing Reactors. *Industrial & Engineering Chemistry Research*, *38*(1), 49–80. https://doi.org/10.1021/ie970504e

Pennemann, H., & Kolb, G. (2016a). Review: Microstructured reactors as efficient tool for the operation of selective oxidation reactions. *Catalysis Today*, *278*, 3–21. https://doi.org/10.1016/j.cattod.2016.04.032

Pennemann, H., & Kolb, G. (2016b). Review: Microstructured Reactors as Efficient Tool for the Operation of Selective Oxidation Reactions. *Catalysis Today*, *278*, 3–21. https://doi.org/10.1016/j.cattod.2016.04.032

Phillips, T. W., Lignos, I. G., Maceiczyk, R. M., deMello, A. J., & deMello, J. C. (2014). Nanocrystal Synthesis in Microfluidic Reactors: Where Next? *Lab on a Chip*, *14*(17), 3172. https://doi.org/10.1039/c4lc00429a

Pileni, M. P. (1993). Reverse Micelles as Microreactors. *The Journal of Physical Chemistry*, *97*(27), 6961–6973. https://doi.org/10.1021/j100129a008

Plouffe, P., Bittel, M., Sieber, J., Roberge, D. M., & Macchi, A. (2016). On the Scale-up of Micro-reactors for Liquid–liquid Reactions. *Chemical Engineering Science*, *143*, 216–225. https://doi.org/10.1016/j.ces.2015.12.009

Plouffe, P., Roberge, D. M., & Macchi, A. (2014). WITHDRAWN: Liquid–liquid Flow Regimes and Mass Transfer in Various Micro-reactors. *Chemical Engineering Journal*. https://doi.org/10.1016/j.cej.2014.10.072

Qiu, J., Charleux, B., & Matyjaszewski, K. (2001). Controlled/Living Radical Polymerization in Aqueous Media: Homogeneous and Heterogeneous Systems. *Progress in Polymer Science*, *26*(10), 2083–2134. https://doi.org/10.1016/s0079-6700(01)00033-8

Ramsey, B. L., Pearson, R. M., Beck, L. R., & Miyake, G. M. (2017). Photoinduced Organocatalyzed Atom Transfer Radical Polymerization Using Continuous Flow. *Macromolecules*, *50*(7), 2668–2674. https://doi.org/10.1021/acs.macromol.6b02791

Razzaq, T., & Kappe, C. O. (2010). Continuous Flow Organic Synthesis under High-Temperature/Pressure Conditions. *Chemistry – An Asian Journal*, *5*(6), 1274–1289. https://doi.org/10.1002/asia.201000010

Rizkin, B. A., Popovic, F. G., & Hartman, R. L. (2019). Review Article: Spectroscopic Microreactors for Heterogeneous Catalysis. *Journal of Vacuum Science & Technology A: Vacuum, Surfaces, and Films*, *37*(5). https://doi.org/10.1116/1.5108901

Roberge, D. (2012). Lonza – Hazardous Flow Chemistry for Streamlined Large Scale Synthesis. *Green Processing and Synthesis*, *1*(1). https://doi.org/10.1515/greenps-2011-0504

Rosenfeld, C., Serra, C., Brochon, C., & Hadziioannou, G. (2007). High-temperature Nitroxide-mediated Radical Polymerization in a Continuous Microtube Reactor: Towards a Better Control of the Polymerization Reaction. *Chemical Engineering Science*, *62*(18–20), 5245–5250. https://doi.org/10.1016/j.ces.2006.12.074

Rossetti, I., & Compagnoni, M. (2016). Chemical Reaction Engineering, Process Design and Scale-up Issues at the Frontier of Synthesis: Flow Chemistry. *Chemical Engineering Journal*, *296*, 56–70. https://doi.org/10.1016/j.cej.2016.02.119

Rubens, M., Latsrisaeng, P., & Junkers, T. (2017). Visible Light-induced Iniferter Polymerization of Methacrylates Enhanced by Continuous Flow. *Polymer Chemistry*, *8*(42), 6496–6505. https://doi.org/10.1039/c7py01157a

Ryu, I., Studer, A., Fukuyama, T., & Kajihara, Y. (2012). Nitroxide-Mediated Polymerization of Styrene, Butyl Acrylate, or Methyl Methacrylate by Microflow Reactor Technology. *Synthesis*, *44*(16), 2555–2559. https://doi.org/10.1055/s-0031-1290780

Schimanski, T., Loos, J., Peijs, T., Alcock, B., & Lemstra, P. J. (2006). On the Overdrawing of Melt-spun Isotactic Polypropylene Tapes. *Journal of Applied Polymer Science*, *103*(5), 2920–2931. https://doi.org/10.1002/app.25128

Su, Y., Hessel, V., & Noël, T. (2015). A Compact Photomicroreactor Design for Kinetic Studies of Gas-liquid Photocatalytic Transformations. *AIChE Journal*, *61*(7), 2215–2227. https://doi.org/10.1002/aic.14813

Su, Y., Straathof, N. J. W., Hessel, V., & Noël, T. (2014). Photochemical Transformations Accelerated in Continuous-Flow Reactors: Basic Concepts and Applications. *Chemistry – A European Journal*, *20*(34), 10562–10589. https://doi.org/10.1002/chem.201400283

Szwarc, M., Levy, M., & Milkovich, R. (1956). polymerization Initiated by Electron Transfer to Monomer. A New Method of Formation of Block polymers. *Journal of the American Chemical Society*, *78*(11), 2656–2657. https://doi.org/10.1021/ja01592a101

Tonhauser, C., Natalello, A., Löwe, H., & Frey, H. (2012). Microflow Technology in Polymer Synthesis. *Macromolecules*, *45*(24), 9551–9570. https://doi.org/10.1021/ma301671x

Urban, P. L., Goodall, D. M., & Bruce, N. C. (2006). Enzymatic Microreactors in Chemical Analysis and Kinetic Studies. *Biotechnology Advances*, *24*(1), 42–57. https://doi.org/10.1016/j.biotechadv.2005.06.001

Wang, K., Li, L., Xie, P., & Luo, G. (2017). Liquid–liquid microflow reaction engineering. *Reaction Chemistry & Engineering*, *2*(5), 611–627. https://doi.org/10.1039/c7re00082k

Wang, P., Wang, K., Zhang, J., & Luo, G. (2015). Preparation of Poly(p-phenylene terephthalamide) in a Microstructured Chemical System. *RSC Advances*, *5*(79), 64055–64064. https://doi.org/10.1039/c5ra10275h

Wang, X., Zhu, J., Bau, H., & Gorte, R. J. (2001). Fabrication of Micro-Reactors Using Tape-Casting Methods. *Catalysis Letters*, *77*(4), 173–177. https://doi.org/10.1023/a:1013236306883

Waterkamp, D. A., Heiland, M., Schlüter, M., Sauvageau, J. C., Beyersdorff, T., & Thöming, J. (2007). Synthesis of Ionic Liquids in Micro-reactors—A Process Intensification Study. *Green Chemistry*, *9*(10), 1084. https://doi.org/10.1039/b616882e

Watts, P. (2004). Chemical Synthesis in Micro Reactors. *ChemInform*, *36*(2). https://doi.org/10.1002/chin.200502258

Weiss, P. (1981). Principles of polymerization, 2nd ed., George Odian, Wiley-Interscience, New York, 1981, 731 pp. *Journal of Polymer Science: Polymer Letters Edition*, *19*(10), 519. https://doi.org/10.1002/pol.1981.130191009

Wenn, B., Conradi, M., Carreiras, A. D., Haddleton, D. M., & Junkers, T. (2014). Photo-induced copper-mediated polymerization of methyl acrylate in continuous flow reactors. *Polymer Chemistry*, *5*(8), 3053–3060. https://doi.org/10.1039/c3py01762a

Wenn, B., & Junkers, T. (2016). Continuous Microflow PhotoRAFT Polymerization. *Macromolecules*, *49*(18), 6888–6895. https://doi.org/10.1021/acs.macromol.6b01534

Wohlgemuth, R., Plazl, I., Žnidaršič-Plazl, P., Gernaey, K. V., & Woodley, J. M. (2015). Microscale Technology and Biocatalytic Processes: Opportunities and Challenges for Synthesis. *Trends in Biotechnology*, *33*(5), 302–314. https://doi.org/10.1016/j.tibtech.2015.02.010

Wojcik, F., O'Brien, A. G., Götze, S., Seeberger, P. H., & Hartmann, L. (2013). Synthesis of Carbohydrate-Functionalised Sequence-Defined Oligo(amidoamine)s by Photochemical Thiol—Ene Coupling in a Continuous Flow Reactor. *Chemistry – A European Journal*, *19*(9), 3090–3098. https://doi.org/10.1002/chem.201203927

Xia, Y., & Whitesides, G. M. (1998). Soft Lithography. *Annual Review of Materials Science*, *28*(1), 153–184. https://doi.org/10.1146/annurev.matsci.28.1.153

Xu, J., Shanmugam, S., Duong, H. T., & Boyer, C. (2015). Organo-photocatalysts for Photoinduced electron Transfer-reversible Addition–Fragmentation Chain Transfer (PET-RAFT) polymerization. *Polymer Chemistry*, *6*(31), 5615–5624. https://doi.org/10.1039/c4py01317d

Xu, W., Su, Y., Song, Y., Shang, M., Zha, L., & Lu, Q. (2018). Process Analysis on Preparation of Cyclobutanetetracarboxylic Dianhydride in a Photomicroreactor within Gas–Liquid Taylor Flow. *Industrial & Engineering Chemistry Research*, *57*(7), 2476–2485. https://doi.org/10.1021/acs.iecr.7b04572

Yang, B., Lu, Y., & Luo, G. (2012). Controllable Preparation of Polyacrylamide Hydrogel Microspheres in a Coaxial Microfluidic Device. *Industrial & Engineering Chemistry Research*, *51*(26), 9016–9022. https://doi.org/10.1021/ie3004013

Yang, S., Wang, Y., Gong, M., & Gong, Y. (2014). A Simple Method to Fabricate Hemocompatible Coating by Crosslinkable Biomimetic Bipolymer. *Asian Journal of Chemistry*, *26*(17), 5745–5749. https://doi.org/10.14233/ajchem.2014.18197

Yokozawa, T., & Yokoyama, A. (2009). Chain-Growth Condensation Polymerization for the Synthesis of Well-Defined Condensation Polymers and π-Conjugated Polymers. *Chemical Reviews*, *109*(11), 5595–5619. https://doi.org/10.1021/cr900041c

Yoon, T.-H., Park, S.-H., Min, K.-I., Zhang, X., Haswell, S. J., & Kim, D.-P. (2008). Novel Inorganic Polymer Derived Microreactors for Organic Microchemistry Applications. *Lab on a Chip*, *8*(9), 1454. https://doi.org/10.1039/b804726j

Yoshida, J., Takahashi, Y., & Nagaki, A. (2013a). ChemInform Abstract: Flash Chemistry: Flow Chemistry That Cannot be Done in Batch. *ChemInform*, *44*(50). https://doi.org/10.1002/chin.201350255

Yoshida, J., Takahashi, Y., & Nagaki, A. (2013b). Flash Chemistry: Flow Chemistry that CANNOT BE DONE in Batch. *Chem. Commun.*, *49*(85), 9896–9904. https://doi.org/10.1039/c3cc44709j

Zebda, A., Innocent, C., Renaud, L., Cretin, M., Pichot, F., Ferrigno, R., & Tingry, S. (2011). Enzyme-Based Microfluidic Biofuel Cell to Generate Micropower. In *Biofuel's Engineering Process Technology*. InTech. https://doi.org/10.5772/17190

Zeng, D., Pan, M., Wang, L., & Tang, Y. (2012). Fabrication and Characteristics of Cube-post Microreactors for Methanol Steam Reforming. *Applied Energy*, *91*(1), 208–213. https://doi.org/10.1016/j.apenergy.2011.08.024

Zhang, B., & Chen, Z. (2019). Screening of Cathepsin B Inhibitors in Traditional Chinese Medicine by Capillary Electrophoresis with Immobilized Enzyme Microreactor. *Journal of Pharmaceutical and Biomedical Analysis*, *176*, 112811. https://doi.org/10.1016/j.jpba.2019.112811

Zhang, J., Wang, K., Teixeira, A. R., Jensen, K. F., & Luo, G. (2017). Design and Scaling Up of Microchemical Systems: A Review. *Annual Review of Chemical and Biomolecular Engineering*, *8*(1), 285–305. https://doi.org/10.1146/annurev-chembioeng-060816-101443

5 Carbon-Based Nanomaterial Synthesis Using Microreactors

Madhubonti Pal, Rahul S. Tanpure, and Sankha Chakrabortty

5.1 INTRODUCTION

Microreactor technology—often referred to as micro process engineering—centers on devices with characteristic channel dimensions below one millimeter that carry out chemical transformations in a continuous, tightly controlled environment. Shrinking the reactor length scales fundamentally reshapes transport phenomena: surface-area-to-volume ratios soar, conductive and convective pathways shorten, and thermal inertia plummets (Kolb & Hessel, 2004). The outcome is a platform that couples rapid heat and mass transfer with exquisite residence-time control, enabling reactions to proceed faster, safer, and with higher selectivity than in conventional vessels. These attributes make microreactors attractive for distributed manufacturing and point-of-use operations in pharmaceuticals, fine and specialty chemicals, energy carriers, photochemical conversions, and bio-based products. From a process-safety perspective, handling small holdups of reactants lowers the risk associated with energetic chemistries, exotherms, or toxic intermediates. From a development standpoint, short diffusion paths and narrow residence-time distributions (RTD) accelerate screening, optimization, and technology transfer from lab to pilot. Crucially, capacity is increased not by scaling the linear dimensions of a single unit but by "numbering-up"—running many identical channels or plates in parallel. This modularity supports flexible production footprints and rapid reconfiguration for new campaigns.

The transport regime in microchannels is dominated by laminar flow (typically $Re \ll 1$), leading to predictable velocity profiles and enabling deterministic control of contact times, temperature histories, and interfacial areas. In single-phase operation, Poiseuille flow governs pressure drop and RTD; in gas–liquid systems, flow patterns such as slug, bubbly, and annular regimes can be engineered via geometry and dosing strategy to enhance interfacial renewal and mass transfer. Because diffusion lengths are tiny, thermal and compositional gradients can be suppressed or intentionally imposed, providing a powerful lever for selectivity in heat- or mass-transfer-limited reactions. At the same time, design must respect micro-scale realities: wall friction is significant, surface roughness and wettability influence slip and wetting, interfacial

DOI: 10.1201/9781003633761-5

tension steers two-phase morphologies, and rarefaction effects may arise at low pressures or high temperatures for gases.

The promise of microreactors is balanced by practical constraints. Small passages are susceptible to clogging by precipitates, particulates, or biofilms; catalyst coatings may deactivate; and uniform flow distribution across multiple channels is non-trivial (Bae et al., 2024; Bugay et al., 2024). Reliability over long on-stream periods, sealing integrity, and mixing in laminar regimes are recurrent design concerns. Economic factors also matter: while footprint and utility demands are modest, microfabrication, precision manifolding, and instrumentation can elevate capital cost. Table 5.1 synthesizes these trade-offs, pairing each benefit—high surface-area-to-volume ratio, compactness, rapid response, improved yields and selectivity, enhanced safety—with

TABLE 5.1
Advantages and Disadvantages of Microreactor Technology (Bae et al., 2024, Bugay et al., 2024)

Advantages	Explanation	Disadvantages/ Limitations	Explanation
High surface-area-to-volume ratio	Enhances heat and mass transfer, improving reaction efficiency	Clogging or fouling	Narrow channels may block easily due to particulates, precipitates, or biofilms
Improved heat and mass transfer rates	Enables precise thermal control and rapid mixing	Scale-up challenges (numbering-up)	Scaling by parallelization requires complex fluid distribution networks
Compact design	Reduced footprint ideal for decentralized and portable systems	Catalyst deactivation	Catalyst-coated channels can degrade or deactivate over time
Short and narrow RTD	Minimizes back-mixing; enhances process uniformity	Distributor malfunction	Maldistribution leads to channel starvation or over-feeding
Enhanced safety	Lower inventory of hazardous chemicals minimizes explosion/ toxicity risks	Reliability during long-term operation	Continuous operation stability remains a challenge
Avoidance of runaway reactions	Efficient heat dissipation prevents thermal runaways	Channel leakage	Microchannel seals can fail causing cross-contamination
Faster system response	Rapid change in operating conditions ensures tight process control	Mixing limitations	In laminar-flow regimes, mixing depends heavily on reactor geometry
Accelerated R&D cycle	Enables rapid screening, process optimization, and scale-down studies	Higher capital cost	Microfabrication and instrumentation costs remain high
Lightweight and modular	Easy relocation and integration in existing units	Short residence time	Suitable only for fast kinetics; slow reactions need redesign or recirculation

corresponding limitations such as fouling, numbering-up challenges, and short residence times that favor intrinsically fast kinetics. Designing effective microreactors therefore requires a thermo-fluid approach. Thermal design targets uniform temperature fields and hot-spot avoidance (especially for exothermic systems) by optimizing channel shapes, coolant strategies, and heat-exchange architectures. Fluid design aims to minimize RTD breadth, equalize maldistribution from manifolds, and set pressure-drop budgets compatible with pumps and seals. Computational fluid dynamics (CFD) has become indispensable for resolving velocity, pressure, and temperature fields, screening channel geometries (e.g., straight, serpentine, herringbone, split-and-recombine), and exploring two-phase pattern maps before hardware is built. In heterogeneous catalysis, one must choose between wall-coated microchannels (low pressure drop, excellent heat removal) and packed or structured beds (high surface area, potential for channeling). Geometry, catalyst location and loading, sensor placement, and materials of construction—metals for robustness, silicon for heat flux and microfabrication compatibility, glass/ceramics for corrosion resistance, polymers for rapid prototyping—are matched to reaction severity and solvent compatibility. Beyond reaction intensification, microreactors are evolving into precision synthesis tools for functional nanomaterials. Continuous-flow (CF) routes for graphene oxide (GO), carbon nanotubes (CNTs), and carbon dots (CDs) leverage the same transport advantages to control nucleation, growth, oxidation/reduction, and assembly with millisecond-to-second residence times. In flow GO synthesis, temperature fields, oxidant delivery, and mixing intensity define the degree and topology of oxidation, defect density, and colloidal stability; micro- and milli-fluidic devices permit safe handling of energetic steps with improved reproducibility. For CNTs, CF chemical vapor deposition (CVD) platforms (tube, fluidized-bed, or microchannel reactors) regulate precursor decomposition, catalyst evolution, and residence time to tune diameter, wall number, alignment, and crystallinity. For CDs, microreactors, hydrothermal flow, microwave-assisted, or ultrasound-assisted continuous processes enable scalable, monodisperse products with tunable photophysics and facile surface functionalization. Across these cases, process parameters—temperature, flow rate and residence time, shear and mixing, activation (light, ultrasound, electric fields), and catalyst selection—interact strongly; microreactors make these interactions measurable and controllable.

This chapter moves from fundamentals to engineering implementation. One section introduces flow dynamics in microchannels, emphasizing laminar transport, two-phase patterning, pressure-drop management, and heat/mass transfer coupling. Another section develops microreactor design principles, covering geometry, materials, catalytic architectures, manifold design, and thermo-fluid optimization assisted by CFD. Subsequent sections outline material choices and fabrication routes and then connect fundamentals to application case studies in synthesis and reaction engineering, with a focus on carbon-based nanomaterials where microreactors deliver distinctive advantages in safety, scalability, and product quality. Table 5.1 serves as a roadmap for recognizing both the strengths and the pitfalls that determine real-world performance, from benchtop prototypes to parallelized production units. By integrating transport science, smart design, and modern control, microreactor technology provides a coherent platform for safer, cleaner, and faster chemistry.

5.1.1 Fundamentals of Microreactor Technology

Microreactors, or microprocess engineering, have gained significant attention in global academic research due to their small, compact designs—typically featuring channel dimensions under 1 mm. These reactors are capable of performing chemical reactions or generating energy in a highly efficient, controlled, and scalable manner. Their design allows them to operate with minimal energy input, making them particularly advantageous for specific applications such as power generation, medical treatments, and other forms of distributed manufacturing.

These small-scale reactors are commonly used in the chemical industry for specific reactions, typically in laboratory or manufacturing settings. They offer high precision and efficiency in reactions that would be challenging to achieve in larger-scale reactors. Additionally, smaller reactors operate with greater efficiency due to enhanced heat and mass transfer. By handling smaller quantities of chemicals, they also minimize the risk of large-scale accidents. Reactions in microreactors occur at faster rates and within more controlled environments, which accelerates the transition from research to production. Furthermore, they can be easily scaled up by integrating multiple reactors in parallel. These benefits open up significant opportunities for microreactor technology across various sectors, including pharmaceutical manufacturing, polymerization, catalytic reactions, and the production of bio-based products like biofuels and biopharmaceuticals. Several advantages, as well as some disadvantages, of using microreactors are summarized in Table 5.1.

5.1.2 Flow Dynamics

Flow chemistry, in this connection, is a related approach of microreactors which bridges these two cutting-edge fields to improve chemical reactions by conducting them in CF systems instead of traditional batch processes. Flow phenomena refer to the practice of conducting chemical reactions in a CF system rather than in batch reactors. Here, reagents are pumped through a reactor, and reactions occur while the reactants are in motion. Devices used in flow chemistry include commercial microcapillaries and fluidic connectors, commonly used in high-pressure liquid chromatography. Due to the smaller hydraulic diameter of these capillaries, the nature of the fluid flow here does not follow the convection-diffusion equations (Navier-Stokes equations) of general flow phenomena. In microreactor design, the parameters like molecular and interfacial tension, wall friction, roughness of surface, viscosity of the fluid, and rarefaction effects play a significant role (Ehrfeld et al., 2000; Kolb & Hessel, 2004; Hessel et al., 2005; Papautsky et al., 2001). When the gases flow in microchannels at high temperature or low pressures, the Maxwell-Boltzmann assumption for the velocity of the particles in a co-moving fluid may break down. In such reactors, due to small hydraulic diameter of the channels, the flow remains laminar, making the Reynolds number (Re) very small. Again, the diffusion paths for both heat and mass transfer are also very small, thus making these designs ideal for heat- and mass-transfer-limited reactions. Likewise, in liquid flow, the Re is usually very low ($Re \ll 1$), indicating laminar flow. This means that fluid flows smoothly without turbulence, and the flow velocity is uniform at any cross-section. In such cases, Poiseuille flow dominates, where fluid moves in layers, with the slowest flow occurring at the walls and the fastest flow in the center of the channel. The transport phenomena (convective and diffusive), microchannel geometries, flow patterns, pressure drop, reaction kinetics, and heat transfer are some key features of fluid flow in microreactors. In microreactors, the flow is driven by pressure differences (e.g., using pumps or capillary forces), and convection plays a major role in transporting reagents through the system. At small scales, diffusion becomes very important for the movement of molecules, especially for the mixing of reactants and products, as the laminar flow does not inherently mix them well. Reactors are typically designed with intricate microchannel networks, which can be in the form of straight channels, serpentine channels, or even multi-phase systems. The geometry impacts both the flow dynamics (e.g., how pressure gradients develop) and heat/mass transfer rates. Due to the small dimensions, the pressure drop in microreactors is significant, but it can also be controlled through channel design and flow rate adjustments. Single-phase flow refers to the flow of a single liquid phase, where the pressure drop is dominated by the viscous resistance of the fluid in the channel. When gas and liquid phases are mixed, such as in gas–liquid microreactors, two-phase flow phenomena, including slug flow, bubbly flow, and annular flow, can occur. The pressure drop is generally determined by the Darcy-Weisbach equation and modified for microflows. In a microreactor, the enhanced heat and mass transfer characteristics facilitate rapid chemical reactions. Heat generation or absorption in reactions can be more effectively managed, and temperature gradients are usually smaller than in traditional reactors (Figure 5.1).

In short, due to small reaction volumes and high surface-area-to-volume ratios, microreactors promote better mixing of reactants. With improved heat and mass transfer, reactions can proceed more rapidly. Smaller quantities of reactants and products are used, making hazardous reactions more manageable. Microreactors allow fine control over reaction parameters, often down to the millisecond or even nanosecond scale. High efficiency leads to reduced consumption of reagents and minimized waste production. These special features make this technology easier to scale up or down compared to batch reactors.

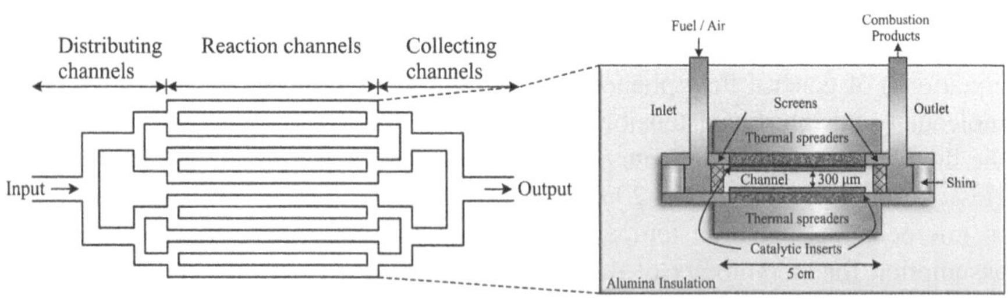

Enhanced Microreactor Assembly Diagram

FIGURE 5.1 (a) Fluid and heat flow in a single-phase microreactor. (b) Single-phase microreactor structure.

(Bugay et al., 2024.)

5.1.3 Microreactor Design

Microreactor design generally refers to the creation of very small-scale reactors, typically with channel dimensions below 1 mm, where chemical reactions or nuclear reactions take place under micro-scale confinement. These designs maximize efficiency, safety, and control by enhancing heat and mass transfer, enabling CF operations, and improving reaction speed and yields. Gokhale et al. (2005) mentioned how proper choice of materials and fabrication methods along with kinetics, transport rates, coating and loading of catalysts, and location of sensors and intrinsic conductivity can help design an efficient microreactor. According to Kashid et al. (2007), CFD is one such method for design and optimization of microreactor parameters. Recent advances in CFD modeling provide relatively precise knowledge on flow, temperature, and pressure distribution, without the need to perform any experiments.

Different types and shapes of microreactors exist for single and multiphase flow systems in this field. These shapes and types of mixers and reactors vary with reaction processes. For example, for single-phase reactions, catalytic wall and packed bed microreactors are considered to be appropriate, whereas capillary and falling film microreactors, as well as micro-contactors are used for multiphase reactions (Kiwi-Minsker and Renken, 2005).Considering the design, the control of two separate flows with their relative velocities through one microchannel, have significant role on the resulting pattern of the multiphase flow. Again, depending upon flow characteristics, microreactors can be classified as liquid-flow and fixed-bed microreactors (Figure 5.2).

Liquid flow microreactors rely on well-defined channels (often T- or Y-shaped) for uninterrupted flow and controlled mixing. They excel in fast reaction kinetics and minimal pressure drops. Fixed-bed microreactors create flow paths through packed beds of particles, which causes pressure drops, channeling, pulsing, and slugging depending on flow rates and gas–liquid ratios. This results in periodic sweeping, sparse slug flow, and sometimes gas CF regimes (Cao et al., 2021). Another key consideration is how to introduce the active phase of the catalysts for a

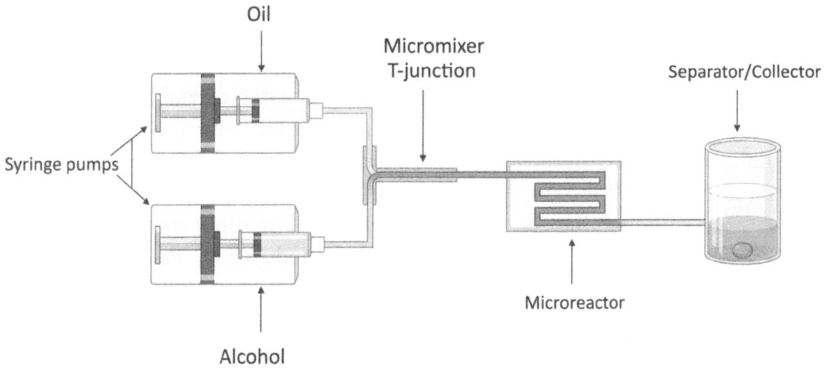

FIGURE 5.2 Microreactor used in biodiesel production.

(Natarajan et al., 2019.)

heterogeneous reaction, where the solid catalyst is either coated onto the wall or placed as a packed bed within the reactor. While packed bed reactors are generally simpler to construct than catalytic wall microreactors, the second option is often preferred due to its lower pressure drop and superior heat transfer capabilities (Kim & Kwon, 2006). Apart from the reactor type, geometry and structure of the microreactors are also crucial factors. For example, microreactors with multiple channels combined with catalytic wall offer several advantages over traditional reactors, addressing limitations related to volumetric flow rates and scaling up microchannel units. The size of a microreactor and its channels must be optimized according to throughput, with the goal of enhancing key performance characteristics. The RTD in a microchannel with a laminar flow profile is significantly influenced by both the diffusion coefficients of the species and the dimensions of the channels (Kolb & Hessel, 2004). A study by Tonomura et al. (2004a) found that an optimal design for a plate-fin microreactor typically includes parallel microchannels with inlet and outlet manifolds. The two primary design parameters for microreactors are thermal and fluid design. In thermal design, the main goal is to control the fluid temperature by optimizing variables such as microchannel shape, flow rate, coolant temperature, and constraints like maximum pressure drop, reaction temperature, and yield/selectivity. In fluid design, the focus is on minimizing the total RTD, optimizing manifold shape, and determining the number of microchannels. Key constraints such as throughput, RTD, and maximum pressure drop must also be accounted for. In a thermo-fluid design approach, adjusting the channel width along the longitudinal direction can help achieve a uniform temperature distribution and prevent hot spots, particularly during exothermic reactions. Therefore, achieving an optimal thermo-fluid design is crucial for the performance of the microreactor system (Tonomura et al., 2004a, 2004b) (Figure 5.3).

There are various construction materials and fabrication methods used for microreactor making. Metals, e.g., stainless steel (SS), Ti (Titanium), are commonly used for high-temperature or high-pressure reactions, as they offer durability and are resistant to corrosion. Polymers (PDMS, PMMA, PEEK) are used in lower-temperature, non-corrosive reactions. They are easier to fabricate but less suitable for harsh conditions. Ceramics and glass materials are often used in microreactors for specific chemical applications due to their excellent resistance to aggressive chemicals and high temperatures (McMullen & Jensen, 2010). Each material offers distinct advantages and disadvantages, and the choice of material for a microreactor depends on factors such as the reaction characteristics as well as the chemical properties and compatibility of the reactants or reagents. For high-temperature reactions and separations, ceramics are a strong option due to their low heat loss and chemical resistance, although they come with a high cost. For applications involving high pressures and temperatures, where superior heat conductivity and high aspect-ratio design are required, silicon and SS microreactors are often the best choices. Plastics, on the other hand, are ideal for fast, cost-effective development, but they are incompatible with organic solvents and unsuitable for high temperatures and pressures. Most microreactors made from silicon wafers are patterned to create microstructures or channels, often integrated with heaters, sensors, and catalytic reaction zones (Figure 5.4).

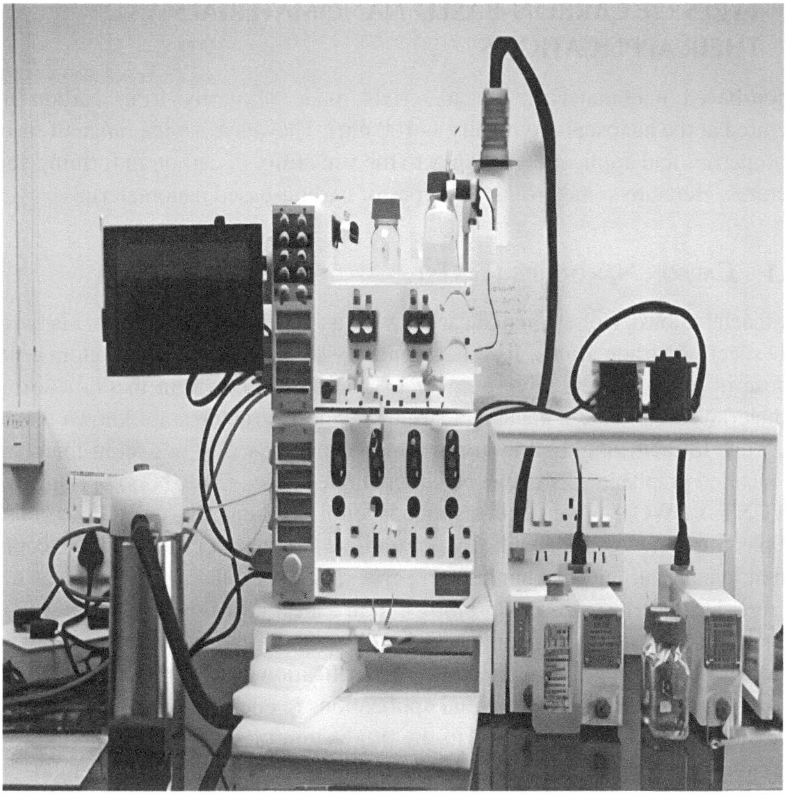

FIGURE 5.3 Example of a Flow reactor system (IACS, Kolkata, India).

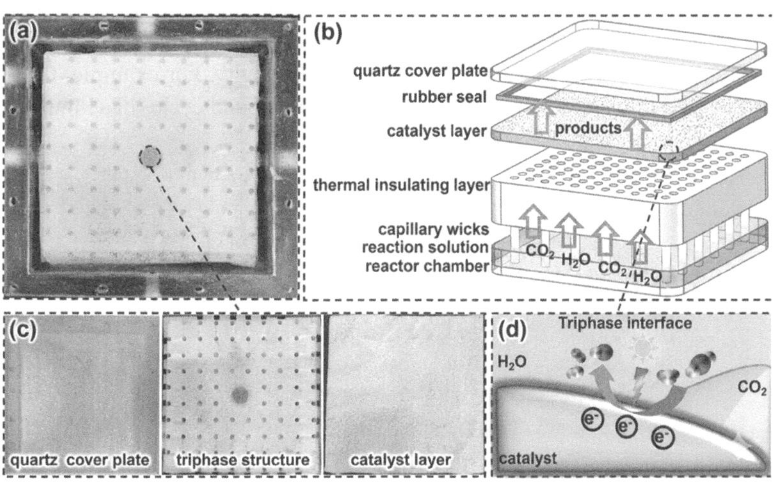

FIGURE 5.4 A compact microreactor design for solar-driven photothermal CO_2 reduction. (Wang et al., 2025.)

5.2 TYPES OF CARBON-BASED NANOMATERIALS AND THEIR APPLICATIONS

Carbon-based nanomaterials are materials made primarily from carbon atoms structured at the nanoscale (typically 1–100 nm). They have a wide range of fascinating properties and applications, thanks to the versatility of carbon in forming various structures. Here are some prominent types of carbon-based nanomaterials:

5.2.1 Carbon Nanotubes (CNTs)

This special allotrope of elemental carbon with a cylindrical nanostructure is basically a 2-D sheet of carbon atoms. It is made entirely of sp^2-bonded carbon atoms, similar to those in graphite. The large number of double bonds within the structure gives the material remarkable mechanical strength. These structures are known as "nanotubes" due to their long, hollow form, which is composed of one-atom-thick carbon sheets called graphene. CNTs can be either single-walled (SWCNTs) or multi-walled (MWCNTs). SWCNTs consist of a single layer of graphene, and MWCNTs consist of multiple graphene layers. Thanks to their unique structure, CNTs exhibit exceptional thermal, electrical, and mechanical properties. The well-defined cylindrical hollow structure, large surface area (~250 m^2/g), high aspect ratios, hydrophobic walls, and easily modifiable surfaces make CNTs highly effective as nano-sorbents and membranes with excellent performance in water purification and desalination applications. CNTs show a wide range of potential applications in energy storage, generation, and conversion. They are widely used in the production of supercapacitors due to their exceptional surface area, electrical conductivity, and mechanical strength. These properties significantly improve the charge-discharge rates and overall performance of supercapacitors. CNTs offer a large surface area for efficient charge storage and are capable of withstanding numerous charge/discharge cycles without degradation. They are also used as electrodes in lithium-ion batteries, lithium-sulfur batteries, and other emerging battery technologies due to their conductivity and structural integrity which help to improve battery efficiency, capacity, and longevity. For example, CNTs can be incorporated into anodes to improve charge capacity and cycle life (Figure 5.5).

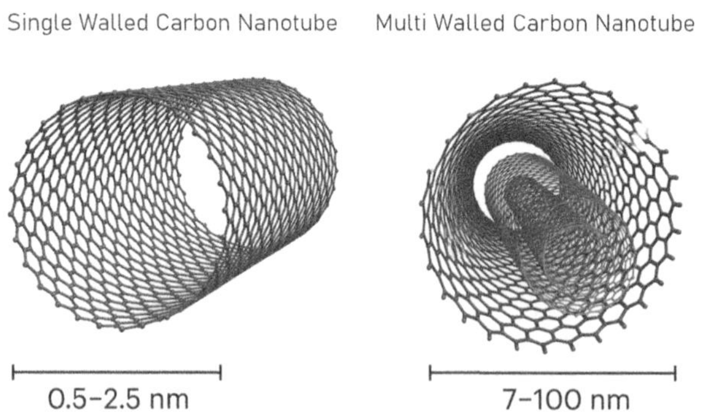

FIGURE 5.5 Single and multi-walled carbon nanotubes.

(Scarselli et al., 2012.)

CNTs are also explored as catalysts in fuel cells, especially in proton exchange membrane (PEM) fuel cells. Besides, they are also used in the fabrication of flexible, lightweight, and cost-effective solar cells to improve charge transport, reduce recombination losses, and enhance the overall efficiency of the cells. These materials can convert heat directly into electricity, thus making them useful for waste heat recovery, providing energy in systems where excess heat is generated, such as in industrial plants or vehicle engines. Due to their exceptional thermal conductivity, CNTs are used in thermal management systems, such as heat exchangers and heat sinks, in energy conversion devices. This helps in improving the efficiency of power plants, engines, and electronic devices. CNTs are explored in triboelectric nano-generators (TENGs), which convert mechanical energy from friction into electricity. They can be used in applications such as wearable devices or small sensors powered by motion or vibrations. CNTs possess an exceptionally high surface area, making them ideal candidates for hydrogen storage applications. Hydrogen molecules can be adsorbed onto the surface of CNTs, offering a compact and efficient solution for hydrogen storage in fuel cells. Additionally, CNTs are utilized in electrocatalysis for water splitting, a process that generates hydrogen from water. When incorporated into electrodes, CNTs enhance the efficiency of the electrolysis process, supporting a cleaner and more sustainable method of hydrogen production. CNTs, especially SWCNTs, are frequently used in drug delivery applications because of their ability to carry a broad spectrum of therapeutic agents, including anticancer drugs, genes, and small molecules. The high surface area of CNTs allows for efficient drug loading, while their nanoscale size facilitates easy penetration of biological membranes. By functionalizing CNTs with specific groups or antibodies, they can be directed to target particular tissues or cells, such as cancer cells, which helps minimize side effects and improve therapeutic efficacy. CNTs can be used in photothermal therapy, where they are loaded with a therapeutic agent and then exposed to near-infrared light. They absorb the light, heat up, and selectively kill cancer cells. Apart from these, they have several medical applications as biosensors, antimicrobial agents and are used in orthopedic surgeries, gene therapy, etc.

5.2.2 Graphene

A monolayer 2D sheet of graphite, known as graphene, consists of covalently bonded sp^2-hybridized carbon atoms arranged in a hexagonal honeycomb lattice. This single-layer structure of graphene is defined by the International Union for Pure and Applied Chemistry (IUPAC) as "a single carbon layer of graphite structure, analogous to a polycyclic aromatic hydrocarbon of quasi-infinite size." In 2004, Novoselov, Geim and their team at the University of Manchester successfully isolated and characterized a mechanically exfoliated graphene monolayer. In the hexagonal lattice of graphene, each carbon atom is bonded to three neighboring atoms from different sublattices, forming three σ bonds (in-plane) and one π bond (out-of-plane), resulting in a C–C bond length of 0.142 nm. The remarkable stability of graphene arises from the tightly packed carbon atoms and the sp^2 orbital hybridization. Additionally, the delocalized π-electrons across the carbon atoms contribute to its excellent electrical conductivity (Figure 5.6).

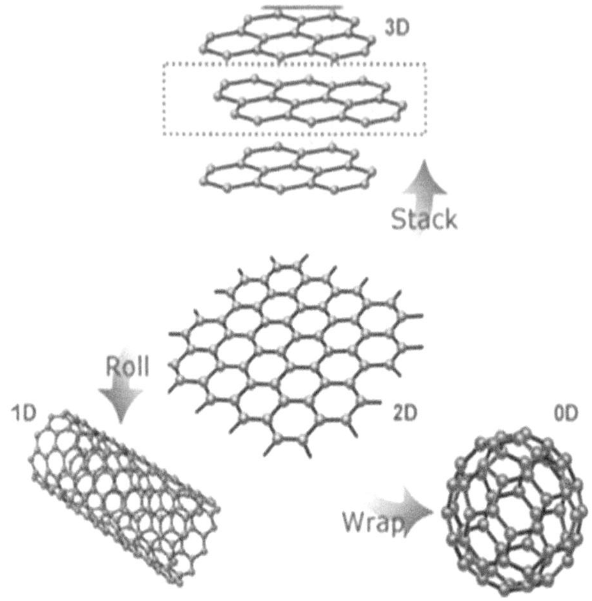

FIGURE 5.6 Graphene family.

(Garg et al., 2014.)

Graphene's unique properties, such as its excellent electrical conductivity, mechanical strength, and large surface area, make it highly valuable for energy-related applications. Graphene-based materials are being used to enhance the performance of supercapacitors, which are devices that store and discharge energy. Graphene allows for faster charging times and higher energy storage capacity, making it an alternative to traditional battery technologies. In particular, graphene can increase the rate of charge and discharge, making batteries more efficient for everything from smartphones to electric vehicles. Graphene can significantly improve the efficiency of both organic and perovskite-based solar cells. Its excellent conductivity enhances charge collection and accelerates electron movement, which could lead to cheaper, more efficient solar panels. Thanks to its porous structure, graphene is well suited for hydrogen gas storage, making it a valuable material for fuel cell technology. Graphene-based materials could provide safer and more efficient methods for storing and transporting hydrogen, contributing to the advancement of clean energy solutions (Figure 5.7).

GO membranes are highly effective at filtering nanoparticles, salts, and other contaminants from water, positioning graphene as a key material in the development of more efficient and cost-effective water filtration systems. These membranes could also play a crucial role in desalination and water purification, particularly in regions with limited access to clean water. Its conductive properties can also be used in electronic waste recycling, making it easier to recover valuable metals like gold, silver, and copper from discarded electronics. Graphene-enhanced materials can be used to reduce the carbon footprint of construction and manufacturing industries. For example, graphene can be integrated into cement or other building materials to enhance their thermal conductivity, reducing energy costs in buildings and infrastructure (Figure 5.8).

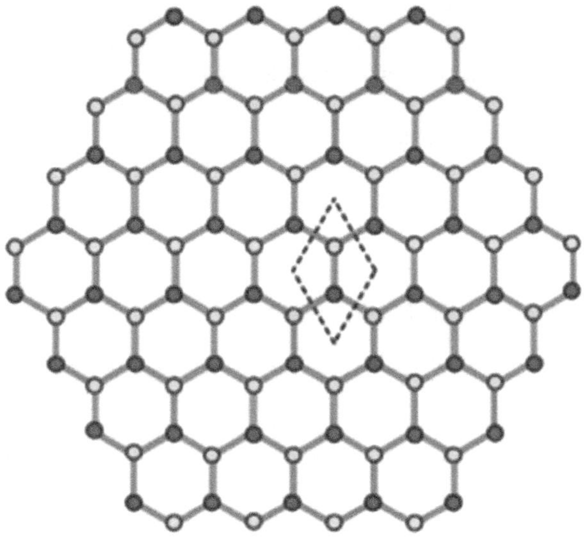

FIGURE 5.7 Graphene unit shell.

(Y. Zhu et al., 2010.)

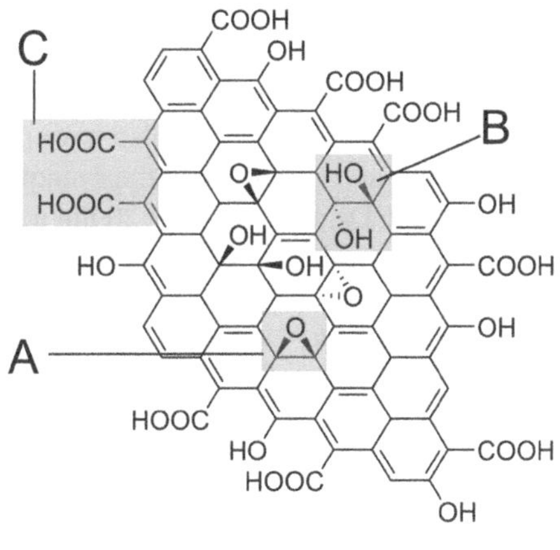

FIGURE 5.8 Structure of GO.

(Paek et al., 2009.)

GO can be functionalized to deliver drugs directly to targeted cells or tissues in the body. This "smart" drug delivery system enhances treatment effectiveness while minimizing side effects, especially in cancer therapy where precision is crucial. Additionally, graphene's electrical properties make it an ideal material for highly sensitive biosensors capable of detecting disease biomarkers, such as those for cancer, diabetes, or infections, at very low concentrations. This enables earlier diagnosis and more personalized treatment plans. Graphene's conductivity also aids in stimulating cell growth and promoting faster healing, particularly in burns or chronic wounds, with some graphene-based dressings already developed to support the healing process. Furthermore, graphene's flexibility and conductivity make it perfect for wearable health devices, such as flexible patches or sensors, which can monitor heart rate, blood glucose levels, and even brain activity in real time, offering vital data to both patients and healthcare providers. Moreover, graphene and GO have demonstrated strong antimicrobial properties, making them ideal for antimicrobial coatings on medical devices, such as catheters, implants, and surgical tools, helping prevent infections and reducing the reliance on antibiotics.

5.2.3 Carbon Dots and Nanospheres

CDs and carbon nanospheres (CNS) are both forms of carbon-based nanomaterials, but they have distinct structural and functional properties (Table 5.2).

CDs consist of carbon atoms arranged in amorphous or graphitic structures, often with oxygen, nitrogen, and other heteroatoms present, contributing to their surface properties. They often have functional groups like hydroxyl, carboxyl, or amino groups, which are responsible for their solubility and stability in aqueous solutions. One of the defining characteristics of CDs is their fluorescence. They exhibit strong, tunable fluorescence across the visible spectrum, which makes them useful for imaging and sensing applications. They are typically non-toxic and biocompatible, which makes them ideal candidates for biomedical applications.

Due to their small size and tunable fluorescence, CDs can be used as imaging agents for live-cell and tissue imaging. They can be used for the detection of various chemical and biological species, including metal ions, small molecules, and biomolecules. Their small size and surface functional groups make them suitable for encapsulating and delivering drugs or other therapeutic agents. CDs also have potential applications in light-emitting devices, solar cells, and LEDs due to their unique optical properties.

CNS are typically larger than CDs, ranging from 20 nm to several micrometers in diameter. They are generally spherical and exhibit a more crystalline structure compared to CDs. Like CDs, CNS are composed of carbon, but they may contain more ordered structures such as graphene-like sheets or fullerene-like arrangements. CNS can exhibit high conductivity, making them useful in electronic and energy storage

TABLE 5.2
Key Differences Between Carbon Dots (CDs) and Carbon Nanospheres (CNS)

Feature	Carbon Dots (CDs)	Carbon Nanospheres (CNS)
Size	Typically <10 nm	Typically >20 nm to several micrometers
Shape	Spherical or quasi-spherical, sometimes irregular	Primarily spherical
Structure	Amorphous or semi-crystalline	More crystalline or ordered
Fluorescence	Strong fluorescence, tunable	Generally, no fluorescence
Conductivity	Low to moderate	High electrical conductivity
Applications	Bioimaging, sensing, drug delivery, solar cells	Energy storage, catalysis, sensors, composites

applications (supercapacitors, lithium-ion batteries, and other energy storage devices). Similar to CDs, CNS can be functionalized with various surface groups to enhance their properties or tailor them for specific applications (e.g., carboxyl, amino, or hydroxyl groups). The feasibility of surface modification makes them suitable for use as catalysts or catalyst supports in chemical reactions, such as hydrogenation or oxidation reactions. CNS are known for their mechanical strength and stability, which make them suitable for various engineering applications. They can be incorporated into polymer composites to enhance mechanical, electrical, and thermal properties.

5.3 SYNTHESIS APPROACHES IN MICROREACTORS

5.3.1 Continuous Flow Methods for Graphene Oxide

GO is a derivative of graphite that contains oxygenated functional groups—such as epoxide, hydroxyl, carbonyl, and carboxyl—that disrupt the sp^2 carbon network. These groups make GO hydrophilic, chemically reactive, dispersible in water, and highly tunable. Key factors such as the degree and type of oxidation, defect density, lateral size, number of layers, and distribution of functional groups greatly influence GO's performance in applications like membranes, sensors, catalysis, composites, energy storage, and biomedicine.

Traditionally, GO is synthesized using batch processes like the Hummers or modified Hummers methods. However, these methods involve harsh chemicals, pose safety risks due to exothermic reactions, and offer limited control over oxidation uniformity, along with long processing times and scale-up challenges. To overcome these limitations, CF and microreactor-based synthesis methods are being investigated. These approaches offer improved control over key reaction parameters—including temperature, flow rate, residence time, mixing, and oxidant delivery—leading to better reproducibility, scalability, and safer operations. This chapter focuses on how precise control of these parameters in CF or microreactor systems can be used to tailor the properties of GO for specific applications (Table 5.3).

Vimalanathan et al. (2022) demonstrates how multiple parameters like oxidant, activation method, flow rate/mixing, and laser power were tuned during a CF fabrication of "green GO" in a vortex fluidic device (VFD) and how it affects the properties of the resultant material. In this work, they used 30% H_2O_2, a greener oxidant compared to permanganate/acids. Initially, they applied UV LED irradiation to generate hydroxyl radicals from H_2O_2, then pulsed near-IR (1064 nm) laser irradiation to oxidize basal plane, combined with high shear from VFD. They varied the rotational speed of the VFD (4000–9000 rpm), the inclination angle (0°–90°), and flow rate under CF. They found the optimal conditions: ~8000 rpm, 45° inclination, flow rate 0.5 mL/min. They also varied NIR laser pulse energy (260, 450, 650 mJ), which impacted the degree of oxidation in the basal plane. At higher power, oxidation approached levels of the conventional Hummers' method; at lower power, it resulted in lower oxidation but still stable colloids. By varying these parameters, they could control where oxidation occurs (edges vs basal plane), the oxidation degree (atomic % oxygen), colloidal stability (lifetime of dispersion), emission properties (optical), maintenance of large intact sp^2 domains for electrical

TABLE 5.3
Key Parameters in Continuous-Flow GO Synthesis

Parameter	What It Affects	Why Critical
Temperature (bulk, local/hot spots, activation)	Rate of oxidation; extent of oxidation; type and distribution of functional groups; defect density; exfoliation; stability of GO; unwanted side reactions (over-oxidation, degradation)	Temperature influences reaction kinetics, activation of oxidants, radical formation, thermal degradation, etc. In microflow reactors local heating (e.g., by photonic or laser excitation) can produce hot spots. The average and peak temperatures both matter
Flow rate/residence time/mixing intensity	Controls exposure time between oxidant and graphite; how uniform the oxidation is across particles; exfoliation; size of flakes; layer number; throughput vs. quality trade-off	Faster flow (shorter residence time) tends to reduce oxidation and defect formation but may give lower yield; slower flow increases oxidation but risks over-oxidation or defects. Mixing intensity ensures uniform contact, avoids concentration gradients
Catalyst/oxidant/activation method	Type of oxidant (e.g., $KMnO_4$, H_2O_2, etc.); presence of catalysts or promoters; whether activation by light, electrochemical, or thermal is used; this influences which functional groups are formed, where (edges vs basal plane), oxidation degree, and how clean the product is	Choice of oxidant/catalyst also influences safety, cost, environmental impact. Activation methods (photonic, catalytic, electrochemical) enable milder conditions or selective oxidation. Catalysts (or catalytic surfaces) may accelerate oxidation or guide its selectivity

or optical applications. Presowitz et al. (2021) explained reduction/exfoliation and graphene gel formation in glycol-thermal CF system. Here they showed how flow rate, temperature, solvent environment, and addition of chemical additives (e.g., base) influence the properties of the gel/reduced GO network. The ethylene glycol solvent was heated to about 150°C to affect thermal deoxygenation and gel formation. They used polytetrafluoroethylene (PTFE) tubing with flow rate ~45 mL/h, giving ~22 min residence time in the heated tubing for the reaction. Addition of NaOH to sequester CO_2 evolved during deoxygenation to prevent blockages (because CO_2 bubbles can disrupt flow or eject the gel prematurely). This chemical additive is a kind of "catalyst"/process enhancer in a broad sense. The resulting gel (or xerogel) had moderate oxidation retained, and after further annealing could be converted to higher conductivity materials. The density, volumetric capacitance, and mechanical stability depend heavily on deoxygenation extent, residence time, and temperature.

5.3.2 How Temperature, Flow Rate, and Catalysts/Activation Interact

These parameters do not operate independently. Below are key interactions and trade-offs to be aware of, plus design considerations for CF GO synthesis:

Temperature

- **Bulk temperature vs local/transient heating**: In microreactors or flow devices, local heating (e.g., from laser irradiation or photonic activation) or hot spots from mixing/shear can give effective temperatures much higher than the bulk. The design must account for uniform heat distribution or controlled local heating depending on whether basal plane activation is desired.
- **Thermal stability**: Too high a temperature or too long an exposure may degrade the carbon backbone, cause over-oxidation, create excessive defects, or fragment sheets.
- **Oxidant decomposition/activation**: Many oxidants require certain activation energies (e.g., H_2O_2 needs UV to generate radicals; $KMnO_4$ reacts more under elevated temperatures). Temperature also affects the rate of oxidation, radical lifetime, and diffusion.
- **Exfoliation assistance**: Higher temperature can assist exfoliation (thermal stress, expansion of interlayer water), but must be balanced to avoid damage.

Flow Rate and Residence Time

- **Residence time** is inversely related to flow rate (for fixed reactor volume). Longer residence time = more oxidation (or reduction, depending on the stage) and more exposure. But beyond a point, more residence does not always improve quality (diminishing returns) and may increase defects or wastage.
- **Flow rate/mixing intensity** also affects how well the oxidant interacts with graphite particles: turbulence or shear helps mixing; laminar flow may cause gradients of concentration.
- **Solid handling**: Graphite/flakes can cause clogging if flow slows too much or if gel forms prematurely. Adjust flow so that the particles stay suspended, avoid settling. Also, flow geometry (e.g., segmented flow, plug flow) helps.
- **Scale vs quality trade-off**: Higher flow rates desired for throughput may compromise oxidation uniformity or consistency, unless reactor design compensates (more mixing, better activation, etc.).

Catalyst/Oxidant/Activation Method

- **Oxidant type**: Classical oxidants (e.g., $KMnO_4$, strong acids) give high oxidation but also produce more defects, waste, safety concerns. Greener oxidants like H_2O_2 are safer; activation (e.g., UV) or cascaded activation (UV + NIR) improves their performance.

- **Catalytic or photonic or electrochemical activation**: Methods like UV or laser irradiation, or electrochemical oxidation, allow greater selectivity (edges vs basal plane), lower defect generation, or the possibility to perform oxidation under milder thermal conditions.
- **Additives/promoters**: For example, bases to capture CO_2 (as in the gel flow case) to manage by-product management, surfactants or stabilizers to prevent aggregation, catalysts or light absorbers that localize energy input.
- **Catalyst surfaces**: In some designs, catalytic surfaces or reactor wall materials may participate (e.g., plasmonic materials, or photocatalysts) to enhance oxidation.

5.4 CONTINUOUS-FLOW METHODS FOR CNTs

CNTs possess exceptional mechanical, electrical, and thermal properties, making them highly desirable for a wide range of applications in nanotechnology, electronics, materials science, and energy systems. Among various synthesis methods, CF-CVD has emerged as a highly scalable and controllable approach, offering advantages in yield, reproducibility, and customization (Sano et al., 2010; Pérez-Cabero et al., 2003). Precise control over temperature, gas flow rate, and catalyst composition can be used to tailor key CNT properties—such as diameter, chirality, alignment, and wall number. Emphasis is placed on the mechanistic understanding of each parameter and how their interplay determines CNT quality and characteristics.

CF-CVD involves a constant and controlled flow of gaseous precursors through a heated reaction zone where catalytic decomposition leads to the formation of CNTs. This approach contrasts with batch processes by enabling steady-state reaction conditions, better scalability, and fine parameter modulation. Horizontal tube furnaces are the most common reactor configurations in CNT synthesis, which offer uniform thermal profiles of synthesized CNTs. Fluidized bed reactors, another variety of reactor configuration, provide dynamic catalyst exposure and are very useful for large-scale production. Microreactors and millifluidic systems allow for rapid heating and quenching, enhancing temporal control (Tran et al., 2007).

Temperature plays an important role in the growth of CNTs. It governs the decomposition of carbon precursors and the activity of catalyst nanoparticles, affecting nucleation and elongation processes. It also influences diameter of CNTs (via catalyst particle size evolution), number of walls (SWCNT vs. MWCNT formation), growth rate and yield, as well as structural quality and defect density in CNTs. A typical temperature range from 700 to 900°C is considered as optimum for synthesis of SWCNTs, whereas a moderately broader range from 600 to 1200°C is suitable for MWCNTs. Another synthetic technique associated with low-temperature synthesis of CNTs is enabled via plasma-enhanced CVD (PECVD).

Controlled thermal gradients can induce aligned CNT growth, especially when combined with substrate patterning. Sharp gradients are also used to spatially tune diameter and chirality along CNT lengths. Gas flow rates also influence CNT length and uniformity, aggregation, and building. Flow rate determines residence time, precursor concentration, and reaction kinetics. High flow rates promote rapid growth, but can lead to poor crystallinity and shorter CNT lengths. Low flow rates, on the other hand, enable longer periods and higher crystallinity but may lead to lower yield. Pulsed flow approaches, in this connection, improve control over nucleation vs elongation stages. Use of inert gases like Ar/N_2 in CVD controls ambient pressure and prevent unwanted reactions, whereas reactive gases (H_2, O_2 etc.) help to modify growth environment of CNTs and etch amorphous carbon.

Again, choice of proper catalysts during synthesis in terms of their types (typically transition metals: Fe, Co, Ni etc.), morphology (nanoparticles/thin films/MOFs), particle size, and distribution as well as catalyst composition (Fe-Co, Ni-Mo) strongly dictate the CNT type (SWCNT vs. MWCNT), chirality, electronic properties, growth yield, and rate. Smaller particles tend to favor SWCNTs with smaller diameters; larger particles promote MWCNTs. Monodispersity of catalysts is crucial for uniform CNT formation. Support materials (e.g., Al_2O_3, SiO_2) influence catalyst mobility and aggregation. Proper catalyst composition (bimetallic catalyst) offers enhanced activity, wider temperature stability, and potential chiral selectivity in CNTs. Recent research focuses on chirality-controlled growth of CNTs using doped or alloyed nanoparticles with specific crystallographic orientations.

5.5 CONTINUOUS-FLOW METHODS FOR CARBON DOTS

CF methods for synthesizing CDs have become increasingly popular due to their scalability, reproducibility, and enhanced control over reaction parameters. These methods are advantageous compared to traditional batch processes, especially for industrial applications and large-scale production (Rao et al., 2017; Somaprakash & Badiya, 2025).

The main CF synthesis methods used for CDs include i) microreactor-based synthesis, where reactions are carried out in microfluidic chips or capillary microreactors where precursors continuously flow through heated zones; high heat and mass transfer efficiency, precise control over reaction time and temperature, and scalability are some advantages of this particular process; ii) millifluidic reactor systems are similar to microreactors but with larger channels (0.1–1 mm), allowing for higher throughput; iii) in hydrothermal flow synthesis, precursors are heated under pressure in a tubular reactor system, mimicking batch hydrothermal synthesis but in continuous mode. This environmentally friendly ("green synthesis") process is compatible with use of biomass/renewable feedstocks for the synthesis of CDs; iv) microwave-assisted flow synthesis is another technique where microwave heating is applied to a CF of reactants in microwave-transparent tubing; v) in ultrasound-assisted flow synthesis, ultrasound energy enhances reaction kinetics during flow synthesis. This technique enhances nucleation and size control and is really useful for synthesis of functionalized or doped CDs.

5.6 CONCLUSIONS

Microreactor technology translates fundamental transport advantages at the sub-millimeter scale into practical gains in safety, selectivity, and speed. High surface-area-to-volume ratios, short diffusion paths, and predictable laminar RTD enable precise thermal histories, efficient interfacial renewal, and accurate residence-time dosing—capabilities that directly improve yield and product quality while shrinking risk and footprint. The same attributes accelerate development cycles and simplify technology transfer via numbering-up. Yet success depends on engineering around micro-scale limitations: preventing fouling, ensuring uniform distribution across channels, maintaining catalyst activity, and balancing pressure drop against pump capacity. A thermo-fluid design framework, supported by CFD, underpins solutions to those challenges. As a synthesis platform, microreactors open controlled, continuous routes to nanomaterials such as GO, CNTs, and CDs, where temperature, flow, mixing, and activation can be orchestrated to tailor structure and properties at scale. Material selections (metal, silicon, glass/ceramic, and polymer) and catalytic architectures (wall-coated vs packed/structured beds) are chosen to match reaction severity and media. Taken together, the principles summarized here position microreactors as versatile building blocks for decentralized and sustainable manufacturing—from pharmaceutical and catalytic processes to advanced materials and bio-based products. The remaining chapters apply these fundamentals to design strategies, case studies, and CF syntheses that illustrate how microreactors move from concept to reliable, high-performance practice.

REFERENCES

Bae, K., Kim, D., Dung, P. A., Lee, D., Hwang, B., Go, K. S., Kim, W., Lee, J. K., Im, J. S., Kang, S. C., & Lee, S. H. (2024). Simultaneous and continuous production of carbon nanotubes and hydrogen by catalytic CH_4 decomposition in a pressurized fluidized-bed reactor. *Industrial & Engineering Chemistry Research*, *63*(2), 930–941. https://doi.org/10.1021/acs.iecr.3c02778

Bugay, C. A., Caballas, M. C., Mercado, J. F., Rubio, P. K., Serote, P. N., Villarte, R. V. C., & Rubi, A. (2024). A review of microreactors for process intensification. *Engineering Proceedings*, *67*(1), 21. https://doi.org/10.3390/engproc2024067021

Cao, Q., Sang, L., Tu, J., Xiao, Y., Liu, N., Wu, L., & Zhang, J. (2021). Rapid degradation of refractory organic pollutants by continuous ozonation in a micro-packed bed reactor. *Chemosphere*, *270*, 128621. https://doi.org/10.1016/j.chemosphere.2020.128621

Ehrfeld, W., Hessel, V., & Löwe, H. (2000). *Microreactors: New technology for modern chemistry*. Wiley-VCH.

Garg, B., Bisht, T., & Ling, Y.-C. (2014). Graphene-based nanomaterials as heterogeneous acid catalysts: A comprehensive perspective. *Molecules*, *19*(9), 14582–14614. http://dx.doi.org/10.3390/molecules190914582

Gokhale, S. V., Tayal, R. K., Jayaraman, V. K., & Kulkarni, B. D. (2005). Microchannel reactors: Applications and use in process development. *International Journal of Chemical Reactor Engineering*, *3*, R2.

Hessel, V., Hardt, S., & Löwe, H. (2005). *Chemical micro process engineering: Fundamentals, modelling and reactions* (pp. 125–147). Wiley-VCH Verlag GmbH.

Kashid, M. N., Agar, D. W., & Turek, S. (2007). CFD modelling of mass transfer with and without chemical reaction in the liquid–liquid slug flow microreactor. *Chemical Engineering Science*, *62*(18–20), 5102–5109. https://doi.org/10.1016/j.ces.2007.01.068

Kim, T., & Kwon, S. (2006). Catalyst preparation for fabrication of a MEMS fuel reformer. *Chemical Engineering Journal*, *123*, 93–102.

Kiwi-Minsker, L., & Renken, A. (2005). Microstructured reactors for catalytic reactions. *Catalysis Today*, *110*(1–2), 2–14.

Kolb, G., & Hessel, V. (2004). Micro-structured reactors for gas phase reactions. *Chemical Engineering Journal*, *98*, 1–38.

McMullen, J. P., & Jensen, K. F. (2010). Integrated microreactors for reaction automation: New approaches to reaction development. *Annual Review of Analytical Chemistry*, *3*, 19–42.

Natarajan, Y., Nabera, A., Salike, S., Tamilkkuricil, V. D., Pandian, S., Karuppan, M., & Appusamy, A. (2019). An overview on the process intensification of microchannel reactors for biodiesel production. *Chemical Engineering and Processing: Process Intensification*, *136*, 163–176. https://doi.org/10.1016/j.cep.2018.12.008

Paek, S. M., Yoo, E., & Honma, I. (2009). Enhanced cyclic performance and lithium storage capacity of SnO_2/graphene nanoporous electrodes with three-dimensionally delaminated flexible structure. *Nano Letters*, *9*(1), 72–75.

Papautsky, I., Ameel, T., & Frazier, A. B. (2001, November 21–26). A review of laminar single-phase flow in microchannels. *Proceedings of ASME International Mechanical Engineering Conference and Exposition*, New York.

Pérez-Cabero, M., Rodríguez-Ramos, I., & Guerrero-Ruiz, A. (2003). Characterization of carbon nanotubes and carbon nanofibers prepared by catalytic decomposition of acetylene in a fluidized bed reactor. *Journal of Catalysis*, *215*(2), 305–316. https://doi.org/10.1016/S0021-9517(03)00026-5

Prestowitz, C.O. L., & Huang, J. (2021). Glycol-Thermal continuous flow synthesis of graphene gel. *ACS Omega*, *6*(29), 18663–18667. https://doi.org/10.1021/acsomega.1c02589

Rao, L., Tang, Y., Li, Z., Ding, X., Liang, G., Lu, H., Yan, C., Tang, K., & Yu, B. (2017). Efficient synthesis of highly fluorescent carbon dots by microreactor method and their application in Fe3+ ion detection. *Materials Science and Engineering: C*, *81*, 213–223. https://doi.org/10.1016/j.msec.2017.07.046

Sano, N., Ishimaru, S., & Tamaon, H. (2010). Synthesis of carbon nanotubes in graphite microchannels in gas-flow and submerged-in-liquid reactors. *Materials Chemistry and Physics*, *122*(2–3), 474–479. https://doi.org/10.1016/j.matchemphys.2010.03.029

Scarselli, M., Castrucci, P., & De Crescenzi, M. (2012). Electronic and optoelectronic nanodevices based on carbon nanotubes. *Journal of Physics: Condensed Matter*, *24*(31). https://doi.org/10.1088/0953-8984/24/31/313202

Somaprakash, N., & Badiya, P. K. (2025). Biodegradable brightness: Green synthesis of carbon dots and their applications. *ACS Applied Engineering Materials*, *3*(7), 1943–1959. https://doi.org/10.1021/acsaenm.5c00294

Tonomura, O., Noda, M., Kano, M., & Hasebe, S. (2004a, October 17–21). Optimal design approach for microreactors with uniform residence time distribution. *10th Asian Pacific Confederation of Chemical Engineering (APCChE)*, 4B-07, Kitakyushu, Japan.

Tonomura, O., Noda, M., Kano, M., & Hasebe, S. (2004b). Thermo-fluid design approach to microreactors with uniform temperature and residence time distribution. *AIChE Annual Meeting*, 430d.

Tran, K. Y., Heinrichs, B., Colomer, J.-F., Pirard, J.-P., & Lambert, S. (2007). Carbon nanotubes synthesis by the ethylene chemical catalytic vapour deposition (CCVD) process on Fe, Co, and Fe–Co/Al_2O_3 sol–gel catalysts. *Applied Catalysis A: General*, *318*, 63–69. https://doi.org/10.1016/j.apcata.2006.10.042

Vimalanathan, K., Scott, J., Pan, X., Luo, X., Rahpeima, S., Sun, Q., … Raston, C. L. (2022). Continuous flow fabrication of green graphene oxide in aqueous hydrogen peroxide. *Nanoscale Advances*, *4*, 3121–3130.

Wang, J., Xuan, Y., & Zhu, Q. (2025). Versatile microreactor for solar-driven photothermal catalytic CO_2 reduction into fuels. *Energy & Fuels*, *39*(6), 3308–3318. https://doi.org/10.1021/acs.energyfuels.4c05586

Zhu, Y., Murali, S., Cai, W., Li, X., Suk, J. W., & Potts, J. R. (2010). Graphene and graphene oxide: Synthesis, properties, and applications. *Advanced Materials*, *22*, 3906–3924.

6 Metal and Metal Oxide Nanoparticle Synthesis Using Microreactors

Mugdha Sinha, Taisia Miroshnichenko, and Priti Sundar Mohanty

6.1 INTRODUCTION

The most important component in making a nanostructure is a nanoparticle (NP). The size of a NP typically falls between 1 and 100 nm. Lowering the melting points, increasing specific surface areas, distinct optical characteristics, mechanical strengths, and distinct magnetizations are just a few of the physical and chemical characteristics that set metallic NPs apart from bulk metals and may make them appealing for a range of industrial uses (Horikoshi and Serpone 2013). NPs, which are three-dimensional, nanoscale particles, include dendrimers, liposomes, nanorods, nanospheres, nanowires, nanobelts, nanorings, nanopores, nanotubes, quantum dots, and nanoshells (Subramani and Ahmed 2012). Drug delivery systems, targeting cancer, dentistry, and other fields like targeted and controlled-release drug delivery use NPs (Khan et al. 2025). NPs can be employed as efficient delivery and transportation systems because they enter cells more readily than bigger microparticles.

NPs are nanomaterials that have all three dimensions at the nanoscale. The four categories of nanomaterials include dendrimers, composites, metal-based materials, and carbon-based materials. Intentionally created nanomaterials that fit into these four primary categories have special qualities that provide them imaging, thermal, mechanical, medicinal, and commercial capabilities.

6.2 OVERVIEW OF MICROREACTOR TECHNOLOGY

The use of tiny, frequently microfabricated reactors to carry out chemical reactions with exact control over reaction conditions is known as microreactor (MR) technology. This novel method facilitates effective mass transport, heat transfer, and mixing at the microscale, which makes it perfect for industrial uses of organometallic catalysts where maximizing reaction parameters is essential to increasing yields and selectivity (Ahmed-Omer, Brandt, and Wirth 2007).

Compact nuclear reactors known as MRs are 100–1,000 times smaller than traditional ones and may produce up to 20 MW of thermal energy for use in industrial processes like desalination and hydrogen production as well as electricity. They may

DOI: 10.1201/9781003633761-6

be quickly deployed to remote locations, military installations, or disaster areas because they are factory-built and portable, providing dependable, carbon-free electricity for years without the need for refueling. They are a versatile option for both on-grid and off-grid requirements due to their portability, scalability, and compatibility with various energy systems. With demonstrations anticipated in the coming years, MRs are gaining traction as a game-changing clean energy technology due to their built-in safety measures, ease of use, and robust backing from both business and the government (Rahimnejad 2023).

6.2.1 Basic Design and Working Principles of Microreactor

A pump, micromixer, microchannels or tubes, and a quench zone make up a basic MR system as shown in (Figure 6.1). Syringes, HPLC, or peristaltic pumps are frequently utilized for continuous operation. Although they have a small capacity (less than 500 mL), syringe pumps provide fine flow control. Because of valve failure, HPLC pumps are not appropriate for viscous fluids, even if they operate at high pressure and high flow rates. Although they can handle viscous, slurry, and shear-sensitive fluids, peristalsis pumps only provide low-rate pulsed flow.

An essential tool for effective mixing is the micromixer. They can be categorized as either passive (using geometry to create chaotic advection) or active (using external energy—such as sound, magnetic, electric, thermal, or pressure fields). Active mixers improve mixing but are more complicated. One example is polydimethylsiloxane (PDMS) devices containing magnetic microneedles (Zhao et al. 2011). Common passive micromixers include split-and-recombine, helical/curved, multi-lamination, and simple contacting structures (T-, Y-, and cross-type).

Although they are simple to construct, simple contacting mixers are less efficient in multiphase flows. While T-type mixers have been utilized to generate paclitaxel-loaded chitosan NPs, cross mixers provide the best particle homogeneity among them. Split-and-recombine mixers have high pressure drops yet perform well with viscous and multiphase fluids.

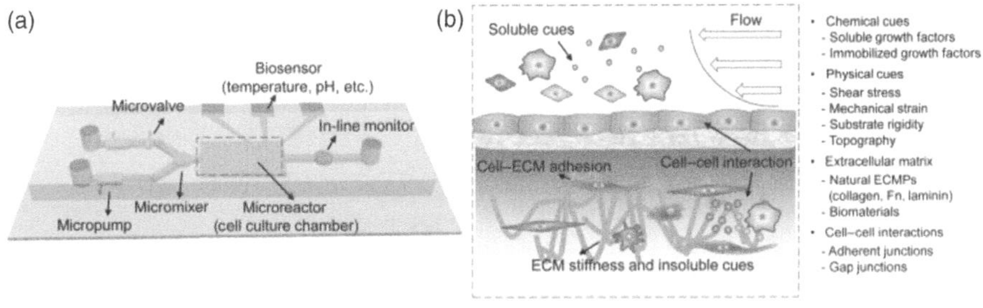

FIGURE 6.1 Few essential MR design concepts. (a) The MR chamber, biosensors, in-line monitor element, and some fluidic control components such as micropump, microvalve, and micromixer are among the various parts of a MR. (b) A thorough diagram showing the variables that affect the MR chamber.

(Shrimal, Jadeja, and Patel 2020.)

6.2.2 Components of Microreactor Design

The structural elements include microchannels and casings composed of silicon, glass, steel, and polymers. MR designs include fuel and moderator for nuclear fission or catalysts and reactants for chemical processes as shown in Figure 6.2. The heat removal/transfer system, which transports heat to an intermediate heat exchanger, is an essential part. It frequently uses heat pipes or coolant loops like helium or molten salt. Additional parts include pumps for constant flow, control systems for temperature and pressure regulation, and multi-component designs for output scaling, like "numbering-up" arrays and parallel microchannels.

For various uses, there are three types of MRs: capillary, plate, and laminar. As a single channel, MRs can deliver more mass and heat transfer rates for a multiphase system. MRs must be scaled up for industrial-scale production in order to attain the necessary throughput range that can be accomplished by numbering-up a single channel. In early works, one of the major predictions of the benefits of MRs was numbering- or scaling-up (i.e., repeated microchannel processing units in parallel) (Schenk et al. 2004, Jensen 2001).

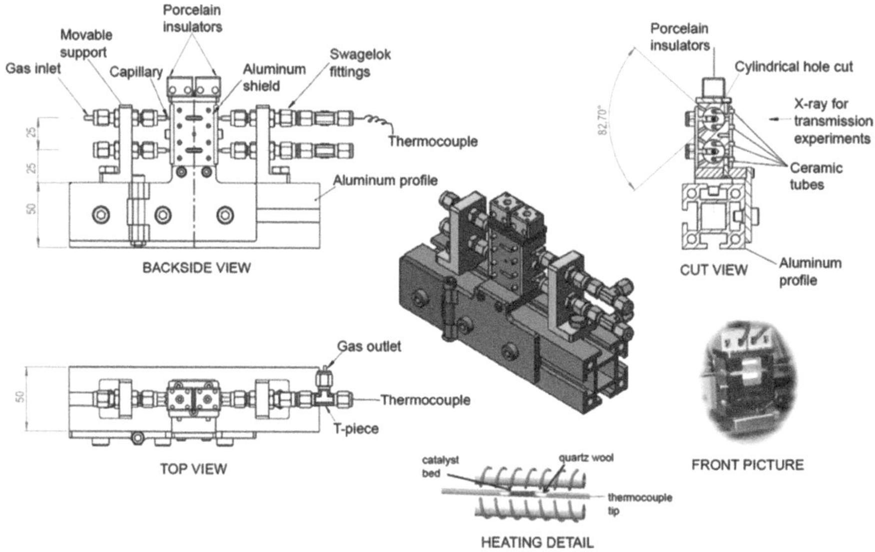

FIGURE 6.2 Technical design and components of a MR. The fully completed flow cell is displayed in the center; the relative positions of the catalyst bed and thermocouple tip between the quartz wool are depicted in the heating detail. The "front picture" shows the flow cell/furnace operating at 1073 K; the Kapton tape is meant to prevent air currents from abruptly altering the temperature.

(Figueroa et al. 2013.)

6.2.3 Microreactor Setup

A MR chip or tube, quenching or collecting zones, and fluid delivery components (such as pumps) are common components of a chemical synthesis MR setup. Reaction channels, micromixers, accurate pumps, and quick cooling systems that stop additional reactions are essential components for chemical applications. A MR configuration for nuclear applications uses a mobile, self-contained unit to generate power, necessitating a specific fuel containment system for radioactive materials.

In order to facilitate controlled synthesis and product collection, a MR system usually comprises a pump, a microchip, and a sample collector. A MR with three inlets for introducing various reagents and a single outlet for collecting the finished product is a popular design. The MR can be separated into different functional zones when seen in cross section. Before moving on to the next phase, Zone I is devoted to reagent mixing, making sure that the introduced solutions mix effectively. The production of stable NPs is the final result of processes including reduction, sluggish nucleation, and rapid autocatalytic growth that take place in Zone II, the principal reaction region as shown in Figure 6.3. Lastly, Zone III denotes the outlet area where the finished product is gathered for analysis or later use (Pach et al. 2024, Murphy et al. 2011).

6.2.4 Continuous and Batch Flow Reactors

A comparison of continuous flow reactors versus batch reactors reveals a number of significant variations in their characteristics. Continuous flow reactors are more sophisticated than batch reactors, which are typically employed during the early stages of research. Batch reactors are simpler to operate than continuous flow reactors, which are comparatively more challenging. Continuous flow reactors minimize the response time to seconds or a few minutes, whereas batch reactors might take minutes or even hours as shown in Figure 6.4. Additionally, the heating period in continuous flow reactors is brief to very short, often only a few seconds, whereas in batch reactors it is longer, frequently requiring several minutes. Continuous flow

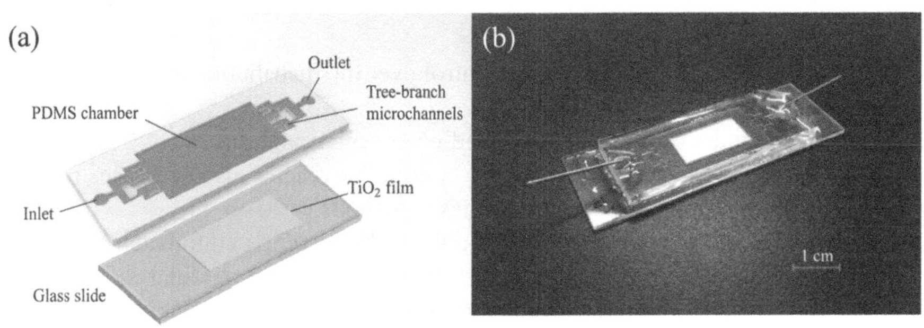

FIGURE 6.3 MR diagram displaying the setup, inlets and outlets, and functional zones for mixing, reaction, and product collection.

(He et al. 2023.)

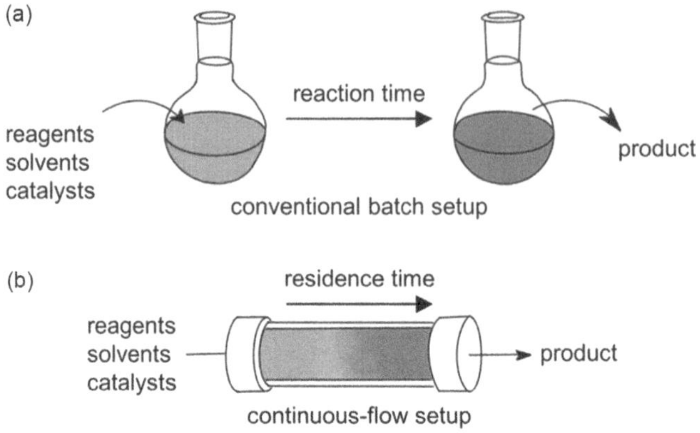

FIGURE 6.4 Difference between (a) batch and (b) continuous flow reactions.

(Laporte et al. 2024.)

reactors can create particles in the nanoscale range, while batch reactors typically produce particles on the order of a few microns. Last but not least, continuous flow reactors enable significant operational control, whereas batch reactors only provide a small operating range (He et al. 2023).

6.2.5 Advantages of Continuous Flow Reactor Over Batch Reactor

In order to synthesize customized nanomaterials, reactor equipment is crucial, both in the lab and for expanding the synthetic processes into large-scale manufacturing. While there are challenges associated with the sequential three-stage scale-up from the conventional process (i.e., lab-scale, pilot-scale, and large-scale) using multipurpose batch reactors, the interface of continuous flow microfluidic chemistry and nanomaterials synthesis processes has shown comparatively better process performance over conventional technologies. As a result, continuous flow microfluidics systems have shown a lot of interest lately as flexible instruments for the production of different structured nanomaterials.

1. Continuous reactor gives better control over the distribution and size of the particle.
2. Heat and mass transfer is proper and gives high scalability.
3. High rate of heating and cooling in continuous reactor also decreases reaction time in high-temperature synthesis methods.
4. It gives a higher production rate and throughput compared to batch reactor.
5. It is possible to build continuous flow synthesis as a modular "plug and play" system that makes it simple to incorporate different synthesis stages, such as nucleation and growth, into an ongoing process.

6.2.6 Microfluidics and Flow Chemistry Concepts

Flow chemistry, sometimes referred to as continuous flow or plug flow chemistry, is the process of conducting reactions in a stream that flows continuously via small channels or MRs. It has many benefits over conventional batch processes,

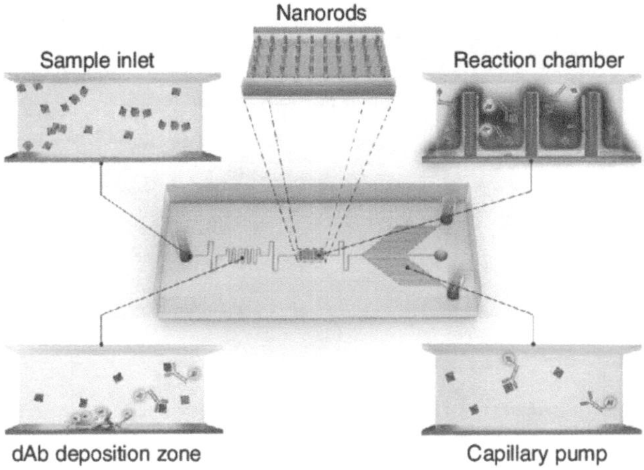

FIGURE 6.5 Flow-through microfluidic device with Au nanorods for one-step protein biomarker immunoassay, showing target capture and enhanced fluorescence detection.

(Wang et al. 2021.)

including better heat and mass transfer, faster response times, increased conversion and selectivity, and enhanced safety. Higher product quality and production are guaranteed when minimal reagent amounts are used with intensive mixing, which also reduces byproducts. Flow chemistry is also very effective for medicines, fine chemicals, and advanced research applications because it makes it easier to scale up, optimizes reaction conditions more quickly, and integrates with inline analytical instruments for automated, self-optimizing processes. Figure 6.5 represents an example of a flow-through microfluidic system that incorporates Au nanorods for protein biomarker immunoassay in a single step. The fluorescent dAb that has been prepatterned in the deposition zone is introduced and coupled with the serum sample that contains the target biomarkers. cAb coupled with dense arrays of embedded Au nanorods captures the complex as it passes through the reaction chamber. A microscope with a CCD camera is used to measure and evaluate the augmented fluorescent signals. The capillary pump determines the flow rate and overall reaction time.

6.3 MECHANISM OF NANOPARTICLE FORMATION IN MICROREACTORS

Controlled chemical synthesis involving the phases of nucleation (first generation of stable particles) and growth (increase in particle size) is necessary for the formation of NPs in MRs as shown in Figure 6.6. Highly effective micromixing of reactants is made possible by MRs, which results in quick and consistent supersaturation, which is essential for regulated nucleation. Monodisperse NPs with specific properties are produced by this exact mixing, which also reduces reaction durations and permits independent control over the nucleation and growth stages.

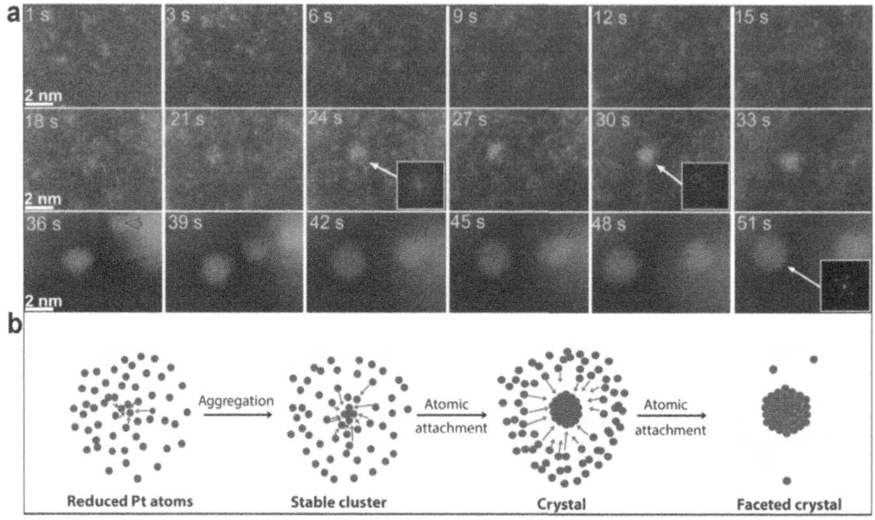

FIGURE 6.6 Two steps of nucleation and atom-by-atom attachment are used to create NPs. (a) A sequence illustrating the formation of Pt NPs with atoms migrating and attaching to a cluster to form a single crystal of around 2 nm. (b) Atom-by-atom nucleation and early growth schematic.

(Yadavalli and Shukla 2017.)

6.3.1 Nucleation and Growth Mechanism

Nucleation is the process where monomers like atoms, ions, or molecules form a new thermodynamic configuration or structure at the atomic or molecular level, further leading to growth during which monomers get incorporated into the surface of the nuclei, which may also aggregate, leading to an increase in size (Figure 6.7).

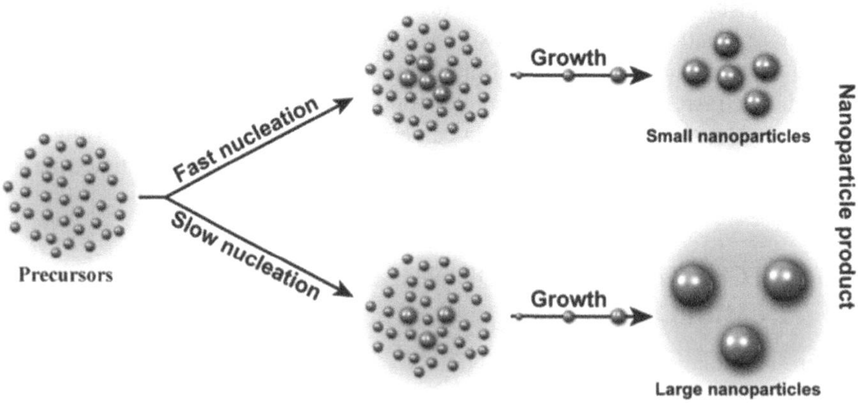

FIGURE 6.7 Schematic representation of the nucleation and growth of NPs.

(Shevchenko et al. 2003.)

The overall decrease in (Gibbs) free energy is the thermodynamic driving force behind both nucleation and growth in solution. The chemical potential gradient between a monomer in solution and in the new phase, the transport "resistance" of the monomers in the solution, and the "concentration" of the monomers are the three competing elements that often define their rates or kinetics. Nucleation and expansion frequently happen at the same time, making them dynamically competitive. The nucleation and growth processes, which are crucial for controlling the precise structural characteristics and attributes of the final solid material, have an effect on every aspect of the industry that relies on a certain set of these materials' capabilities (Wu et al. 2022).

6.3.2 Some Factors Which Affect Biosynthesis of Nanoparticles

There are different methods for synthesizing NPs, ranging from physical techniques using mechanical procedures to chemical or biological protocols using various organic or inorganic chemicals and living organisms. Each procedure has its special benefits and drawbacks. However, biological methods for the synthesis of NPs use nontoxic and environmentally benign materials in conjunction with green technology and are therefore eco-friendly and more acceptable than traditional methods. Some factors that affect NP biosynthesis are described below.

1. **Particular Method**: There are a number of ways of preparing or synthesizing NPs from chemical to mechanical or biological processes with the help of different organic or inorganic chemicals or organisms. Biological methods use nontoxic and eco-friendly materials with green technology for the synthesis of NPs, so they are more acceptable than traditional methods (Kharissova et al. 2013).
2. **pH**: In a solution, pH affects the size and texture of the NP which is synthesized. Through green technology, pH affects the synthesis of NPs. We can therefore alter the size of a NP by changing the pH of the media (Gardea-Torresdey et al. 1999).
3. **Temperature**: The synthesis of NPs is affected by temperature, where the physical method needs a maximum temperature of >350°C and the chemical methods need a temperature which is less than 350°C. The reaction medium's temperature determines the nature of the NP formed (Armendariz et al. 2004).
4. **Pressure**: The shape and size of the NP are affected by the pressure. At optimum pressure the rate of reduction of metal ions has been faster (Rai et al. 2006).
5. **Time**: The length of time affects the quality and type of the NP produced by green synthesis. The synthesis procedure, exposure to light, and storage conditions all have a significant impact on the properties of the produced NPs, which also change over time. Aeration in time can happen in different ways, which includes particle aggregation from prolonged storage, particle growth or shrinkage over long time of storage, shelf life, and other factors that impact their potential (Rai et al. 2006).

6. **Preparation cost**: Although the chemical method of synthesis yields a high cost, the cost-effectiveness of the production procedure is a significant factor that influences NP synthesis. This is because the costs associated with the synthesis of NPs must be regulated and controlled to facilitate their potential application in modern uses.
7. **Particle Shape and Size**: The characteristics of NPs are significantly affected by their size. It has been shown that when NPs get smaller, their melting point drops. Because of their comparable energy, NPs with various configurations can easily change shape (Kuchibhatla et al. 2012). The form of the NP changes depending on the energy type that is commonly used in NP analysis. The form and dynamic nature of the generated NPs have a major impact on their chemical properties.

6.4 SYNTHESIS OF METAL NANOPARTICLE BY MICROREACTOR

A distinct class of nanomaterials, metal NPs are distinguished by their exceptional optical, electrical, and catalytic characteristics that set them apart from their bulk counterparts. They are very adaptable in areas like medical, imaging, sensing, catalysis, and environmental applications due to their small size and high surface-to-volume ratio. Due to their strong surface plasmon resonance (SPR), which enables improved optical responses and applications like surface-enhanced Raman scattering (SERS) and targeted drug delivery, noble-metal NPs such as gold (Au) and silver (Ag) are particularly significant among them (Yadavalli and Shukla 2017, Dachraoui et al. 2021).

6.4.1 Gold Nanoparticle (Au)

Gold NPs are stable and biocompatible. They frequently have dielectric shells (such as oxides) applied to them to avoid aggregation and further adjust their characteristics, or they are joined with other metals to create core-shell structures (Hassan et al. 2022).

6.4.2 Silver Nanoparticle (Ag)

This process includes reduction of silver ions by a reducing agent in an alkaline environment for better electron transfer and particle formation.

6.4.3 Chemical Synthesis of Gold or Silver Nanoparticle

The manufacture of CH-AgNPs uses 0.75 mM silver nitrate and 38 mM trisodium citrate dehydrate (Na3C6H5O7·2H2O). 50 mL of AgNO3 and 10 mL of trisodium citrate were combined in a dropwise manner and heated to 90°C. CH-AgNPs were produced by chemical degradation and erosion over time, and the translucent light hue changed to brownish black to show the existence of AgNPs. However, for CH-AuNPs, 1.0 mM $HAuCl_4$, a 1% trisodium citrate dehydrate solution, and the chemical production of AuNPs were employed. 20 mL of chloroauric acid combined with 2 mL of 1% trisodium citrate dihydrate solution were continuously rolled and heated with a magnetic stir bar. Trisodium citrate is used in this method to decrease the gold solution, which results in a mild reddish violet coloration. This reddish violet color indicates AuNPs (Tyagi et al. 2021).

6.4.4 Biological Synthesis of Silver and Gold Nanoparticle

As a precursor to create Bio-AgNPs, 5 mL of 0.75 mM $AgNO_3$ aqueous solution was added to 95 mL of plant leaf extract (supernatant) in conical flasks. For 72 hours, this solution was shaken at 30°C and 150 rpm in an incubator. The extract of this solution functions as a stabilizing and reducing agent during the incubation period, and the silver ions may transform into a colloidal silver solution. The production of Bio-AgNPs is shown by the color changing from pale yellow to blackish brown during this process. The synthesis of gold NPs followed the same procedure. 95 mL of plant leaf extract (supernatant) was placed in conical flasks, and 5 mL of an aqueous tetrachloroauric acid solution was added as a precursor to create purple-reddish Bio-AuNPs (Pal, Rai, and Pandey 2019).

6.4.5 Au–Ag Synthesis in Microreactor

MRs allow for regulated mixing and reaction periods, offering a scalable and accurate synthesis approach. When Au (III) is reduced with ascorbic acid in Au–Ag core–shell synthesis, for example, Au nuclei are produced. When silver nitrate is added, these nuclei serve as seeds for Ag deposition as shown in (Figure 6.8). A low flow rate only produces Au colloids, whereas an optimal flow rate of 42–58 µL/min guarantees effective diffusion of silver ions, resulting in homogenous Au–Ag NPs (Pal, Rai, and Pandey 2019).

A premade mixture of oleylamine, octadecene, gold chloride, and silver nitrate (Au3+/Ag+ = 20) was added to a MR that was kept at 140 °C. At this temperature, the thermally assisted diffusion-reduction of Ag inside the Au cores was preceded by the rapid nucleation of gold into the primary core materials. The gold seeds were approached by more silver ions as the residence duration grew from 30 to 320 seconds. The Au–Ag alloy's silver content consequently rose proportionately. Particles of the Au–Ag alloy that were highly monodisperse and composition-defined, with an average diameter of 2.7 nm, were created after three minutes of residence. Au–Ag NPs with a wide size distribution were created under similar reaction conditions in a batch reactor, but it took longer (Figure 6.9).

6.5 IMPORTANCE OF METAL AND METAL OXIDE NANOPARTICLES

Metal and metal oxide NPs have important qualities and various applications in industries including biomedicine, electronics, conversion and storage of energy, catalysis, and environmental technologies are all made possible by the nanoscale materials. These NPs due to their small size and precise morphology have high surface area to volume ratios, optical and magnetic behavior, and increased reactivity.

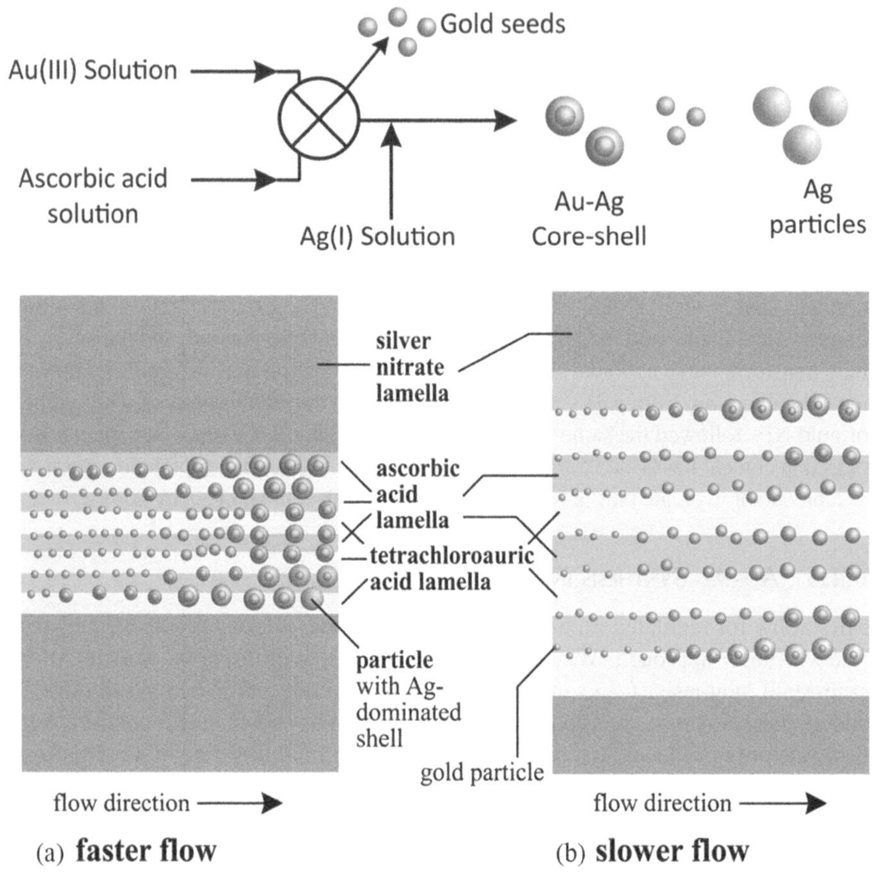

FIGURE 6.8 Upper panel: schematic of Au–Ag core-shell synthesis in a MR and possible products. Lower panel: multilamination and mixing efficiency at different flow rates (a: typical flow rate 42–58 μL/min, while b: 25 μL/min).

(Pal et al. 2019.)

Although traditional batch synthesis techniques are widely used, they frequently have built-in drawbacks. Batch-to-batch variability, irregular reaction conditions, insufficient control over particle size and shape, and scaling up issues are still common problems (Sun et al. 2012, Hakke et al. 2021).

Metals and their oxide NPs have been recently used in virus-targeted formulations for therapeutic activity in enhancing targeted drug delivery. NPs vary in shape and size, composition, and surface properties, and they have been used in humans and tested for clinical diagnosis and viral infections were found relatively less in number (Watanabe and Miyahara 2025).

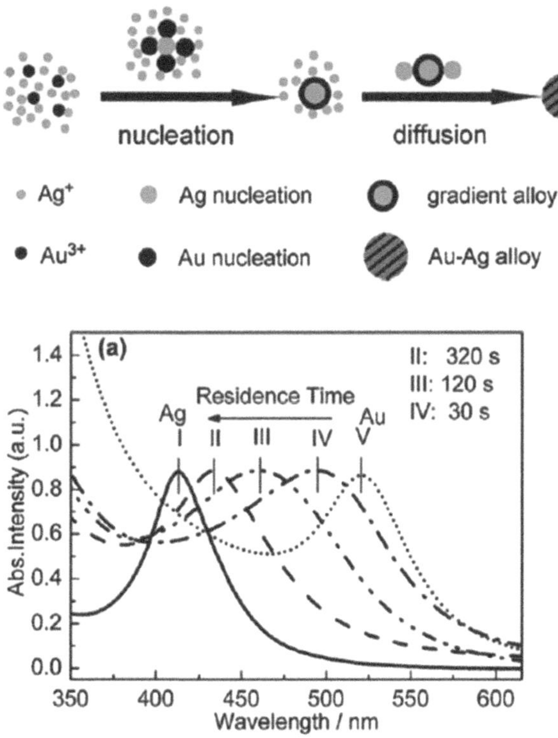

FIGURE 6.9 Schematics of the mechanism for Au–Ag alloy synthesis. Change of product composition (colloidal Ag or Au and Au–Ag alloy) with residence time inside the MR (variation of plasmonic absorption with residence time is shown in the UV-Vis trace of (a)).

(Sun et al. 2012.)

6.5.1 Common Oxide Formation

Improved physicochemical and electrical properties are demonstrated by iron (Fe) and titanium (Ti) oxide NPs that are produced by physical (such as laser ablation), chemical (such as precipitation, sputtering), or green (using plant or microbial extracts) approaches. By employing biomolecules for stabilization, green synthesis—such as the reduction of Fe salts mediated by algae extract—offers a sustainable substitute. Because of their high surface area, reactivity, and electrically programmable characteristics, these NPs are widely used in sensors, electronics, biological sectors (drug delivery, MRI imaging, antibacterial agents), and catalysis. Although they have a lot of potential, particularly for biomedical and sustainable applications, further study is required to completely guarantee their scalability and safety.

6.5.2 Continuous Setup Better Than Batch Setup for Metal Oxide Nanoparticle Synthesis

In order to synthesize customized nanomaterials, reactor equipment is crucial, both in the lab and for expanding the synthetic processes into large-scale manufacturing. While there are challenges associated with the sequential three-stage scale-up from the conventional process (i.e., lab-scale, pilot-scale, and large-scale) using multipurpose batch reactors, the interface of continuous flow microfluidic chemistry and nanomaterials synthesis processes has shown comparatively better process performance over conventional technologies. As a result, continuous flow microfluidics systems have shown a lot of interest lately as flexible instruments for the production of different structured nanomaterials (Watanabe and Miyahara 2025).

1. Continuous reactor gives better control over the distribution and size of the particle.
2. Heat and mass transfer is proper and gives high scalability.
3. High rate of heating and cooling in continuous reactors also decreases reaction time in high temperature synthesis methods.
4. It gives higher production rate and throughput compared to batch reactors.
5. It is possible to build continuous flow synthesis as a modular "plug and play" system that makes it simple to incorporate different synthesis stages, such as nucleation and growth, into an ongoing process.

6.6 MICROREACTOR-BASED SYNTHESIS OF METAL OXIDE

One of the most extensively studied types of magnetic nanoparticles (MNPs) is metal oxide nanostructures. Because of their diverse electrical architectures, metal oxide nanostructures can be either metallic, semiconductor, or insulator. Numerous magnetic metal oxides, such as MnO_3, Co_3O_4, Fe_2O_3, and NiO_2, have been created utilizing an easy, affordable, and repeatable method. Fe_2O_3 and Fe_3O_4 are examples of magnetic iron oxide NPs as shown in (Figure 6.10), a promising class of metal sulfide/oxide nanomaterials for a range of uses. Magnetic iron oxide NPs are frequently mixed with other photothermal agents (PTAs) to create composite nanomaterials in the field of photothermal therapy and antibacterial treatments, which allows for efficient therapeutic results (Watanabe and Miyahara 2025).

6.6.1 Synthesis of Metal Oxide (Fe_2O_3) Nanoparticle

In (Figure 6.11) arrangement, an affordable spiral copper tube MR (800 μm ID, 1240 mm length) is wound into six loops with a diameter of 200 mm. Using syringe pumps, reagents like iron (III) nitrate and NaOH are injected through a Y-type intake, enabling complete mixing as they move through the heated spiral tube in a water bath. A colloidal solution of iron oxide NPs is created by this

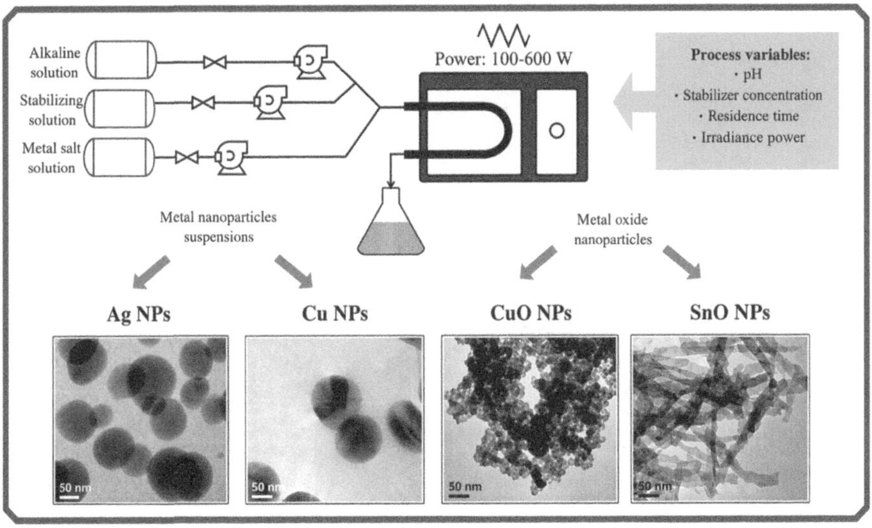

FIGURE 6.10 Experimental setup with a fabricated MR for the synthesis of magnetic iron oxide NPs.

(Długosz and Banach 2020.)

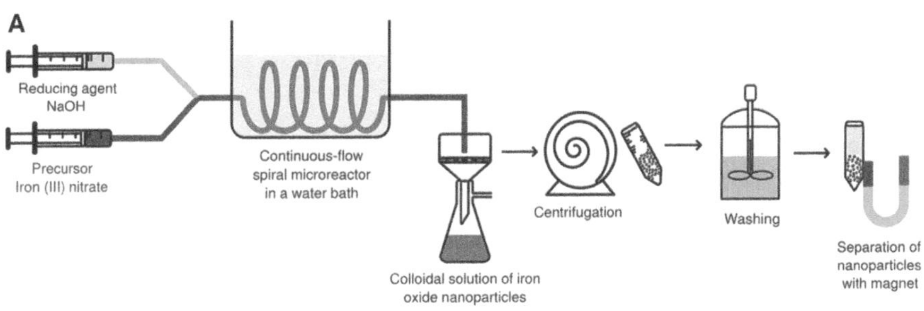

FIGURE 6.11 The schematic representation of the experimental setup based on spiral copper tube MR has been shown.

(Suryawanshi et al. 2018.)

continuous flow technique, then centrifugation, washing, and magnetic separation are used to purify it.

6.6.2 Bottom Up and Top Down Approach for Synthesis of Metal Oxide Nanoparticle

The bottom up technique builds NPs from smaller atomic or molecular blocks, whereas the top down process breaks down the larger molecules into NPs by physical or chemical disintegration (Figure 6.12)

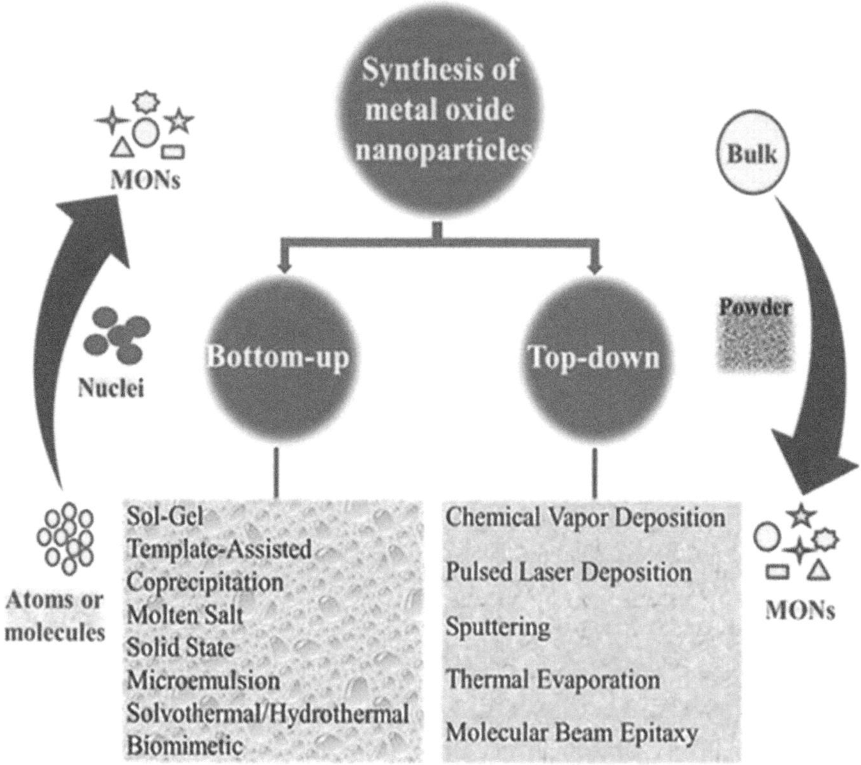

FIGURE 6.12 Representation of top down and bottom up approaches for synthesis of metal oxide NPs.

(Zadehnazari 2025, Glace et al. 2023.)

6.6.3 Green Synthesis Using Microreactor

Using harsh reducing agents to exploit physical and chemical channels results in high energy consumption, low yield, high expense, and environmental harm. The biological processes entail the employment of plants or microorganisms as in (Figure 6.13) (fungi, bacteria, yeast, algae, etc.); the use of microorganisms is riskier due to the pathogenicity issue; vast cultures must also be maintained. Consequently, it is preferable to synthesize NPs using greener techniques.

6.7 MICROREACTORS ARE NECESSARY FOR LARGE-SCALE PRODUCTION

With the ability to produce low-carbon energy, MRs are becoming a new nuclear energy technology (Pal, Rai, and Pandey 2019, Aparicio 2023). Their potential for expansion is enhanced by their capacity to penetrate new markets where

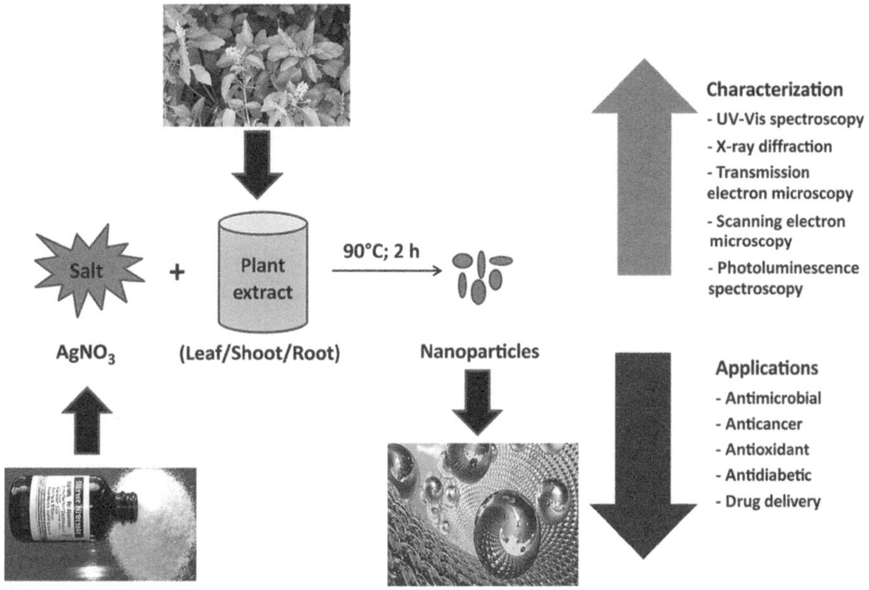

FIGURE 6.13 Green synthesis of NPs and its applications.

(Pal, Rai, and Pandey 2019.)

large-scale energy technologies are inappropriate (Testoni, Bersano, and Segantin 2021, Black et al. 2023, Apolinar-Hernández et al. 2023, Ingersoll 2009). These extremely tiny reactors can produce electricity utilizing both advanced light water reactor (LWR) and non-LWR designs, usually producing less than 1 MWe to 20 MWe (Thomas and Ramana 2022, Petti et al. 2017). 50 MWe is the maximum power output that can be achieved. They can process heat for immediate use, produce electricity, or do both. MRs have the ability to meet unmet demands in energy portfolios where fossil fuel limits are clear, provided that the difficulties associated with production and operating costs, as well as institutional and regulatory barriers, can be overcome (Dittmeyer et al. 2017, Pennemann et al. 2004, Kolb 2013). MRs can also become competitive in areas without large electrical grids, where fuel delivery is cumbersome or expensive, where economies of scale do not apply, or where space constraints or specific weather conditions make renewables unfeasible.

6.7.1 Conventional Reactors Show Problem in Large-Scale Production and Why is Microreactor Used Instead

Conventional reactors have certain limitations for large scale production such as during heat and mass transfer. Temperature and concentration gradients can often negatively affect the reaction yield and quality of the product; therefore, larger reactors have difficulty in heat and mass transfer.

There can be safety issues while using large scale reactors when they deal with hazardous or exothermic reactions and toxic metal-organic precursor. So, scaling up is necessary for more complex safety measures.

High temperature and pressure lead to more energy consumption and material degradation in conventional methods. Maintaining uniform NP size and distribution can be problematic in conventional reactors. Conventional reactors are big in size, therefore may cause problems in proper mixing and heating and may need further post-synthesis processing.

Conventional reactors often take a long synthesis time which is problematic for industrial applications. Therefore, MR should be used for large scale production of NPs.

6.7.2 Microreactors Are Important for Large-Scale Production of Metal Oxide Nanoparticle

Microfluidic technology has advanced quickly in recent years and has several uses in the production of NPs and the execution of different chemical reactions. The reaction occurs in channels with an inner diameter of less than 1 mm in a MR. Higher heat and mass transfer results from the improved surface area to volume ratio caused by the narrower channel dimensions. This method provides accurate control over process variables (temperature, concentration, and reaction time) in contrast to traditional batch reactors.

Monodispersed NPs, temperature and concentration homogeneities, and repeatable composition, structure, and physicochemical properties are all made possible by high surface area to volume ratios and mixing qualities. MR technology is currently regarded as the preferred method for creating NPs due to all of these features (Zhao et al. 2011).

6.7.3 Why Do We Use Microreactors for Nanoparticle Synthesis

Numerous research teams have demonstrated in recent years that microfluidic devices present a viable method for producing high-quality NPs. Mass and heat transport are significantly more efficient in MRs than in the traditional round bottom flask because of their compact size and higher surface to volume ratio. Heat exchange between the reaction medium and reaction vessel is extremely efficient, and reagents are mixed by diffusion very quickly. Consequently, a continuous flow microchannel's response conditions are uniform and controllable.

A MR can easily be used to harness highly exothermic and even explosive processes. A reaction's yield, purity, and selectivity can all be improved by carefully controlling the reaction temperature and residence time. In order to produce monodisperse particles, the reaction must be set up so that the entire nucleation process occurs quickly and that more material is added very gradually so that it can reach the nuclei while maintaining the solution concentration below a point at which additional nucleation could occur (Singh, Malek, and Kulkarni 2010).

6.8 CHARACTERIZATION OF SYNTHESIZED NANOPARTICLE

The following are important characteristics of NPs: size, shape, size distribution, aggregation, surface charge, surface area, surface chemistry, crystal structure, and chemical composition. These elements significantly affect the qualities and uses of NPs. Additional challenges include monitoring waste and effluents, analyzing in complicated matrices, and measuring concentrations in situ and online during scale-up.

Strong and dependable techniques are necessary for both regulatory compliance and commercial adoption. The following are important characteristics of NPs: size, shape, size distribution, aggregation, surface charge, surface area, surface chemistry, crystal structure, and chemical composition. These elements significantly affect the qualities and uses of NPs. The absence of standardized procedures, calibrated reference materials, difficult sample preparation, and problems with data interpretation, however, continue to be obstacles. In situ and online concentration measurements during scale-up, analysis in complex matrices, and waste and effluent monitoring present additional challenges. Both commercial adoption and regulatory compliance depend on solid and dependable methodologies. It is necessary to characterize the NP core and surface ligands using both traditional and contemporary in situ/operando approaches in order to investigate formation kinetics and defect control (Mourdikoudis, Pallares, and Thanh 2018).

6.8.1 Common Methods for Nanoparticle Characterization

XRD, TEM, and SEM: A crystal lattice's atomic arrangement can be studied using X-ray diffraction (XRD), which also yields data on unit cell characteristics, crystal phases, phase purity, crystallite size, and lattice strain. It is frequently used to guarantee material quality and validate the successful production of NPs. By detecting secondary electrons, scanning electron microscopy (SEM) creates surface pictures that enable examination of particle size and shape, surface morphology, and elemental composition using energy dispersive X-ray spectroscopy (EDX). It is frequently used to investigate the elemental distribution and surface characteristics of NPs. By passing electrons through a small sample, transmission electron microscopy (TEM) gives high-resolution imaging that can reveal internal structure, crystal lattice, and defects (via SAED), as well as the size, shape, and distribution of NPs.

6.8.2 Inline and Real Time Monitoring in Microreactors

The USFDA's introduction of Process Analytical Technology (PAT) is essential for effective continuous microflow reaction monitoring and control. Because it uses probes or flow cells that are directly attached to the process stream, inline analysis is particularly significant among its instruments. This allows for real-time, in-situ monitoring of reactant consumption, product production, and changing reaction conditions without interfering with flow or concentration. This makes it possible to integrate into sophisticated process control methods, handle hazardous or air-sensitive intermediates safely, and make quick modifications. In contrast to offline techniques, which frequently result in artifacts, delays, and higher expenses because of the need for additional reagents, inline analytics enhance flow chemistry's accuracy, productivity, and overall product quality.

6.8.2.1 Raman Spectroscopy

The Raman effect, an inelastic light scattering caused by molecular contact, is the foundation of Raman spectroscopy. A monochromatic laser light source, often in the visible or near-infrared spectrum, is used as the excitation source in Raman spectrometers. Using the spectral fingerprint (300–1900 cm^{-1}) and data analysis, inline Raman spectroscopy offers a real-time insight of the process. Unlike infrared spectroscopy, water and alcohols do not produce any discernible signal in Raman spectra because they are weak Raman scatterers. In the infrared study, the O-H bonds produce notable absorption signs. Frequently, it obscures the signal of interest. Raman spectroscopy is therefore very useful in alcoholic or watery reaction settings. It primarily describes how functional groups alter throughout the course of a reaction (Glace et al. 2023, Hengoju et al. 2022, Esmonde-White, Cuellar, and Lewis 2022, Miyai et al. 2021, Su et al. 2014, Mainini and Eccles 2020).

6.8.2.2 Advantages and Disadvantages of Using Microreactors for Nanoparticle Synthesis

Advantages of using MRs for synthesis of NPs are:

- Heat and mass transfer are well controlled
- Properties of NPs are controlled
- Repeatability is high
- Reagent consumption is less
- Residence time is controlled
- It is cost-effective
- Production of hazardous waste is reduced
- Reaction system is compact

Disadvantages of using MRs for synthesis of NPs are:

- There is a chance of clogging of the channels
- Compatibility of some base materials with reagents is poor
- The resistance is low for the base material at high temperature
- There may be problems for large-scale production
- This may not eliminate the formation of undesirable products

6.8.3 Reproducibility and Scalability

- Improved reproducibility: Excellent heat and mass transport allow for precise control of reaction conditions.
- Consistent product quality: Batch-to-batch consistency is ensured by a smaller particle size dispersion and less polydispersity.
- Effective mixing: Compared to batch reactors, integrated micromixers enable quicker and more consistent mixing.

6.8.4 Improved Safety and Environment Friendly Synthesis

Enhanced Security

- Decreased hazardous materials: Accident risks are decreased by small reactor volumes.
- Improved thermal control: In exothermic reactions, a high surface-to-volume ratio avoids hot patches.
- Safe intermediate handling: Unstable intermediates can be handled under ambient settings due to short residence durations.

Eco-Friendliness

- Reduced waste: Byproducts are reduced by high yields and selectivity.
- Greener or no solvents can be used, thanks to safer solvents.
- A higher atom economy means fewer auxiliary functions, such as guarding groups, are required.
- On-demand production: On-site synthesis saves energy and transportation.

6.9 APPLICATIONS OF SYNTHESIZED NANOPARTICLES

The multifunctional NPs are widely used in various applications such as pharmaceutical and biomedical applications, catalysis, electronics, enhance photoluminescence, create photonic crystals, etc. Mostly in the biomedical field, these NPs are used for bioimaging, controlled drug release, targeted drug delivery, cell labeling, and tissue engineering applications. For instance, core–shell NPs have played important roles in clinical and therapeutic applications. While core-shell NPs that are effective scatterers can be employed in imaging applications, those that are powerful absorbers can be used in photothermal therapy. By altering their core-to-shell ratio, silica core–gold shell nanoshells—a unique core–shell nanostructure—can be created that efficiently scatters or absorbs light. Core-shell NPs can be coupled with particular antibodies for tumors or damaged tissues for use in imaging applications (Bao, Mitragotri, and Tong 2013) (Figure 6.14).

6.9.1 Biomedical Applications

Management of Inflammation

- Chronic inflammation is associated with a number of chronic diseases, including Alzheimer's, cancer, COPD, diabetes, and cardiovascular disease.
- Traditional anti-inflammatory medications have poor side effects and bioavailability.
- NPs boost solubility, improve site-specific delivery, and adjust to the elevated ROS, acidic pH, and enhanced permeability of inflammatory tissue microenvironments.

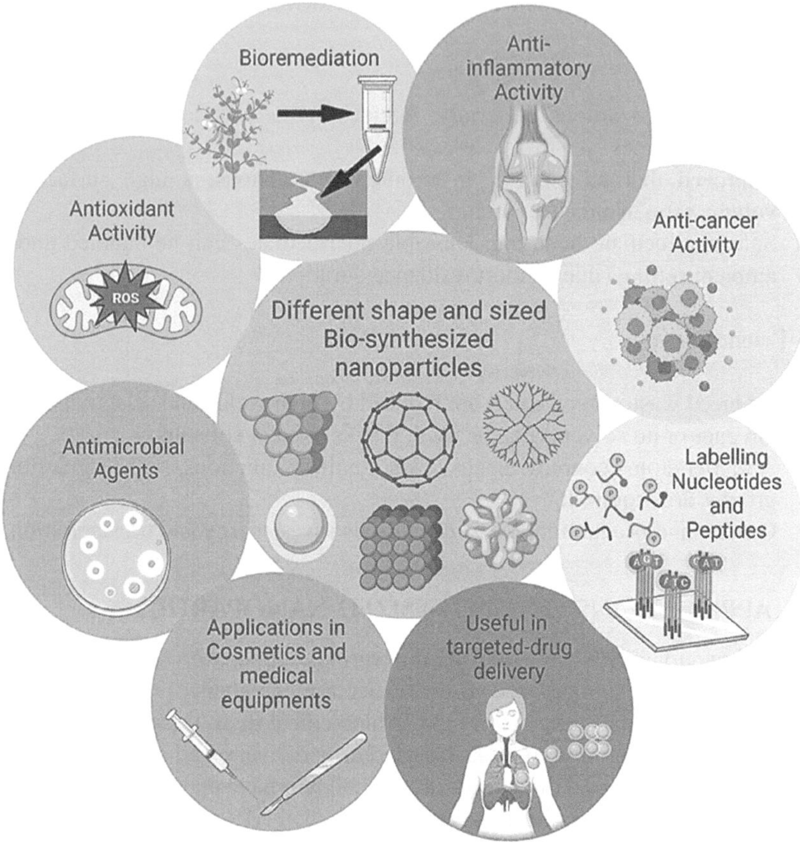

FIGURE 6.14 Various applications of NPs.

(Singh et al. 2023.)

Cancer Treatment

- Certain characteristics of cancer cells include an acidic environment, elevated temperature, hypoxia, reactive oxygen species, and certain surface molecules or enzymes.
- The size and surface of polymeric NPs (10–200 nm) can be adjusted for both passive and aggressive tumor targeting.
- NPs boost medication bioavailability, absorption, and localization.
- For instance, compared to free curcumin, curcumin-loaded NPs (PLGA, chitosan, lecithin, and silk fibroin) exhibit much higher bioavailability and cancer-cell adhesion.

Management of Contagious Illnesses

- Take action against the growing antibiotic resistance, a serious health risk recognized by the WHO.
- For prophylactic antimicrobial action against resistant infections, NPs can be added to medical materials such as bandages, catheters, and hospital linens (Lopez-Sanchez et al. 2011, Hodges et al. 2021, Van Hal et al. 2012, Kalinichenko et al. 2024)

6.9.2 Environmental Applications

- Remediation of Water and Soil: NPs, especially iron-based ones such as nanoscale zero-valent iron (nZVI), are very good in eliminating pollutants from soil and groundwater. They have the ability to both adsorb and accelerate reactions that break down contaminants such synthetic colors, insecticides, and arsenic.
- Using Nanotechnology to Treat Water: Dendrimers can be effective and non-toxic water treatment agents, and engineered polymeric NPs are employed for the bioremediation of hydrophobic pollutants. Compared to previous techniques, nanofiltration membranes use less energy to clean water by using nanomaterials.
- Environmental Monitoring: To develop nanosensors that can identify pollutants, NPs are used. The ability of NPs to recognize biological species and contaminants is improved by functionalizing them with particular ligands or enzymes.

6.10 CURRENT CHALLENGES AND FUTURE PROSPECTS

Agglomeration, toxicological issues, and cost-effectiveness are recognized as obstacles. Potential avenues for development include future developments in the synthesis of nanomaterials, their incorporation with other materials, and their influence on environmental regulations. To sum up, metal oxide NPs have a great deal of promise for lowering dye pollution, but more study and development are needed to fully understand their function in long-term environmental management.

Environmental Toxicity: Harsh conditions and toxic reagents are used in conventional synthesis, which puts both human health and the environment at risk.Agglomeration: NPs readily group together, which lowers their efficacy and stability.

Cost and Scalability: It is expensive and challenging to scale production while maintaining a high standard of quality.

6.10.1 Future Prospects

Eco-friendly NP production is possible through green and biogenic synthesis that uses microorganisms, algae, and plants. For increased efficiency, hybrid synthesis techniques seek to integrate chemical and biological methods. The goal of morphological control advancements is to customize size, shape, and structure for particular uses. It is anticipated that integration with additional materials, including catalysts or polymers, will result in multifunctional hybrid nanostructures with improved characteristics.

Rather than using single, large batch reactors, we can use many MRs together in parallel, which would help in scaling up the production process efficiently. Therefore, continuous flow systems used in MRs give better throughput synthesis along with less reaction times.

6.10.2 Integration with AI and Automation

AI-driven optimization: Data is analyzed by machine learning to adjust parameters (temperature, flow rate, and concentration).

Automated experimentation: High-throughput screening and methodical data collecting are made possible by robotic MR configurations.

Closed-loop systems: AI continuously improves synthesis by proposing new parameters based on experimental results.

Advantages of Integration

Faster discovery: The best synthesis conditions are found quickly.
Precise control: Accurate management of yield, size, and shape of NPs.
NPs with adjustable structures and characteristics for particular uses are known as tunable NPs.
Inverse design: Based on the desired attributes of the NPs, AI forecasts the synthesis conditions.

Real-World Examples

Automated synthesis: Effective investigation of process and reagent parameters.
Rapid testing of several synthesis conditions is known as high-throughput screening.AI-derived kinetic parameters are used in process modeling to improve the process.

6.10.3 Emerging Trends in Microreactor-Based Nanoparticle Synthesis

Enhanced Functionality and Control

- Accurate regulation of nucleation and growth results in customized NP composition, size, and shape.
- Consistent, scalable, and on-demand production are made possible by continuous flow.
- Complex shapes and consistent sizes are achieved via hybrid systems, such as those that use microwaves.
- Low polydispersity is achieved through improved mixing in 3D vortex MRs.
- Permits the use of functionalized NPs, such as MNPs coated with salicylic acid.

- Used in biomedicine such as wearable/implantable controlled release devices, drug encapsulation.
- Tissue engineering: Promotes the migration, growth, and differentiation of cells.
- Environmental remediation: Photocatalytic water treatment using nanomaterials.
- Sensing: Nanosensors based on microfluidics.

Benefits Compared to Conventional Techniques

- Improved safety: Less dangerous chemical handling.
- Reduced waste and expense: Fewer chemicals and reagents are needed.
- Improved monodispersity and selectivity: consistent production of NPs.
- Improved process control: Nanomaterials that are precisely specified and repeatable.

6.11 CONCLUSION

The synthesis of metal and metal oxide NPs has emerged as a cornerstone in advancing nanotechnology, with MR-based continuous flow systems offering significant advantages over traditional batch methods. By enabling superior heat and mass transfer, precise process control, and reproducibility, MRs allow the fabrication of NPs with tailored size, morphology, and functionality. Strategies such as numbering-up, hybrid integration, and advanced reactor designs further enhance scalability, making them suitable for industrial translation. Polymeric and functionalized NPs have potential uses in drug delivery, tissue engineering, diagnostics, biomedicine, environmental cleanup, electronics, and sensing. A paradigm change toward intelligent and adaptive synthesis is signaled by the incorporation of AI and automation into MR platforms, which speed up discovery, provide closed-loop optimization, and facilitate inverse design. Additionally, the creation of NPs is in line with economic and environmental factors through the use of sustainable techniques and green synthesis procedures.

REFERENCES

Ahmed-Omer, Batoul, Johan C Brandt, and Thomas Wirth. 2007. "Advanced organic synthesis using microreactor technology." *Organic & Biomolecular Chemistry* 5 (5):733–740.

Aparicio, Santiago Andrade. 2023. *Technical and Commercial Feasibility Assessment of Nuclear Microreactors as a Clean Energy Source for Data Centers and Mining Sites*: Massachusetts Institute of Technology.

Apolinar-Hernández, Jesús Efraín, Savio Leandro Bertoli, Humberto Gracher Riella, Cintia Soares, and Natan Padoin. 2023. "An overview of low-temperature fischer–tropsch synthesis: Market conditions, raw materials, reactors, scale-up, process intensification, mechanisms, and outlook." *Energy & Fuels* 38 (1):1–28.

Armendariz, Veronica, Isaac Herrera, Jose R Peralta-Videa, Miguel Jose-Yacaman, Horacio Troiani, Patricia Santiago, and Jorge L Gardea-Torresdey. 2004. "Size controlled gold nanoparticle formation by Avena sativa biomass: Use of plants in nanobiotechnology." *Journal of Nanoparticle Research* 6 (4):377–382.

Arostegui, D. A., Holt, M., & Library of Congress. Congressional Research Service, Issuing Body. (2018). *Advanced nuclear reactor: technology overview and current issues* ([Library of Congress public edition]). Congressional Research Service.

Bao, Gang, Samir Mitragotri, and Sheng Tong. 2013. "Multifunctional nanoparticles for drug delivery and molecular imaging." *Annual Review of Biomedical Engineering* 15 (1):253–282.

Black, Geoffrey, David Shropshire, Kathleen Araújo, and Aliki van Heek. 2023. "Prospects for nuclear microreactors: A review of the technology, economics, and regulatory considerations." *Nuclear Technology* 209 (sup1):S1–S20.

Dachraoui, Walid, Trond R Henninen, Debora Keller, and Rolf Erni. 2021. "Multi-step atomic mechanism of platinum nanocrystals nucleation and growth revealed by in-situ liquid cell STEM." *Scientific Reports* 11 (1):23965.

Dittmeyer, R, T Boeltken, P Piermartini, M Selinsek, M Loewert, F Dallmann, H Kreuder, M Cholewa, A Wunsch, and M Belimov. 2017. "Micro and micro membrane reactors for advanced applications in chemical energy conversion." *Current Opinion in Chemical Engineering* 17:108–125.

Długosz, Olga, and Marcin Banach. 2020. "Continuous synthesis of metal and metal oxide nanoparticles in microwave reactor." *Colloids and Surfaces A: Physicochemical and Engineering Aspects* 606:125453.

Esmonde-White, Karen A, Maryann Cuellar, and Ian R Lewis. 2022. "The role of Raman spectroscopy in biopharmaceuticals from development to manufacturing." *Analytical and Bioanalytical Chemistry* 414 (2):969–991.

Figueroa, Santiago JA, Dean Gibson, Trevor Mairs, Sebastien Pasternak, Mark A Newton, Marco Di Michiel, Jerome Andrieux, Konstantinos C Christoforidis, Ana Iglesias-Juez, and Marcos Fernández-García. 2013. "Innovative insights in a plug flow microreactor for operando X-ray studies." *Applied Crystallography* 46 (5):1523–1527.

Gardea-Torresdey, JL, KJ Tiemann, G Gamez, K Dokken, and NE Pingitore. 1999. "Recovery of gold (III) by alfalfa biomass and binding characterization using X-ray microfluorescence." *Advances in Environmental Research* 3 (1):U7–93.

Glace, Matthew, Wei Wu, Harrison Kraus, David Acevedo, Thomas D Roper, and Adil Mohammad. 2023. "Development of a continuous synthesis process for carbamazepine using validated in-line Raman spectroscopy and kinetic modelling for disturbance simulation." *Reaction Chemistry & Engineering* 8 (5):1032–1042.

Hakke, Vikas, Shirish Sonawane, Sambandam Anandan, Shriram Sonawane, and Muthupandian Ashokkumar. 2021. "Process intensification approach using microreactors for synthesizing nanomaterials — A critical review." *Nanomaterials* 11 (1):98.

Hassan, Homa, Pradakshina Sharma, Mohd Rahil Hasan, Shiwani Singh, Deepanshi Thakur, and Jagriti Narang. 2022. "Gold nanomaterials – The golden approach from synthesis to applications." *Materials Science for Energy Technologies* 5:375–390.

He, Meng-Qi, Yongjian Ai, Wanting Hu, Liandi Guan, Mingyu Ding, and Qionglin Liang. 2023. "Recent advances of seed-mediated growth of metal nanoparticles: From growth to applications." *Advanced Materials* 35 (46):2211915.

Hengoju, Sundar, Oksana Shvydkiv, Miguel Tovar, Martin Roth, and Miriam A Rosenbaum. 2022. "Advantages of optical fibers for facile and enhanced detection in droplet microfluidics." *Biosensors and Bioelectronics* 200:113910.

Hodges, Lisa M, Eduardo N Taboada, Adam Koziol, Steven Mutschall, Burton W Blais, G Douglas Inglis, Daniel Leclair, and Catherine D Carrillo. 2021. "Systematic evaluation of whole-genome sequencing based prediction of antimicrobial resistance in Campylobacter jejuni and C. coli." *Frontiers in Microbiology* 12:776967.

Horikoshi, Satoshi, and Nick Serpone, eds. 2013. *Microwaves in Catalysis: Methodology and Applications*. Weinheim: Wiley-VCH Verlag GmbH & Co. KGaA. https://doi.org/10.1002/9783527648122

Ingersoll, Daniel T. 2009. "Deliberately small reactors and the second nuclear era." *Progress in nuclear energy* 51 (4–5):589–603.

Jensen, Klavs F. 2001. "Microreaction engineering—Is small better?" *Chemical Engineering Science* 56 (2):293–303.

Kalinichenko, Dimitri, Lucas Wodrich, Alvin JH Lee, Tomasz Kozlowski, and Caleb S Brooks. 2024. "Analysis of nuclear microreactor efficacy with hydrogen production methods." *Progress in Nuclear Energy* 168:104994.

Khan, Hafeez Ullah, Nadeem Raza, Safirah Maheen, Hanzala Abbas, Mostafa E Salem, Ikramullah Khan, Manal Naseem, and Syed Salman Shafqat. 2025. "Advances in targeted drug delivery systems for cancer treatment: Current trends and future prospects." *Journal of Drug Delivery Science and Technology*:107402.

Kharissova, Oxana V, HV Rasika Dias, Boris I Kharisov, Betsabee Olvera Pérez, and Victor M Jiménez Pérez. 2013. "The greener synthesis of nanoparticles." *Trends in Biotechnology* 31 (4):240–248.

Kolb, Gunther. 2013. "Microstructured reactors for distributed and renewable production of fuels and electrical energy." *Chemical Engineering and Processing: Process Intensification* 65:1–44.

Kuchibhatla, Satyanarayana VNT, Ajay S Karakoti, Donald R Baer, Saritha Samudrala, Mark H Engelhard, James E Amonette, S Thevuthasan, and S Seal. 2012. "Influence of aging and environment on nanoparticle chemistry: Implication to confinement effects in nanoceria." *The Journal of Physical Chemistry C* 116 (26):14108–14114.

Laporte, Annechien AH, Tom M Masson, Stefan DA Zondag, and Timothy Noël. 2024. "Multiphasic continuous-flow reactors for handling gaseous reagents in organic synthesis: Enhancing efficiency and safety in chemical processes." *Angewandte Chemie International Edition* 63 (11):e202316108.

Lopez-Sanchez, Jose A, Nikolaos Dimitratos, Ceri Hammond, Gemma L Brett, Lokesh Kesavan, Saul White, Peter Miedziak, Ramchandra Tiruvalam, Robert L Jenkins, and Albert F Carley. 2011. "Facile removal of stabilizer-ligands from supported gold nanoparticles." *Nature chemistry* 3 (7):551–556.

Mainini, Francesco, and Michael R Eccles. 2020. "Lipid and polymer-based nanoparticle siRNA delivery systems for cancer therapy." *Molecules* 25 (11):2692.

Miyai, Yuma, Anna Formosa, Cameron Armstrong, Brian Marquardt, Luke Rogers, and Thomas Roper. 2021. "PAT implementation on a mobile continuous pharmaceutical manufacturing system: Real-time process monitoring with in-line FTIR and Raman spectroscopy." *Organic Process Research & Development* 25 (12):2707–2717.

Mourdikoudis, Stefanos, Roger M Pallares, and Nguyen TK Thanh. 2018. "Characterization techniques for nanoparticles: Comparison and complementarity upon studying nanoparticle properties." *Nanoscale* 10 (27):12871–12934.

Murphy, Catherine J, Lucas B Thompson, Davin J Chernak, Jie An Yang, Sean T Sivapalan, Stefano P Boulos, Jingyu Huang, Alaaldin M Alkilany, and Patrick N Sisco. 2011. "Gold nanorod crystal growth: From seed-mediated synthesis to nanoscale sculpting." *Current Opinion in Colloid & Interface Science* 16 (2):128–134.

Pach, Adrianna, Aleksandra Szot, Krzysztof Fitzner, and Magdalena Luty-Błocho. 2024. "Opportunities and challenges in the synthesis of noble metal nanoparticles via the chemical route in microreactor systems." *Micromachines* 15 (9):1119.

Pal, Gaurav, Priya Rai, and Anjana Pandey. 2019. "Green synthesis of nanoparticles: A greener approach for a cleaner future." In Ashutosh Kumar Shukla and Siavash Iravani (Eds.) *Green Synthesis, Characterization and Applications of Nanoparticles*, 1–26. Elsevier. https://doi.org/10.1016/B978-0-08-102579-6.00001-0

Pennemann, Helmut, Paul Watts, Stephen J Haswell, Volker Hessel, and Holger Löwe. 2004. "Benchmarking of microreactor applications." *Organic Process Research & Development* 8 (3):422–439.

Petti, David Andrew, R Hill, J Gehin, Hans David Gougar, Gerhard Strydom, F Heidet, J Kinsey, Christopher Grandy, A Qualls, and Nicholas Brown. 2017. Advanced demonstration and test reactor options study. Idaho National Lab.(INL), Idaho Falls, ID.

Rahimnejad, Mostafa. 2023. *Biological fuel cells: Fundamental to applications*: Elsevier.

Rai, Akhilesh, Amit Singh, Absar Ahmad, and Murali Sastry. 2006. "Role of halide ions and temperature on the morphology of biologically synthesized gold nanotriangles." *Langmuir* 22 (2):736–741.

Schenk, Rainer, Volker Hessel, Christian Hofmann, Janos Kiss, Holger Löwe, and Athanassios Ziogas. 2004. "Numbering-up of micro devices: A first liquid-flow splitting unit." *Chemical Engineering Journal* 101 (1–3):421–429.

Shevchenko, Elena V, Dmitri V Talapin, Heimo Schnablegger, Andreas Kornowski, Örjan Festin, Peter Svedlindh, Markus Haase, and Horst Weller. 2003. "Study of nucleation and growth in the organometallic synthesis of magnetic alloy nanocrystals: The role of nucleation rate in size control of CoPt3 nanocrystals." *Journal of the American Chemical Society* 125 (30):9090–9101.

Shrimal, Preena, Girirajsinh Jadeja, and Sanjaykumar Patel. 2020. "A review on novel methodologies for drug nanoparticle preparation: Microfluidic approach." *Chemical Engineering Research and Design* 153:728–756.

Singh, Akanksha, Chantal Khan Malek, and Sulabha K Kulkarni. 2010. "Development in microreactor technology for nanoparticle synthesis." *International Journal of Nanoscience* 9 (01n02):93–112.

Singh, Harjeet, Martin F Desimone, Shivani Pandya, Srushti Jasani, Noble George, Mohd Adnan, Abdu Aldarhami, Abdulrahman S Bazaid, and Suliman A Alderhami. 2023. "Revisiting the green synthesis of nanoparticles: Uncovering influences of plant extracts as reducing agents for enhanced synthesis efficiency and its biomedical applications." *International Journal of Nanomedicine*:4727–4750.

Su, Hui, Kuo Sun, Zhixiang Yao, Peixian Huang, and Liu Liu. 2014. "Raman Spectrum Online Monitoring in Aspirin Synthesis Process." *Asian Journal of Chemistry* 26 (2).

Subramani, Karthikeyan, and Waqar Ahmed. 2012. *Nanobiomaterials in clinical dentistry*: William Andrew.

Sun, Li, Weiling Luan, Yuejin Shan, and Shan-tung Tu. 2012. "One-step synthesis of monodisperse Au–Ag alloy nanoparticles in a microreaction system." *Chemical Engineering Journal* 189:451–455.

Suryawanshi, Prashant L, Shirish H Sonawane, Bharat A Bhanvase, Muthupandian Ashokkumar, Makarand S Pimplapure, and Parag R Gogate. 2018. "Synthesis of iron oxide nanoparticles in a continuous flow spiral microreactor and Corning® advanced flow™ reactor." *Green Processing and Synthesis* 7 (1):1–11.

Testoni, Raffaella, Andrea Bersano, and Stefano Segantin. 2021. "Review of nuclear microreactors: Status, potentialities and challenges." *Progress in Nuclear Energy* 138:103822.

Thomas, Stephen, and MV Ramana. 2022. "A hopeless pursuit? National efforts to promote small modular nuclear reactors and revive nuclear power." *Wiley Interdisciplinary Reviews: Energy and Environment* 11 (4):e429.

Tyagi, Pankaj Kumar, Cristina Quispe, Jesús Herrera-Bravo, Shruti Tyagi, D Barbhai Mrunal, Manoj Kumar, Anas S Dablool, Saad Alghamdi, Gaber El-Saber Batiha, and Javad Sharifi-Rad. 2021. "Synthesis of silver and gold nanoparticles: Chemical and green synthesis method and its toxicity evaluation against pathogenic bacteria using the toxtrak test." *Journal of Nanomaterials* 2021 (1):3773943.

Van Hal, Sebastian J, Slade O Jensen, Vikram L Vaska, Björn A Espedido, David L Paterson, and Iain B Gosbell. 2012. "Predictors of mortality in Staphylococcus aureus bacteremia." *Clinical microbiology reviews* 25 (2):362–386.

Wang, Ye, Jiongdong Zhao, Yu Zhu, Shurong Dong, Yang Liu, Yijun Sun, Liling Qian, Wenting Yang, and Zhen Cao. 2021. "Monolithic integration of nanorod arrays on microfluidic chips for fast and sensitive one-step immunoassays." *Microsystems & Nanoengineering* 7 (1):65.

Watanabe, Satoshi, and Minoru T Miyahara. 2025. "Synthesis of Functional Nanoparticles Using a Microreactor." *KONA Powder and Particle Journal* 42:156–169.

Wu, Ke-Jun, CM Edmund, Congxiao Shang, and Zhengxiao Guo. 2022. "Nucleation and growth in solution synthesis of nanostructures – From fundamentals to advanced applications." *Progress in Materials Science* 123:100821.

Yadavalli, Tejabhiram, and Deepak Shukla. 2017. "Role of metal and metal oxide nanoparticles as diagnostic and therapeutic tools for highly prevalent viral infections." *Nanomedicine: Nanotechnology, Biology and Medicine* 13 (1):219–230.

Zadehnazari, Amin. 2025. "Chemical synthesis strategies for metal oxide nanoparticles: A comprehensive review." *Inorganic and Nano-Metal Chemistry* 55 (6):734–773.

Zhao, Chun-Xia, Lizhong He, Shi Zhang Qiao, and Anton PJ Middelberg. 2011. "Nanoparticle synthesis in microreactors." *Chemical Engineering Science* 66 (7):1463–1479.

7 Biomaterial Synthesis and Biofunctionalization Using Microreactors

Jayato Nayak and Sourab Paul

7.1 INTRODUCTION

The synthesis and biofunctionalization of biomaterials represent a rapidly evolving frontier at the interface of chemistry, materials science, and biotechnology. Biomaterials, encompassing nanoparticles, nanocrystals, biopolymers, and engineered hybrid structures, play a central role in diverse fields ranging from drug delivery and tissue engineering to biosensing, catalysis, and environmental remediation (Holghoomi et al., 2024). A fundamental challenge in the fabrication of these materials lies in achieving reproducibility, scalability, and tunability of physicochemical properties, such as size, morphology, surface chemistry, and functional activity (Krasley et al., 2024). Conventional batch processes for nanoparticle or biomaterial synthesis often suffer from uncontrolled mixing, uneven heat and mass transfer, and broad size distributions, which limit their translational potential. To address these challenges, microreactor technology has emerged as a transformative platform that enables continuous, precise, and scalable synthesis and functionalization of biomaterials under well-defined and controllable conditions (Hameedat et al., 2024).

Microreactors are miniaturized reaction systems, typically comprising channels or chambers with dimensions ranging from micrometers to millimeters, that offer exceptional control over reaction environments. Their small characteristic dimensions confer high surface-to-volume ratios, leading to enhanced mass and heat transfer compared to conventional reactors. This ensures uniform reaction conditions, rapid mixing of reactants, and tight control over residence times, which are critical for synthesizing materials with narrow size distributions and reproducible functional properties. Importantly, microreactors facilitate the integration of multiple sequential or parallel processes within a single platform, such as synthesis, purification, surface modification, and biofunctionalization, thereby streamlining the fabrication of complex biomaterials (Han et al., 2022; Li et al., 2024; Silva et al., 2024). In recent years, microreactors have been widely employed for the synthesis of nanomaterials and biomaterials due to their ability to provide precise reaction control. One prominent example is the synthesis of nanocrystals using solvent–antisolvent methods in microreactors. In microchannels, the presence of strong cohesion significantly mitigates the impact of high slip velocity or drag between particles and fluid, thereby effectively

 DOI: 10.1201/9781003633761-7

suppressing the resuspension of deposited particles and ultimately contributing to clogging (Shao et al., 2021). Yuan et al. (2025) developed a scale-up microreactor system coupled with binary surfactants to prepare dasatinib nanocrystals, a poorly water-soluble anticancer drug. By optimizing parameters such as solvent-to-antisolvent ratio, flow rate, and surfactant composition, they achieved nanocrystals with uniform size (85 nm), high stability (zeta potential 47.5 mV), and improved dispersibility. The confined microreactor environment facilitated controlled nucleation and growth, while enhanced mixing dynamics promoted reproducibility and scalability. Such studies underscore the superiority of microreactor-based approaches over traditional crystallization or ball-milling methods, which often lead to heterogeneous products and low yields (Yuan et al., 2025).

Microreactors have also proven effective in nanoparticle synthesis for biomedical applications. Wang et al. (2024) reported the ultrasound-assisted synthesis of superparamagnetic Fe_3O_4 nanoparticles in a vortex microreactor. Conventional nanoparticle synthesis often faces clogging and agglomeration due to poor mixing, but the application of ultrasound intensified micromixing and mitigated channel blockage. The resulting Fe_3O_4 nanoparticles exhibited improved size uniformity and enhanced magnetic responsiveness, highlighting the ability of microreactors to integrate external fields (e.g., ultrasound, magnetic, or electric) for process intensification. Such advances are particularly relevant in the context of biomaterials, where precise control over nanoparticle properties determines their performance in drug delivery, imaging, or biosensing. Moreover, microreactors have facilitated biosynthetic routes to biomaterials through engineered microbial systems. In a recent study, water-in-water Pickering emulsion microreactors were used to encapsulate engineered *Bacillus amyloliquefaciens* for enhanced biosynthesis of 4-methyl-5-thiazoleethanol (4M5TE), a key thiazole compound. Encapsulation within compartmentalized microenvironments reduced product volatility and improved yields, while maintaining cell viability and metabolic activity. This demonstrates the potential of microreactors not only for chemical synthesis but also for biological and hybrid bio–chemo systems aimed at sustainable biomaterial production (Ding et al., 2025).

Beyond synthesis, microreactors are increasingly used for biofunctionalization, which involves modifying biomaterial surfaces with functional groups, ligands, or biomolecules to impart specific biological activity. Biofunctionalization is critical for tailoring biomaterials toward biomedical applications such as targeted drug delivery, cell adhesion, antimicrobial action, and biosensing. Microreactor environments provide uniform reaction conditions and high interfacial areas that are advantageous for surface modification processes. For instance, surface modification using surfactants, polymers, or ligands can be integrated within microreactor flows to stabilize nanoparticles and improve their biocompatibility. In the case of dasatinib nanocrystals, the incorporation of binary surfactants (sodium dodecylbenzene sulfonate and BS-12) not only improved stability but also tailored the surface characteristics to enhance drug performance (Yuan et al., 2025). Similarly, Tuntithavornwat et al. (2025) designed a cost-effective poly(methyl methacrylate) (PMMA) photocatalytic microreactor coated with immobilized TiO_2 or ZnO nanoparticles. By further modifying the microreactor surface with cetyltrimethylammonium bromide (CTAB), they reduced the water contact angle, improved wettability, and enhanced photocatalytic efficiency. This demonstrates how microreactor-enabled surface functionalization can be employed not only for biomedical but also for environmental bioapplications, such as pollutant degradation and antibacterial activity.

The advantages of microreactors in biomaterial synthesis and functionalization lie in their ability to overcome many of the inherent limitations of conventional batch processes. Their small characteristic dimensions ensure precise reaction control by enabling uniform mixing and short diffusion distances, which in turn allow fine-tuned control of reaction kinetics. Continuous-flow operation enhances reproducibility by minimizing batch-to-batch variability, a critical requirement for biomedical applications where consistency in product quality is essential. Microreactors are also resource-efficient, as they significantly reduce the consumption of reagents, catalysts, and solvents, thus aligning with eco-friendly and sustainable manufacturing practices. Another important advantage is their integration capability, since synthesis can be seamlessly coupled with downstream processes such as separation, purification, and biofunctionalization within the same platform. Finally, microreactors offer scalability through numbering-up strategies and the design of scale-up systems, which have been successfully demonstrated in the high-throughput production of nanocrystals. Compared to conventional reactors, these features not only yield higher-quality biomaterials but also enable systematic exploration of process parameters using minimal quantities of valuable starting materials, a major advantage for high-value, low-volume products (Asano et al., 2024; Jaiswal and Afzal, 2023; Xiong et al., 2023). The applications of microreactor-enabled synthesis and biofunctionalization extend across multiple domains, reflecting their versatility and industrial relevance. In the pharmaceutical field, they have been employed for the high-throughput preparation of drug nanocrystals, such as dasatinib, where control over particle size and morphology translates to enhanced solubility and bioavailability. In biomedical imaging and therapy, microreactors facilitate the synthesis of uniform superparamagnetic nanoparticles that can function as MRI contrast agents or in targeted hyperthermia treatments. Synthetic biology has also benefited from microreactor technology, particularly through the encapsulation of engineered microbes in Pickering emulsion systems, which improve biosynthetic productivity and product stability. Environmental remediation represents another area of application, where photocatalytic microreactors functionalized with TiO_2 or ZnO have been used for the degradation of dyes, recovery of toxic metal ions, and antibacterial disinfection. Collectively, these examples highlight the broad utility of microreactors in producing and tailoring biomaterials, underscoring their potential impact across healthcare, biotechnology, and sustainability-oriented industries (Ding et al., 2025; Shao et al., 2021; Wang et al., 2024; Yuan et al., 2025).

7.2 BIOMATERIALS IN SCALE-DOWN APPROACH IN MICROREACTORS

7.2.1 Enzymatic Microbioreactor

Although micro-bioreactor systems involving enzymatic treatments are currently at a stage (Hessel et al., 2014), in the current times, many researchers believe that ongoing progress in the design, optimization, and application of microenzymatic reactors will mark the beginning of a new phase in the bio-based economy (Miyazaki and Maeda, 2006; Tamborini et al., 2018; Wohlgemuth et al., 2015; Žnidaršič-Plazl, 2017). Broadly, the use of enzymatic microreactors can be categorized into two major directions: one concerning biotransformation processes and the other dealing with molecular screening and kinetic parameter assessment (Posthuma-Trumpie et al., 2007). Enzymes, which serve as the driving agents in biotransformations, can be employed either in soluble form or immobilized. Immobilization in such reactors is most commonly achieved by covalent binding, adsorption, or co-polymerization techniques (Urban et al., 2006). A straightforward demonstration of this principle was provided by Jurinjak Tušek et al. (2013), who carried out enzymatic oxidation of phenolic compounds, catechol and l-DOPA, using commercial laccase derived from *Trametes versicolor* within a two-aqueous-phase microflow reactor. They reported catechol and l-DOPA conversion rates of 41% and 45% at a residence time of $\tau = 72$ s. The effectiveness of the system was validated by comparing oxidation rates: catechol oxidation was 18–167 times faster than in a conventional macroreactor, while kinetic studies indicated that the maximum reaction rate in the microflow setup was approximately double that in a cuvette. This improvement stems from the fact that smaller channel dimensions reduce diffusion resistance for enzymes and substrates, as well as product removal, thereby enhancing the overall reaction rate (Bolivar et al., 2013; Žnidaršič-Plazl, 2014). Another noteworthy case involves the enzymatic oxidation of hexanol to hexanal, a fragrance molecule, using an NAD^+-dependent alcohol dehydrogenase from *Saccharomyces cerevisiae* (Šalić & Zelić, 2014; Tušek et al., 2012). These experiments were carried out in four tubular microreactors of varying internal volumes, with and without micromixers, and with surfaces differing in roughness (a smooth channel with 1% relative roughness and a rough channel with about 10% roughness). Process performance was analyzed in terms of conversion and volumetric productivity. In a 6-mm^3 microreactor, 30% hexanol conversion was achieved within 36 s, whereas in a macroreactor only 5.3% conversion was observed after 180 s (53), underscoring the microreactor's efficiency. Comparisons between smooth and rough surfaces showed that smooth microchannels doubled the conversion rate at identical residence times. This effect is linked to interfacial mass transfer: smooth walls provided a larger effective interfacial area (1.66×10^{-5} m^2) than rough walls, enhancing conversion. Additionally, smooth channels promoted stable, parallel fluid flows, enabling effective phase separation and the integration of multiphase systems. Kinetic analysis of the same setup revealed that the enzymatic process could be described as a pseudo-homogeneous reaction following a double-substrate Michaelis–Menten model (Matosevic et al., 2011).

Whole-cell systems in microreactors have also been explored. For example, permeabilized baker's yeast cells were employed to regenerate NAD^+ via acetaldehyde reduction to ethanol, using alcohol dehydrogenase activity. This setup achieved a 65.3% NADH conversion within $\tau = 36$ s. Despite such successes, whole-cell applications introduce complications, as multiple enzymes may catalyze side reactions, leading to undesired byproducts. A case in point is hexanol oxidation with permeabilized yeast cells, which generated both hexanal and the byproduct hexanoic acid (Šalić et al., 2013a; Šalić et al., 2013b). Covalent immobilization, through multipoint attachment, can sometimes distort the enzyme's active site and reduce activity, making structural analysis crucial before immobilization (Meller et al., 2017). To simplify the process, one-step immobilization methods are being developed. For example, polycationic tags like Zbasic2 can be fused with enzymes to facilitate electrostatic interactions with negatively charged microchannel surfaces (Bolivar et al., 2017). Miložič et al. (2017, 2018) demonstrated that fusing ω-transaminase with Zbasic2 improved enzyme loading and productivity compared with covalent immobilization. Similarly, Valikhani et al. (2016) and Bolivar et al. (2017) showed higher effectiveness and longer operational lifetimes using Zbasic2-fused enzymes, including a 16-day continuous production of αGG. Microreactors with immobilized enzymes can be categorized into systems with direct surface immobilization, bead-packed beds, monoliths, and membranes. Packed bed reactors are versatile platforms, used for screening, scale-up studies, and diverse reactions, including transesterification, amine synthesis, and production of specialty chemicals like ABT (Bajić et al., 2017; Laurenti et al., 2015; Strniša et al., 2018; Wohlgemuth et al., 2015).

7.2.2 Liposomes in Microreactors

Microfabricated bioreactor flow systems offer tightly controlled reaction environments that suit the self-assembly of liposomes. Within synthetic biology and biomedicine, this control is leveraged to build liposomes as modules for artificial cells/organelles and as drug delivery carriers, alongside other microfluidic products like microgels, fibers, and nanoparticles (Lee et al., 2016; Ma et al., 2017). Among microreactor strategies, microfluidic hydrodynamic focusing (MHF) is a popular platform for nanoscale liposomes. In a typical cross-junction, an alcohol stream carrying dissolved lipids is pinched between aqueous side streams; at the well-defined interface, solvent exchange drives spontaneous self-assembly into closed, spherical liposomes (Cheng et al., 2017; Choi et al., 2017). Liposome size and distribution can be tuned by the flow-rate ratio (FRR) and the total flow rate (TFR), with experiments showing that FRR is the dominant lever for size control and can even modulate in vitro transfection efficiency. Comparative studies of mixing modes (bulk injection, stream mixing, laminar microfluidic mixing) highlight that liposome size correlates strongly with microconvective mixing; only certain modes show dependence on lipid concentration in ethanol, underscoring how controlled diffusion/advective timescales shape nucleation/growth during assembly. Device wetting by the organic stream can introduce polydispersity; to mitigate this, coaxial three-dimensional MHF designs have produced highly uniform nanoscale liposomes (Hood & DeVoe, 2015; Molinaro et al., 2018). MHF-produced liposomes also meet functional requirements for delivery. For instance, reactors have generated vesicles with multifunctional surfaces for tumor targeting and demonstrated stability over weeks with high encapsulation efficiencies; release kinetics and drug–membrane interactions can be tuned by operating conditions and composition. Beyond simple carriers, the same flow-defined assembly can fabricate biomimetic nanovesicles that display leukocyte-like functions (NA-Leuko). These showed selective inflammatory targeting in vivo and reduced macrophage uptake compared to controls, highlighting how microreactor fabrication can align physicochemical cues with biological performance. While MHF enables robust, reproducible production with control over size and lamellarity, the article notes that translation to industrial-scale manufacturing remains an open challenge despite the inherent high-throughput features of continuous flow (Correia et al., 2017; Ran et al., 2016). A second major strategy is droplet-based microreactors (double emulsions and phase transfer), which excel at giant unilamellar vesicles (GUVs) and architected liposomes. Droplet systems use T-junction or flow-focusing geometries to produce monodisperse aqueous droplets in a lipid-containing oil continuous phase. Double emulsions (water-in-oil-in-water) act as templates for bilayer formation, with size and encapsulation efficiency governed by flow-rate and concentration ratios. These platforms enable controlled co-encapsulation of cells, proteins, and functional materials (Deng et al., 2016). A microfluidic "assembly line" can convert lipid-stabilized droplets into unilamellar vesicles via phase transfer across a stable oil/water interface, yielding highly uniform GUVs (≈20–70 μm) (Deng et al., 2017). Combining double-emulsion templating with solvent extraction offers continuous, low-dispersity fabrication of liposomes or polymersomes, though residual oil in bilayers is a known concern for membrane proteins and permeability (Deng & Huck, 2017). To resolve this, octanol-assisted liposome assembly (OLA) separates residual solvent during formation, producing unilamellar, cell-sized liposomes (≈5–20 μm) with high encapsulation and biocompatibility, well-suited as nanoreactors for protein expression. Building on these principles, surfactant-assisted double-emulsion routes (e.g., using Pluronic F-68) create solvent-free single- or multicompartment liposomes (≈20–200 μm) in minutes to hours and support cell-free gene expression, extending to vesosomes (liposomes-in-liposome architectures) where the number of inner vesicles and inter-membrane volumes can be precisely tuned by flow control. These structured vesicles can host distinct cargoes in separate sub-compartments, incorporate membrane proteins like melittin to form nanopores for regulated transport, and even support simultaneous transcription/translation in independent chambers—capabilities that make droplet microreactors powerful engines for bottom-up cell models (Cristaldi et al., 2018).

Pushing further toward modular artificial cells, droplet-stabilized GUVs (dsGUVs) can be produced at high throughput in automated microfluidic platforms. Sequential picoinjection loads biomolecules without compromising membrane integrity; after reconstitution, lipid compartments can be released into physiological media, preserving cargo biofunctionality. This bottom-up route yields bespoke vesicles assembled from intracellular modules, with enhanced mechanical/chemical stability and cell-like dimensions. Researchers have also encapsulated coacervate organelles—non-membrane-bound, phase-separated sub-compartments—inside cell-sized liposomes. These coacervates remain centrally positioned and exhibit thermally reversible coacervation/dissolution, illustrating how microreactors can integrate multiple, orthogonal compartment types within a single synthetic cell chassis. MHF platforms tune size/lamellarity through FRR/TFR and geometry, achieving stable, functional nanoparticles with potential for targeted delivery (Hood & DeVoe, 2015; Molinaro et al., 2018). Droplet-based reactors generate architected GUVs, vesosomes, and multicompartment systems, with solutions to residual solvent (OLA) and methods for sequential, high-throughput cargo loading (picoinjection). Meanwhile, pragmatic device-fabrication strategies (surface-treated polydimethylsiloxane (PDMS), 3D-printed molds, commercial hardware) address cost and scalability. The overarching message is that by controlling micromixing, interfaces, and flow histories, microreactors enable liposome products with predictable structure–function relationships, advancing both drug delivery and bottom-up synthetic biology toward manufacturable, application-ready systems (Deng & Huck, 2017; Deng et al., 2016, 2017).

The field of microfluidics has rapidly advanced into an enabling platform for biotechnology and medicine, offering precise manipulation of fluids at the micrometer scale and introducing new possibilities in the design of biomaterials, diagnostics, and therapeutic systems. This article emphasizes how microfluidic systems or lab-on-a-chip devices can transform biomedical research by integrating sample preparation, analysis, and functional material synthesis into miniaturized, efficient, and highly controlled platforms (Qi et al., 2015). Microfluidics in biomedical diagnostics is one of the most impactful areas in the development of diagnostic devices, particularly point-of-care testing platforms. Microfluidic biosensors integrate sample preparation, reaction, and signal detection in a single chip, enabling rapid and sensitive identification of pathogens, genetic markers, or metabolites. Examples include devices for nucleic acid amplification, immunoassays, and cell sorting. The laminar flow regime allows precise spatial control of reagents, while droplet microfluidics offers high-throughput compartmentalization for parallel reactions (Deng et al., 2016). Moreover, microfluidic diagnostic tools have demonstrated high sensitivity and specificity with minimal sample volumes, which is particularly critical in clinical contexts such as neonatal testing or infectious disease monitoring. The possibility of integrating these devices with smartphones and portable readers further strengthens their potential for decentralized healthcare delivery (Deng et al., 2017).

Another major application of microfluidics is in drug discovery, where miniaturization and automation streamline high-throughput screening. Microfluidic systems can mimic physiological environments more effectively than conventional multiwell plates, enabling better prediction of in vivo drug responses. For example, microfluidic chips can recreate tumor microenvironments or organ-specific conditions, facilitating testing of anticancer agents under more realistic scenarios. Droplet-based microfluidics has proven especially valuable in screening libraries of compounds or enzymes, where each droplet serves as an isolated microreactor. These approaches increase throughput while reducing costs, making them attractive for pharmaceutical development. Furthermore, coupling microfluidics with advanced detection systems, such as mass spectrometry or fluorescence microscopy, enhances real-time monitoring of biochemical interactions and cellular responses (Correia et al., 2017; Deng & Huck, 2017; Ran et al., 2016).

Microfluidic platforms also play a critical role in biomaterial synthesis, especially for nanoparticles, polymers, and vesicles used in drug delivery and regenerative medicine. In microreactors, reactions occur under continuous-flow conditions with well-defined mixing profiles, leading to materials with narrow size distributions and improved reproducibility. Liposomes and polymeric nanoparticles produced in microfluidic devices exhibit superior control over size, lamellarity, and encapsulation efficiency compared with bulk methods. This is vital for biomedical applications, as particle characteristics directly influence biodistribution, cellular uptake, and therapeutic efficacy. For instance, MHF allows the formation of liposomes with tunable diameters by adjusting FRRs, ensuring high encapsulation and stability.

Similarly, double emulsion techniques in droplet microfluidics enable fabrication of GUVs for synthetic biology applications (Hood & DeVoe, 2015; Molinaro et al., 2018). Beyond liposomes, microreactors have been employed for synthesizing polymeric micelles, dendrimers, and inorganic nanoparticles with tailored functionalization. Integration of downstream processes such as surface modification, ligand conjugation, or purification further enhances the potential for continuous, on-chip biomaterial production.

A particularly transformative direction is the development of organ-on-a-chip systems, which aim to replicate the microarchitecture and function of human tissues within microfluidic devices. These platforms combine engineered biomaterials with living cells in dynamic microenvironments, allowing the study of organ-level physiology and disease mechanisms in vitro. For example, lung-on-a-chip models simulate alveolar-capillary interfaces, enabling studies of gas exchange, mechanical stretching, and inflammatory responses. Similarly, heart-on-a-chip systems integrate cardiomyocytes in microfluidic scaffolds to investigate electrophysiology and drug cardiotoxicity. By recapitulating organ-specific microenvironments, these devices bridge the gap between conventional cell cultures and animal models, providing more predictive platforms for preclinical testing. Moreover, coupling multiple organ chips into "body-on-a-chip" configurations supports systemic studies of pharmacokinetics and pharmacodynamics, offering insight into whole-body drug responses. These integrated systems may revolutionize personalized medicine by tailoring drug regimens based on patient-specific organ models (Qi et al., 2015).

7.3 CONCLUSION

Overall, microfluidic technology is rapidly evolving from a laboratory tool to a transformative platform for biomedical applications. Its ability to precisely manipulate fluids, create reproducible biomaterials, and simulate organ-level functions opens vast opportunities in diagnostics, drug discovery, biomaterial synthesis, and personalized medicine. While challenges remain in scaling, standardization, and cost, the continuous innovation in device design and integration suggests that microfluidics will play a pivotal role in the future of biotechnology and healthcare. Despite the advances, several challenges remain in translating microreactor-based biomaterial synthesis to industrial practice. Scale-up, although addressed through numbering-up and novel designs, still requires optimization to balance throughput with product quality. Fouling and clogging, especially during nanoparticle synthesis, pose operational challenges, though approaches such as ultrasound assistance offer promising solutions. Moreover, integrating multiple unit operations synthesis, purification, and biofunctionalization within a seamless microreactor platform remains an ongoing area of research. Despite significant progress, several challenges must be addressed before widespread clinical and industrial adoption of microfluidic technologies. Scalability remains a critical limitation, as devices designed for small-scale laboratory experiments often face difficulties in mass production and large-volume operation. Fouling, clogging, and surface interactions also complicate long-term use. Standardization and regulatory approval are additional hurdles, particularly for diagnostic devices and therapeutic material production. Ensuring reproducibility across different fabrication methods and laboratories will be essential for translation. Cost reduction is another priority, as current microfabrication methods can be expensive and resource-intensive. Nevertheless, emerging strategies such as 3D printing, low-cost polymer fabrication, and modular device design offer promising solutions. Advances in automation, integration with machine learning, and coupling with wearable or implantable systems could further extend the reach of microfluidics in healthcare and biotechnology.

Future developments are likely to focus on combining microreactor technology with advances in microfabrication, synthetic biology, and machine learning-driven process optimization. For instance, machine learning models could predict optimal reaction conditions in real time, while synthetic biology approaches may integrate living systems with microreactors for continuous biosynthesis of complex biomaterials. Additionally, the development of low-cost, modular, and optically accessible reactors, such as PMMA-based platforms, will expand accessibility and adoption across laboratories and industries.

REFERENCES

Asano, S., Kudo, S., & Hayashi, J.-I. (2024). Chaotic-flow-driven mixing in T- and V-shaped micromixers. *Chemical Engineering Journal*, *489*, 151183.

Bajić, M., Plazl, I., Stloukal, R., & Žnidaršič-Plazl, P. (2017). Development of a miniaturized packed-bed reactor with ω-transaminase immobilized in LentiKats®. *Process Biochemistry*, *52*, 63–72. https://doi.org/10.1016/j.procbio.2016.09.021

Bolivar, J. M., Luley-Goedl, C., Leitner, E., Sawangwan, T., & Nidetzky, B. (2017). Production of glucosyl glycerol by immobilized sucrose phosphorylase in microscale flow format. *Journal of Biotechnology*, *257*, 131–138. https://doi.org/10.1016/j.jbiotec.2017.01.019

Bolivar, J. M., & Nidetzky, B. (2013). Multiphase biotransformations in microstructured reactors: Opportunities for biocatalytic process intensification and smart flow processing. *Green Processing and Synthesis*, *2*(6), 541–559. https://doi.org/10.1515/gps-2013-0091

Cheng, J., Jun, Y., Qin, J., & Lee, S.-H. (2017). Electrospinning versus microfluidic spinning of functional fibers. *Biomaterials*, *114*, 121–143.

Choi, A., Seo, K. D., Kim, D. W., Kim, B. C., & Kim, D. S. (2017). Engineering microparticles for biomedical delivery and diagnostics. *Lab on a Chip*, *17*, 591–613.

Correia, M. G. S., Briuglia, M. L., Niosi, F., & Lamprou, D. A. (2017). Microfluidic manufacturing of phospholipid nanoparticles. *International Journal of Pharmaceutics*, *516*, 91–99.

Cristaldi, D. A., Yanar, F., Mosayyebi, A., García-Manrique, P., Stulz, E., Carugo, D., & Zhang, X. (2018). Easy-to-perform and cost-effective fabrication of continuous-flow reactors and their application for nanomaterials synthesis. *New Biotechnology*, *47*, 1–7. https://doi.org/10.1016/j.nbt.2018.02.002

Deng, N.-N., Yelleswarapu, M., & Huck, W. T. S. (2016). Monodisperse uni- and multicompartment liposomes. *Journal of the American Chemical Society*, *138*(24), 7584–7591. https://doi.org/10.1021/jacs.6b02107

Deng, N.-N., Yelleswarapu, M., Zheng, L., & Huck, W. T. S. (2017). Microfluidic assembly of monodisperse vesosomes as artificial cell models. *Journal of the American Chemical Society*, *139*(2), 587–590. https://doi.org/10.1021/jacs.6b10977

Deng, N.-N., & Huck, W. T. S. (2017). Microfluidic formation of monodisperse coacervate organelles in liposomes. *Angewandte Chemie International Edition*, *56*(33), 9736–9740. https://doi.org/10.1002/anie.201703145

Ding, X., Huang, W., Cao, H., Ao, S., Zou, D., Wei, X., & Liu, S. (2025). Microreactors derived from water-in-water Pickering emulsions for Bacillus amyloliquefaciens proliferation and synthesis of 4-methyl-5-thiazoleethanol. *Bioresource Technology, 438*, 133231. https://doi.org/10.1016/j.biortech.2025.133231

Hameedat, F., Mendes, B. B., Conniot, J., Di Filippo, L. D., Chorilli, M., Schroeder, A., Conde, J., & Sousa, F. (2024). Engineering nanomaterials for glioblastoma nanovaccination. *Nature Reviews Materials*. https://doi.org/10.1038/s41578-024-00684-z

Han, J. Y., La Fiandra, J. N., & DeVoe, D. L. (2022). Microfluidic vortex focusing for high-throughput synthesis of size-tunable liposomes. *Nature Communications, 13*, 6997.

Hessel, V., Tibhe, J., Noël, T., & Wang, Q. (2014). Biotechnical micro-flow processing at the EDGE – Lessons to be learnt for a young discipline. *Chemical and Biochemical Engineering Quarterly, 28*(2), 167–188. https://doi.org/10.15255/CABEQ.2014.1939

Holghoomi, R., Kharab, Z., Rahdar, A., Pandey, S., & Ferreira, L. F. R. (2024). Harnessing the power of green synthesis of nanomaterials for anticancer applications: A review. *Coordination Chemistry Reviews, 513*, 215903.

Hood, R. R., & DeVoe, D. L. (2015). High-throughput continuous-flow production of nanoscale liposomes. *Small, 11*, 5790–5799.

Jaiswal, A. K., & Afzal, A. (2023). Design optimization of prismatic rib turbulators in a rectangular channel based on a multi-objective criterion. *International Journal of Thermal Sciences, 185*, 108091.

Jurinjak Tušek, A., Tišma, M., Bregović, V., Ptičar, A., Kurtanjek, Ž., & Zelić, B. (2013). Enhancement of phenolic compounds oxidation using laccase from *Trametes versicolor* in a microreactor. *Biotechnology and Bioprocess Engineering, 18*(4), 686–696. https://doi.org/10.1007/s12257-012-0688-8

Krasley, A. T., Li, E., Galeana, J. M., Bulumulla, C., Beyene, A. G., & Demirer, G. S. (2024). Carbon nanomaterial fluorescent probes and their biological applications. *Chemical Reviews, 124*, 3085–3185.

Laurenti, E., & dos Santos Vianna Jr., A. (2015). Enzymatic microreactors in biocatalysis: History, features, and future perspectives. *Biocatalysis, 1*(1), 148–165. https://doi.org/10.1515/boca-2015-0008

Lee, T. Y., Choi, T. M., Shim, T. S., Frijns, R. A. M., & Kim, S.-H. (2016). Microfluidic production of multiple emulsions and functional microcapsules. *Lab on a Chip, 16*, 3415–3440.

Li, Z., Lin, L., Shen, J., Li, C.-D., & Ostrikov, K. K. (2024). Microflow synthesis of fluorescent carbon dots for selective Co^{2+} detection. *Industrial & Engineering Chemistry Research, 63*, 4420–4429.

Ma, J., Lee, S. M.-Y., Yi, C., & Li, C.-W. (2017). Controllable synthesis of functional nanoparticles by microfluidic platforms. *Lab on a Chip, 17*, 209–226.

Matosevic, S., Szita, N., & Baganz, F. (2011). Fundamentals and applications of immobilised microfluidic enzymatic reactors. *Journal of Chemical Technology and Biotechnology, 86*(3), 325–334. https://doi.org/10.1002/jctb.2564

Meller, K., Szumski, M., & Buszewski, B. (2017). Microfluidic reactors with immobilized enzymes – Characterization, dividing, perspectives. *Sensors and Actuators B: Chemical, 244*, 84–106. https://doi.org/10.1016/j.snb.2016.12.021

Miložič, N., Lubej, M., Lakner, M., Žnidaršič-Plazl, P., & Plazl, I. (2017). Enzyme kinetics in a microreactor system with surface-immobilized biocatalyst. *Chemical Engineering Journal, 313*, 374–381. https://doi.org/10.1016/j.cej.2016.12.030

Miložič, N., Stojkovič, G., Vogel, A., Bouwes, D., & Žnidaršič-Plazl, P. (2018). Development of microreactors with surface-immobilized biocatalysts for transamination. *New Biotechnology, 47*, 18–24. https://doi.org/10.1016/j.nbt.2018.05.004

Miyazaki, M., & Maeda, H. (2006). Microchannel enzyme reactors and their applications for processing. *Trends in Biotechnology, 24*(10), 463–470. https://doi.org/10.1016/j.tibtech.2006.08.002

Molinaro, R., Evangelopoulos, M., Hoffman, J. R., Corbo, C., Taraballi, F., Martinez, J. O., Hartman, K. A., Cosco, D., Costa, G., Romeo, I., et al. (2018). Biomimetic nanovesicles using a microfluidic approach. *Advanced Materials*, *30*, e1702749.

Posthuma-Trumpie, G. A., Venema, K., van Berkel, W. J. H., & Korf, J. (2007). A low perfusion rate microreactor for continuous monitoring of enzyme characteristics: Application to glucose oxidase. *Analytical and Bioanalytical Chemistry*, *389*(6), 2029–2033. https://doi.org/10.1007/s00216-007-1596-1

Qi, C., Yan, X., Huang, C., Melerzanov, A., & Du, Y. (2015). Biomaterials as carrier, barrier and reactor for regenerative medicine. *Protein & Cell*, *6*(9), 638–653. https://doi.org/10.1007/s13238-015-0179-8

Ran, R., Middelberg, A. P. J., & Zhao, C.-X. (2016). Microfluidic synthesis of multifunctional liposomes. *Colloids and Surfaces B: Biointerfaces*, *148*, 402–410.

Šalić, A., Faletar, P., & Zelić, B. (2013a). NAD^+ regeneration in a microreactor using permeabilized baker's yeast cells. *Biochemical Engineering Journal*, *77*, 88–96. https://doi.org/10.1016/j.bej.2013.05.007

Šalić, A., Pindrić, K., & Zelić, B. (2013b). Bioproduction of food additives hexanal and hexanoic acid in a microreactor. *Applied Biochemistry and Biotechnology*, *171*(8), 2273–2284. https://doi.org/10.1007/s12010-013-0495-5

Šalić, A., & Zelić, B. (2014). ADH-catalysed hexanol oxidation with fully integrated NADH regeneration performed in microreactors connected in series. *RSC Advances*, *4*(79), 41714–41721. https://doi.org/10.1039/C4RA05421K

Shao, Y., Ruan, X., & Li, S. (2021). Mechanism for clogging of microchannels by small particles with liquid cohesion. *AIChE Journal*, *67*, e17288.

Silva, J. L., Celestino, M. S. C., Taranto, O. P., & Santana, H. S. (2024). Smart scale-up of micromixers for efficient continuous biodiesel synthesis: A numerical study for process intensification. *Chemical Engineering and Processing: Process Intensification*, *196*, 109664.

Strniša, F., Bajić, M., Panjan, P., Plazl, I., Sesay, A. M., & Žnidaršič-Plazl, P. (2018). Characterization of an enzymatic packed-bed microreactor: Experiments and modeling. *Chemical Engineering Journal*, *350*, 541–550. https://doi.org/10.1016/j.cej.2018.05.028

Tamborini, L., Fernandes, P., Paradisi, F., & Molinari, F. (2018). Flow bioreactors as complementary tools for biocatalytic process intensification. *Trends in Biotechnology*, *36*(1), 73–88. https://doi.org/10.1016/j.tibtech.2017.09.005

Tuntithavornwat, S., Saengpitak, K., Phitaksidchareon, P., Sdayuruch, T., & Hunsom, M. (2025). Cost-effective PMMA photocatalytic microreactor for nanocatalyst screening and environmental research. *Chemical Engineering Journal Advances*, *23*, 100815. https://doi.org/10.1016/j.ceja.2025.100815

Tušek, A., Šalić, A., Kurtanjek, Ž., & Zelić, B. (2012). Modeling and kinetic parameter estimation of alcohol dehydrogenase-catalyzed hexanol oxidation in a microreactor. *Engineering in Life Sciences*, *12*(1), 49–56. https://doi.org/10.1002/elsc.201100020

Urban, P. L., Goodall, D. M., & Bruce, N. C. (2006). Enzymatic microreactors in chemical analysis and kinetic studies. *Biotechnology Advances*, *24*(1), 42–57. https://doi.org/10.1016/j.biotechadv.2005.06.001

Valikhani, D., Bolivar, J. M., Pfeiffer, M., & Nidetzky, B. (2016). Multivalency effects on immobilization of sucrose phosphorylase in flow microchannels. *ChemCatChem*, *9*(1), 161–166. https://doi.org/10.1002/cctc.201601019

Wang, S., Zhang, J., Ma, K., Zhang, W., Gao, Y., Yu, P., Zhao, S., Feng, Y., Yang, J., Sun, R., Li, Y., Zhu, N., He, W., & Guo, K. (2024). Design and optimization of novel vortex microreactors for ultrasound-assisted synthesis of high-performance Fe_3O_4 nanoparticles. *Chemical Engineering Journal*, *501*, 157672. https://doi.org/10.1016/j.cej.2024.157672

Wohlgemuth, R., Plazl, I., Žnidaršič-Plazl, P., Gernaey, K. V., & Woodley, J. M. (2015). Microscale technology and biocatalytic processes: Opportunities and challenges for synthesis. *Trends in Biotechnology*, *33*(5), 302–314. https://doi.org/10.1016/j.tibtech.2015.02.010

Xiong, M., Yang, J., Ding, X., Li, H., & Zhang, H. (2023). Topology optimization design of micromixer based on the principle of Tesla valve: An experimental and numerical study. *Chemical Engineering and Processing: Process Intensification*, *193*, 109560.

Yuan, F., Sun, Y., Guo, J., Wei, Y., Wang, S., Zhao, S., Sun, R., He, W., Zhu, N., Li, Y., Guo, K. (2025). Scale-up microreactor coupling binary surfactants for high-throughput preparation of flaky dasatinib nanocrystals. *Journal of the Taiwan Institute of Chemical Engineers*, *167*, 105823. https://doi.org/10.1016/j.jtice.2024.105823

Žnidaršič-Plazl, P. (2014). Enzymatic microreactors utilizing non-aqueous media. *Chimica Oggi*, *32*(1), 54–61.

Žnidaršič-Plazl, P. (2017). Biotransformations in microflow systems: Bridging the gap between academia and industry. *Journal of Flow Chemistry*, *7*(3–4), 111–117. https://doi.org/10.1556/1846.2017.00021

8 Quantum Dot Synthesis Using Microreactors

Sarojini Nanthakumar, Bhargav Midathani, Priyadharshini Sivalingam, Subash Baskaran, and Subhendu K. Panda

8.1 INTRODUCTION

The dawn of the 21st century has been characterized by a growing mastery over matter at its most fundamental scale: the nanoscale. Within this realm, materials often exhibit properties dramatically different from their bulk counterparts, leading to unprecedented opportunities in diverse fields. Among these fascinating materials, nanoparticles, generally described as particles with dimensions varying from 1 to 100 nanometer, have attracted significant attention. Their high surface area-to-volume ratio and size-dependent quantum mechanical effects give rise to unique optical, electronic, magnetic, and catalytic characteristics. Harnessing these novel properties for technological advancements requires the potential to synthesize nanoparticles with exquisite control over their size, shape, composition, crystal structure, and surface properties (Kulkarni and Goel 2020).

In response to the limitations of batch processing for quantum dot synthesis, microfluidic reactor technology has emerged as a transformative platform, offering unparalleled control over reaction conditions (Kulkarni and Goel 2020; Martínez-Cisneros et al. 2012). Microreactors are characterized by channel dimensions typically ranging from tens to hundreds of micrometers (Rašljić Rafajilović et al. 2020). This small scale fundamentally alters the reaction environment compared to macroscopic reactors (Kulkarni and Goel 2020; Rašljić Rafajilović et al. 2020). A key advantage of these small dimensions is the significantly high surface area-to-volume ratio. This property is critical for efficient heat and mass transfer within the microchannels (Martínez-Cisneros et al. 2012; Rašljić Rafajilović et al. 2020) permits the reaction mixture to be quickly heated or cooled, enabling precise control over reaction temperature and minimizing temperature gradients that can affect nucleation and growth (Kulkarni and Goel 2020). Similarly, efficient mass transfer ensures that reagents are quickly and uniformly dispersed throughout the reaction volume (Kulkarni and Goel 2020).

Significantly improved mixing is another major benefit offered by microreactors (Kulkarni and Goel 2020). Due to the low Reynolds numbers characteristic of flow in microchannels, the fluid flow is typically laminar (Rašljić Rafajilović et al. 2020). While seemingly counterintuitive, laminar flow in appropriately designed microchannels (e.g., serpentine or spiral designs) or utilizing techniques like segmented flow

 DOI: 10.1201/9781003633761-8

(droplet-based microfluidics) promotes mixing through diffusion over short distances or by including chaotic advection (Kubendhiran et al. 2019; Kulkarni and Goel 2020). Various passive and active mixing principles are employed in microreactors to ensure rapid and uniform mixing (Kulkarni and Goel 2020; Sulaxna Pandey and Bodas 2022). Rapid and uniform mixing is essential for controlled QDs synthesis, as it helps to prevent local supersaturation and ensures that all precursor molecules experience similar reaction conditions, which is vital for achieving narrow particle size distributions and consistent properties (Kulkarni and Goel 2020; S. Pandey et al. 2021).

Beyond heat transfer and mixing, microreactors facilitate precise control over reaction time, also known as residence time (Rao et al. 2017; Tang et al. 2018). The channel volume and flow rate of a continuous flow microreactor dictate how long the reaction mixture remains within the heated channel (Jeong et al. 2016; Rao et al. 2017; Sonawane et al. 2015; Tang et al. 2018). By simply adjusting the flow rate or the channel length, the reaction time can be precisely controlled from milliseconds to minutes, allowing fine-tuning of particle size and morphology (Jeong et al. 2016; Rao et al. 2017; Sonawane et al. 2015; Tang et al. 2018). This level of temporal control is often difficult to achieve reproducibility in batch systems (Kulkarni and Goel 2020; S. Pandey et al. 2021).

Furthermore, recent advancements in microreactor design have introduced multi-temperature zone microreactors. These devices allow sequential heating or cooling within different sections of the microchannel, providing independent control over the nucleation and growth stages of nanoparticle synthesis. This capability is particularly valuable for synthesizing highly monodisperse quantum dots, as nucleation can be initiated and controlled at a specific temperature, followed by particle growth at a different, optimized temperature within the same continuous flow system (Rašljić Rafajilović et al. 2020). The continuous flow nature of microreactors also presents a distinct advantage for scalability. Instead of relying on traditional scale-up methods that increase reactor size, production capacity in microfluidics is typically increased by operating multiple identical microchannels in parallel – a strategy known as scale-out. This approach maintains the optimized reaction conditions achieved in a single microchannel across multiple channels, ensuring consistent product quality regardless of the production volume. This transition from laboratory-scale development in a single channel to potential industrial-scale production via parallelization is a significant aspect of the technology (Kubendhiran et al. 2019; Lin et al. 2021).

The robustness and suitability of microreactors for various synthesis conditions are also addressed through material selection and fabrication techniques (Berenguel-Alonso et al. 2019; Martínez-Cisneros et al. 2012; Rašljić Rafajilović et al. 2020). Materials like silicon-Pyrex glass with integrated heaters are used for high-temperature operations (Rašljić Rafajilović et al. 2020). Low-Temperature Co-fired Ceramic (LTCC) microreactors offer robustness for high-temperature and pressure hydrothermal synthesis (Berenguel-Alonso et al. 2019; Lin et al. 2021). Other materials like PDMS, PTFE, and stainless-steel tubes have also been employed, depending on the specific chemical requirements and desired operating conditions (Jeong et al. 2016; S. Pandey et al. 2021; Rao et al. 2017; Sonawane et al. 2015; Tang et al. 2018). The development includes monolithically integrated heaters and robust fluidic connections (Rašljić Rafajilović et al. 2020).

In summary, microfluidic reactor technology offers a powerful and controllable environment for the synthesis of high-quality nanomaterials, including QDs (Lin et al. 2021; Martínez-Cisneros et al. 2012; S. Pandey et al. 2021; Rašljić Rafajilović et al. 2020). The amplified heat and mass transfer, superior mixing capabilities, precise temperature and residence time control, the possibility of multizone temperature profiles, and the advantages of scale-out address many of the limitations associated with conventional batch synthesis (Kubendhiran et al. 2019; Kulkarni and Goel 2020; Lin et al. 2021; Martínez-Cisneros et al. 2012; Rao et al. 2017; Rašljić Rafajilović et al. 2020; Tang et al. 2018). This has opened the path for the reproducible fabrication of QDs and other nanomaterials with tailored properties for diverse applications (Kulkarni and Goel 2020; Lin et al. 2021).

8.2 MICROREACTOR TYPES

Chemical synthesis and processing have usually mostly used batch reactors, in which all reactants are mixed together in a vessel and given a set amount of time to react before the result is collected. Batch processes are useful for many applications, but they are limited by issues like poor heat and mass transfer, uneven product quality, safety hazards when handling dangerous reactions, and limited scalability (Jähnisch et al. 2004). In chemical engineering, microreactors have become a game-changing technique to tackle these issues. These are miniature flow-based systems that enable continuous processing under strictly regulated conditions. They are usually characterized by channels with dimensions in the micrometer range. Microreactors provide substantial advantages over batch reactors, exhibiting improved heat and mass transfer, precise reaction control, reduced reagent consumption, accelerated reaction times, and improved safety, particularly for exothermic or hazardous reactions (M. Wojnicki and Hessel 2022). The relative benefits and drawbacks of microreactors versus batch reactors (see Figure 8.1).

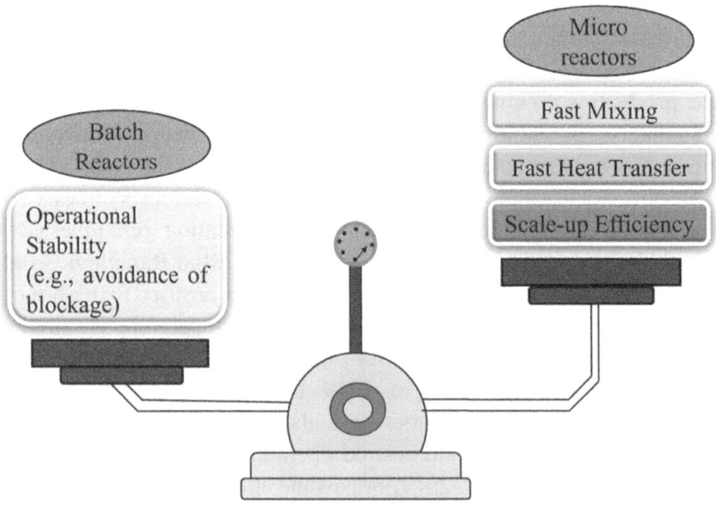

FIGURE 8.1 Balance of advantages and disadvantages of the batch reactors and microreactor.

In microreactors, a steady and ideal environment for QDs preparation can be easily established (Cheng et al. 2021). Microreactors may be divided into three varieties based on the liquid flow pattern in a microchannel reactor: droplet-based microreactors, segmented flow microreactors, and continuous laminar flow microreactors. Continuous laminar flow microreactors make manipulation and scaling up easier and are comparatively simple to construct. But they naturally have problems with the residence time distributions and the parabolic velocity profile. These issues can be mitigated to some extent by employing segmented flow microreactors. Nonetheless, segmented flow microreactors are prone to physical contact between reagents and the reactor walls, potentially leading to product contamination. In contrast, droplet-based microreactors can circumvent cross-contamination issues. However, they necessitate meticulous attention to geometry design for droplet generation, which may impede the scale-up of thc synthesis process (Hao, Nie and Zhang 2018).

8.2.1 Continuous Laminar Flow Microreactor

The most popular way of creating QDs is the continuous laminar flow microreactor, which works well with single-phase liquid flows. These microreactors are distinguished by their great degree of controllability and easy accessibility. Continuous laminar flow microreactors can be constructed using stainless steel or silica glass capillary tubes. With this setup, a water bath or an oil bath can be used to precisely and readily regulate the reaction temperature inside the microtube. Controlling the feed pump makes it simple to adjust the volume or flow velocity of miscible solvents that are pumped into the reactor. In laminar flow streams and microstructures like T-junctions, reagents are combined by diffusion. For greater mixing efficiency, membrane dispersion can be used. However, the tiny route can cause obstruction and fouling, which is a problem for continuous laminar flow reactors. A microfiltration membrane dispersion microreactor was created by the Luo group to minimize obstruction by achieving rapid and uniform mixing.

Single-phase liquid flows are used in continuous laminar flow microreactors (see Figure 8.2a), in which distinct liquid reagents (LR1 and LR2) with different concentrations and compositions are added to the microfluidic systems via the inlets. Mixing is a crucial step in maximizing production in these laminar-dominant microreactors. The channel dimensions needed for mixing under various reaction circumstances can be estimated using the Peclet number (Pe), a dimensionless metric that is the ratio of convection to diffusion. Distributive mixing-based micromixers and wrapping microchannels are common mixing modules for 2D micromixer design (see Figure 8.2a) (Luo et al. 2011).

8.2.2 Segmented Flow Microreactor

Typically, segmented flow microreactors use a gas phase flow to create gas ‘bubbles’ that separate distinct reagent segments and several liquid phase flows for the reagents (see Figure 8.2b). These gas ‘bubbles’ are crucial for containing reactions inside liquid segments and form as a result of surface tension variations between the gas and liquid phases. Two or more liquid-phase reagents (LR1, LR2, etc.) are added to

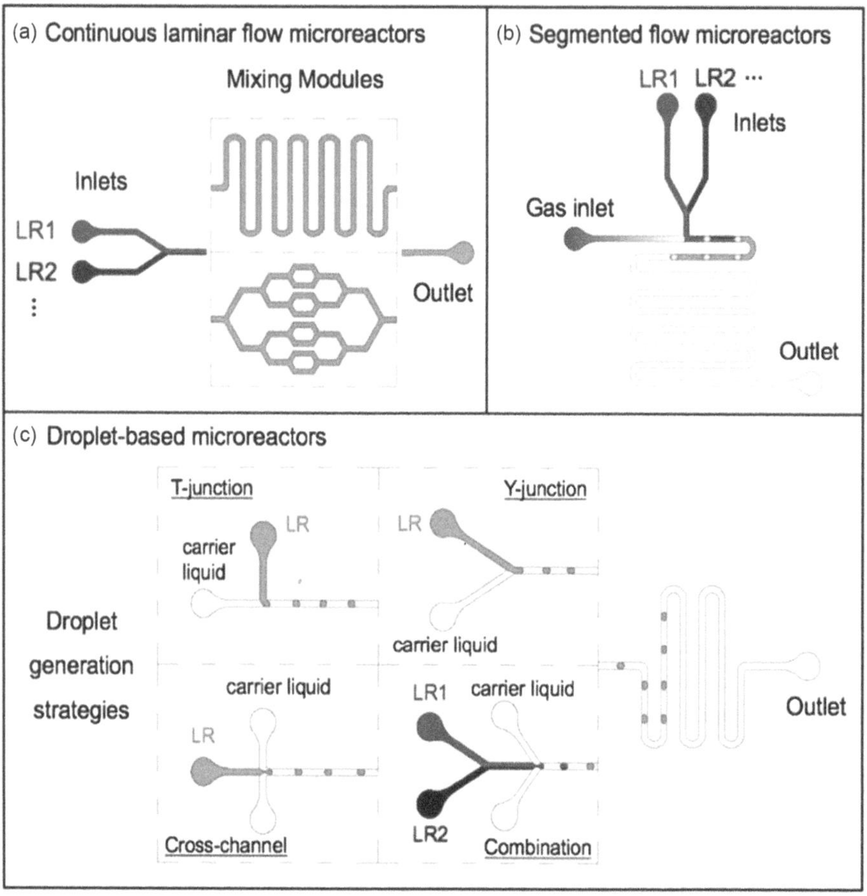

FIGURE 8.2 There are three primary types of microreactors: (a) continuous laminar flow microreactors, which inject single-phase flows of various liquid reagents (LR1, LR2, etc.) into the systems after passing through later mixing modules like distributive mixing-based mixers and winding microchannels. (b) Segmented flow microreactors, which typically consist of multiple liquid phase flows for the introduction of specific reagents (LR1, LR2, etc.) and gas phased flow to produce 'bubbles' for the separation of reagent segments. (c) Microreactors that are droplet-based, which allow for the development of various droplet production techniques, such as T-junction, Y-junction, cross-channel, and a combination of multiple simple designs, like the Y-junction and the cross channel for more complex reactions.

the microreactor through the inlets in a standard configuration (see Figure 8.2b). To satisfy certain response needs, more mixing sections might be added to the design. Because reagents and channel walls come into physical contact, segmented flow microreactors may have contamination problems. By carefully taking into account the hydrophobicity of the channel walls, these problems can be largely resolved (Chen et al. 2004, 2004)

Often, segmented flow microreactors use two Non-mixing flows (liquid–liquid system or gas–liquid system). The solution stretches and folds when a fresh phase is introduced, which improves fluid mixing efficiency by creating recirculation movement. A segmented flow reactor, on the other hand, separates continuous flows into individual columns. Each and every column can be thought of as a separate microreactor since it is a unique included volume that is usually on the fL-nL range. This eliminates the axial fluid's back-mixing in the single-phase flow microreactor and allows for precise control of the residence period. This might make it easier to keep the reagent composition in each reaction unit consistent. Furthermore, at the junction of various flows, intense segment-interval convective mixing may be generated, which would hasten the nucleation time interval and cause nucleation to begin quickly (Kawasaki et al. 2010).

8.2.3 Droplet-Based Microreactors

Another kind of multiphase flow reactor is a droplet-based microreactor, which typically consists of several liquid phase reagent flows and an additional immiscible fluid phase flow serving as a transport liquid for droplet formation. Because droplet-based microreactors may compartmentalize, they have a major advantage over segmented-flow and laminar-flow microreactors. This feature allows for rapid mixing, precise control over time and reagents, and a contamination-free microenvironment. An immiscible liquid phase flow that serves as a carrier for droplet formation and many liquid phase flows of reagents are common features of these microreactors, a kind of multiphase flow system (see Figure 8.2c). The creation of droplets requires consideration of the dimensionless capillary number $Ca = \eta V / \gamma$, which represents the balance between surface tension and viscous forces. Droplet break-off occurs when Ca exceeds a critical value. As depicted (see Figure 8.2c, common geometrical designs for droplet generation include T-junction, Y-junction, cross-channel flow-focusing, and combinations of simple strategies like Y-junction and cross-channel for more complex reactions. T-junction and Y-junction designs are relatively straightforward, with droplet size influenced by channel width and flow rate, allowing for easy manipulation of the reaction process in these microreactors. In cross-channel flow-focusing microreactors, optimizing geometry and flow conditions is essential to produce uniformly sized droplets for chemical synthesis. Another strategy for droplet generation is the electrohydrodynamic method, which encompasses dielectrophoresis-driven and electrowetting on dielectric techniques. Droplet-based microreactors are adaptable microsystems for chemical synthesis, where droplet manipulation, such as fission, fusion, mixing, and sorting, can significantly aid in the production of micro/nanoparticles under various reaction conditions. However, when generating droplets, the hydrophobicity of channel walls must be taken into account. Ensuring that either the continuous phase or the dispersed phase does not adhere to the channel walls is crucial for the optimal performance of droplet-based microreactors (Teh et al. 2008).

8.3 QUANTUM DOTS SYNTHESIS USING MICROREACTORS

8.3.1 Noble Metal Quantum Dots

Noble metal quantum clusters (NMQCs) serve as a transitional form between isolated noble metal atoms and nanoparticles. These clusters, measuring less than 1 nm, consists of a core group of atoms (Jeong et al. 2016; Song et al. 2017; Wallace and

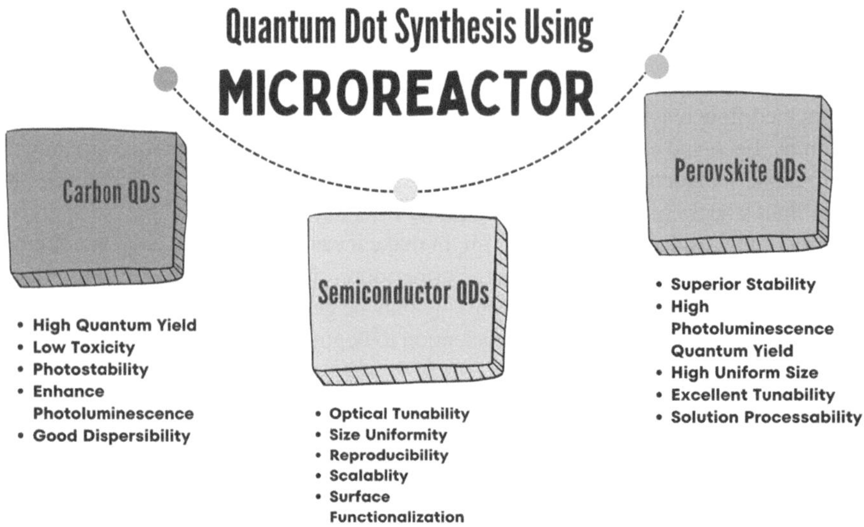

FIGURE 8.3 Various types of QDs using microreactors.

Whetten 2002). Their size is comparable to an electron's Fermi wavelength, which is around 0.5 nm for gold or silver, leading to properties akin to molecules, such as a distinct electronic structure and photoluminescence based on HOMO–LUMO transitions. These materials exhibit intriguing photonic, chemical, and physical characteristics. Bulk metals do not display photoluminescence due to nonradiative decay and the absence of an energy gap (Peyser, Lee, and Dickson 2002). However, in noble metals, the emission color can be adjusted by altering both the size and composition of NMQCs (Figure 8.3).

Unlike smaller structures like citrates used in the Turkevich approach, proteins are big molecules that may effectively mask contacts between metal clusters, which is why they are frequently used in the manufacture of these clusters. Potential uses in biology and medicine are therefore made possible. These materials have potential for bioimaging and targeted cancer treatment since their characteristics are regulated by both the metal atoms and the protein structure. The application of protein-gold nanoclusters in bioimaging on cells that mimic the human immune system is a noteworthy example (Czyżowska et al. 2021; Xavier et al. 2012). In this synthesis, the Au nanohybrids were stabilized and reduced by lysozyme (LYZ), gamma globulin (γG) immunoproteins, bovine serum albumins (BSA), and human serum albumins (HSA). The resulting cluster emits at $\lambda_{em} = 560$ nm with a quantum yield of 5%, although the size of the NMQCs was not given. Au-NMQCs may be synthesized with ease. Ungor et al. created red-emitting gold nanoclusters for the rapid fluorescence detection of tryptophan metabolites using a one-step manufacturing procedure. γ-globulin was used as a stabilizing and reducing agent in the synthesis of red-emitting ($\lambda_{em} \sim 645$ nm) NMQCs. γ-globulin was used as a stabilizing and reducing agent in the synthesis of red-emitting ($\lambda_{em} \sim 645$ nm) NMQCs (Carlander et al. 2019; Ungor et al. 2019). The production of red-emitting ($\lambda_{em} \sim 645$ nm) NMQCs involved the use of γ-globulin as a stabilizing and decreasing agent. The core has a QY of 4.4% and a dimension of $d = 1.5 \pm 0.3$ nm. High sensitivity to tryptophan metabolites was shown by these Au-NMQCs. By varying the molar ratio of Au(III) and L-tryptophan, Csapó et al. investigated the regulation of optical characteristics in sub-nanometric sized ($d < 2$ nm) Au NCs exhibiting blue or green photoluminescence ($\lambda_{em} = 470 - 500\,nm$) (Csapó et al. 2017). Zheng et al. synthesized fluorescent and size-tunable gold QDs by reducing gold salts (e.g., $HAuCl_4$ or $AuBr_3$) in aqueous PAMAM solutions.

Au-NMQCs were explored as particular fluorescent sensors for various metal ions, including Fe^{3+} (Ungor et al. 2017), Hg^{2+} (Xie et al. 2010), and Cu^{2+} (Shibu and Pradeep 2011). Balu et al. detailed a one-pot green synthesis method for Pt-NMQCs, resulting in highly stable, blue and green fluorescent, water-soluble materials. These Pt-NMQCs were produced using H_2PtCl_6 as a precursor and Rec1-resilin as both a reducing and stabilizing agent, achieving a quantum yield of approximately 7%. A mixture of atoms can also be used since, as was already said, the qualities are affected by the metal atoms used. Au/Ag bimetallic nanoclusters that produce yellow light at $\lambda_{em} = 560\,nm$ with a quantum efficiency of 7.3% were manufactured by Ungor et al (Ungor et al. 2021). This kind of quantum material was used to find iron in biological samples because of its high QY. As Pb^{2+} ion sensors, copper quantum clusters in a protein matrices were assessed (Goswami et al. 2011). Similarly, photoluminescent metal mixtures can be applied. $Ag_{10}Cu_6$ clusters stabilized by a PNNP ligand and phenylacetylene were examined for the selective and reversible detection of ammonia in air and water (Y. Wang et al. 2021). The studies mentioned share a common feature: all syntheses were performed in a batch reactor. One advantage of continuous microreactors is that, as Jin et al. showed with a microreactor, they may produce highly monodispersed phosphine-stabilized gold nanoclusters at a pace of around 11.8 (mg s^{-1})(Y. Wang et al. 2021). This rate is approximately 500 times greater than conventional batch syntheses based on production rate per reactor volume. Bulk synthesis in batch reactors faces challenges like poorly controlled mixing, heat and mass transfer issues, and limited control over nucleation and growth processes. Consequently, Wu et al. employed a droplet-based microreactor for synthesizing Au nanoclusters (AuNCs). In this experiment, we used Au(III) chloride complex as the precursor, ascorbic acid as the reducing agent, and ficin as the stabilizing agent for AuNCs. Perfluorodecalin was used as the continuous oil phase (Wu et al. 2019). Ascorbic acid is known to be a slower reducer than sodium borohydride, which is a fast-reacting agent (Carlander et al. 2019; Luty-Błocho et al. 2013). As a result, the process was carried out at high temperatures (65 °C) to increase the reaction rate. Haesuwannakij et al. looked at using a microflow reactor to prepare gold nanoclusters in a size-controlled manner. In their investigation, gold (III) ions were reduced using sodium borohydride. Both reactor types maintained the same size distribution, however the particles produced by the batch reactor were around 0.5 nm larger than those acquired in the flow reactor (Haesuwannakij et al. 2014).

Sandeep group presented a method for using microdroplets to produce luminous noble metal clusters (Bose et al. 2021). They used silver (I) nitrate and gold (III) chloride complex as precursors. The NMQCs were stabilized by the proteins of lysozyme and bovine serum albumin, and the metal ions were reduced by sodium borohydride. An electrospray technique was used to create the microdroplets.

The response rate is greatly accelerated by this technique of creating clusters inside microdroplets. Furthermore, these clusters have significantly higher luminescence than those made using conventional solution-phase techniques. In order to help with diagnosis and the identification of associated brain abnormalities, the NMQCs produced from the droplet-based technique were used in this study for in vitro imaging of retinoblastoma NCC-RbC-51 cells (Bose et al. 2021).

8.3.2 Non-Metal Quantum Dots

Non-metal quantum dots are a new class of nanomaterials that can be further divided into three categories: carbon, semiconductor, and perovskite (see Figure 8.4). These substances make suitable alternatives to traditional metal-based quantum dots because to their high biocompatibility, chemical stability, and photoluminescence.

8.3.2.1 Carbon Quantum Dots

The distinct characteristics of carbonic nanomaterials, including fluorescent carbon nanoparticles or carbon quantum dots (CQDs), fullerenes, carbon nanotubes, graphene sheets, and nanodiamonds, have sparked a significant deal of research because of their immense potential for a wide range of technical applications. Regarding Xu et al. (2004) unintentional discovery of CQDs while purifying and separating single-walled carbon nanotubes (SWCNTs), more research was conducted to take use of CQDs fluorescence characteristics and develop a new class of practical fluorescent nanomaterials (X. Xu et al. 2004). Sun et al. named fluorescent carbon nanoparticles

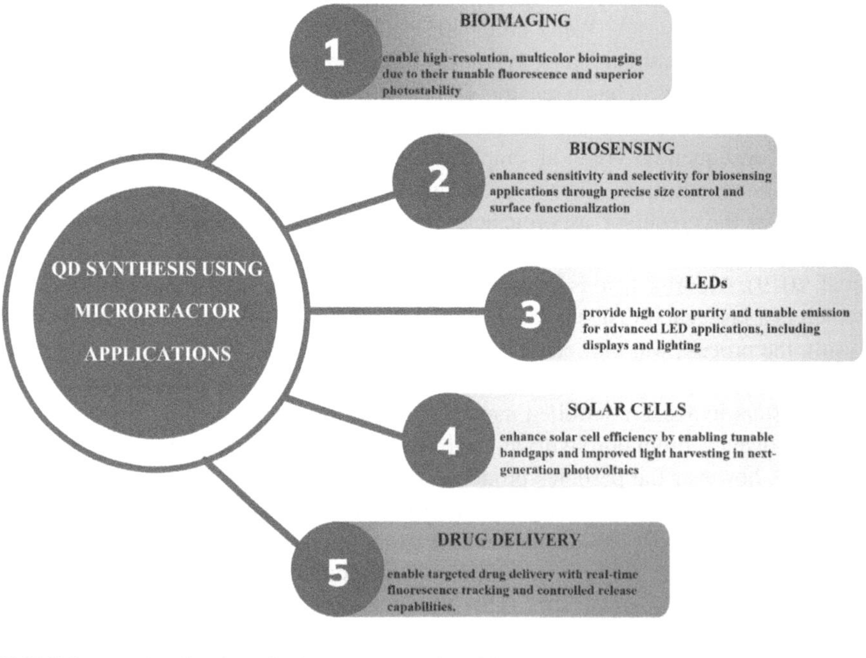

FIGURE 8.4 Application of microreactors using QDs.

"carbon quantum dots" in 2006 (Y.-P. Sun et al. 2006). Recently, these types of QDs are emerged as superior and versatile one. There are three different types of QDs. They are: Polymer dots (PDs), Carbon nanodots (CNDs), and Graphene quantum dots (GQDs) (Lim et al. 2015).

Similar to carbon quantum dots (CQDs), graphene quantum dots (GQDs) are a type of zero-dimensional (0D) nanomaterial belonging to the graphene family. They combine the properties of both graphene and carbon dots (CDs) and are typically anisotropic, with lateral dimensions usually less than 20 nm. In contrast, carbon nanodots (CNDs) are fluorescent carbon-based nanomaterials that can be synthesized from various carbon sources, including graphene, fullerenes, graphite, and carbon nanotubes (CNTs), and generally have sizes around 10 nm. Polymeric dots (PDs) represent a subclass of CDs characterized by a polymer/carbon hybrid structure, featuring numerous carbon-based surface linkages and a dense polymeric network within their core. There are two ways to create CQDs: top-down and bottom-up approaches. To create carbon dot (CD) nanoparticles, top-down methods break down bigger carbon precursor materials. Arc discharge, laser ablation, and ultrasonic treatment are typical techniques in this field. Conversely, bottom-up procedures start with small molecular precursors that go through pyrolysis, microwave irradiation, electrochemical processes, solvothermal or hydrothermal treatment, and polymerization and carbonization (Elugoke et al. 2024). For making CQDs, the electrochemical method is a popular and very effective technique. To create CQDs, carbon-based materials are usually electrochemically exfoliated or electrooxidized.

Carbon dots (CDs) have diverse applications, including serving as markers in medical settings (Molaei 2019). Because they have low biotoxicity, CDs have attracted a lot of attention as possible alternatives to semiconductor quantum dots (Lin et al. 2021). Furthermore, the utility of CDs has been substantiated in the fields of photocatalysis (Heng et al. 2021) and as components in light-emitting devices (F. Wang et al. 2011), with their use documented on multiple occasions (Kurian and Paul 2021). The source also notes their application as corrosion inhibitors (J. Li et al. 2020).

An exemplary application of steady flow microreactors in process chemistry and engineering is demonstrated in the research conducted by Lu et al (Lu et al. 2014). The choice of solvents, precursors, and conditions for CD synthesis was methodically examined in this work. The solubility of the precursors served as the basis for the first screening stage. After that, fifteen mixtures of solvents and carbon precursors were selected for more optimization. In order to obtain the required photoluminescence (PL) characteristics of CDs, more than 300 reaction conditions were subsequently assessed. With roughly 15 minutes for each condition, the microreactor device allowed for the efficient and rapid screening of reaction scenarios. The method was straightforward: syringe pumps were connected to a capillary heated in an oil bath with an internal diameter of 1 mm. Excellent PL quantum outputs were obtained from analysis of the screening conditions, which showed tunable PL emission maxima with a range of roughly 330–550 nm. During the CD manufacturing process, the chemical undergoes thermal breakdown in the flow-through microreactor. For the synthesis of noble metal nanoparticles and semiconductors, the Watzke and Finke (W-F) model, in its original or altered version, has been smoothly applied to the formation of nano silver (Amirjani and Haghshenas 2018), gold (M. Luty-Błocho et al. 2017), palladium (Marek Wojnicki, Fitzner, and Luty-Błocho 2016), platinum (Marek Wojnicki et al. 2016), and transition-metal nanoclusters (Finney and Finke 2008). However, the nucleation mechanism for CDs remains unclear, prompting numerous studies in this domain. Initially, individual free nuclei form, which subsequently aggregate into clusters and ultimately develop into nanoparticles or QDs. By ensuring that the separate processes of particle creation are carried out by a flow system, highly narrow nanoparticle size distributions are produced (Xia et al. 2019).

Recently, the introduction of carbon quantum dot (QD) nanomaterials has led to subsequent investigations utilizing microreactors. Microreactors are particularly well-suited for the synthesis of carbon QD nanomaterials, even more so than for metal QDs, due to the significant transmission of heat required. On average, organic molecules are thermally broken down into carbon and residues to create carbon QDs. Monodisperse QD manufacturing is made possible by effective heat exchange throughout the reactor volume, which is necessary to support rapid reaction kinetics. Microreactors uniquely provide this capability. Furthermore, the synthesis often involves temperatures significantly exceeding the solvent's boiling point, necessitating high-pressure conditions. In microreactors, pressures exceeding 10 bar are technically feasible. While PTFE microreactors can function at temperatures up to 300 °C and pressures up to 50 bar, glass microreactors can tolerate pressures exceeding 15 bar, depending on their design. The range of uses is further increased by the use of other building materials. Prior studies have frequently relied on the color of produced light and the yield of quantum light (QY). The QY of carbon QDs is significantly higher than that of metal QDs, according to a comparison. However, the control over the emitted light color remains limited. Numerous studies in microreactors demonstrate that the emitted light color can be modulated by tailoring experimental parameters, specifically flow rate. This represents a distinct advantage of microflow reactors over batch reactors (M. Wojnicki and Hessel 2022).

8.3.2.2 Semiconductor QDs

Semiconductor nanoparticles, specifically quantum dots, are nanocrystals capable of electron transport, it makes it easier to use them in situations where excellent photostability, resistance to photobleaching, programmable surface properties, and adjustable emission spectra are required (William et al. 2006). These semiconductor nanoparticles are utilized, among other applications, for the bio-labeling of DNA, proteins, and cells. Microreactors have proven effective in the efficient synthesis of these QDs (Vannoy et al. 2011). Cadmium selenide (CdSe) QDs are the most commonly synthesized, owing to both historical and practical reasons (Amreen and Goel 2021). In the late 1970s, Russian physicist Alexei Ekimov first synthesized nanocrystals of copper chloride and subsequently cadmium selenide within a molten glass matrix, laying the groundwork for the study of semiconductor-doped glasses and the development of theories to explain their properties (Efros and Brus 2021). The synthesis of II-VI systems, such as CdSe, is also comparatively straightforward (Nightingale and De Mello 2009). In the context of CdSe quantum material synthesis, continuous flow, (Park et al. 2016) supercritical processing (Marre et al. 2008), and droplet-based systems have been employed. Marre et al. were the first to utilize supercritical conditions in a microflow system for the synthesis of CdSe QDs, which significantly narrowed the QDs range of sizes (2008). In a microreactor, Park et al. created core (CdSe) shell (ZnS) colloidal nanoparticles, showing that the precursor flow rate regulates the produced QDs' particle size (M. Wojnicki and Hessel 2022). The QD size decreases and the bandgap energy increases as the precursor flow rate increases. The maximum luminous wavelength (λ_{max}) gradually shifted to shorter wavelengths (blue shift) due to the significant reliance of photoluminescence characteristics on the CdSe precursor's flow rate. After synthesizing CdSe in a continuous flow system, Mirhosseini et al. investigated the impacts of temperature, residence time, and flow rate on the size and optical characteristics of CdSe QDs (Mirhosseini Moghaddam et al. 2013). In the temperature range of 240–270 °C, the connection between a rise in synthesis temperature and a decrease in quantum yield (QY) is linear. Nevertheless, it was discovered that there was a nonlinear relationship between the synthesis temperature and emission wavelength. The maximum wavelengths of light emitted by QDs synthesized at 240, 250, 260, and 270 °C are 500, 530, 560, and 560 nm, respectively. Using a high-pressure, high-temperature microreactor, Marre et al. additionally produced CdSe QDs and examined how temperature affected their size and optical characteristics. A nonlinear relationship between reaction temperature and particle size was found in experiments carried out between 180 and 350 °C. Higher temperatures enabled the synthesis of bigger particles with distinct emission maxima. It was also shown that the kind of solvent had an impact on photoemission maxima. Two primary effects are typically visible in the photoluminescence (PL) curves of QDs produced in various solvents (Marre et al. 2009). The initial variation is the change in maximum emission that is seen when moving from solvents with a long carbon chain that are viscous to solvents with smaller molecules. A decrease in the full width at half maximum indicates the second factor, which is the narrowing of the photoluminescence (PL) curves. Yang et al. looked into using a microreactor to scale up the manufacture of CdSe QDs. A significant amount of 0.8 g of dry CdSe NCs powder (3.5 nm, PSD 8.2%) was produced in just one hour, indicating that the scaled-up synthesis of CdSe nanocrystals (NCs) was successfully shown based on the data obtained (Yang et al. 2011). This development could make it easier for microreactors to be used in the synthesis of nanomaterials in the future.

However, it is crucial to observe that QDs containing cadmium present potential toxicity concerns (Murray et al. 1993). As cadmium is toxic, Zinc was used to exchange cadmium to form non-cadmium quantum dots. As a result, various research groups have previously reported the synthesis of ZnO (Bahnemann et al. 1987), ZnS (Mahamuni et al. 1993), ZnSe, and ZnTe (Liao et al. 1999) QDs. In contrast, the complex and heterogeneous surface of CdX QDs often leads to the formation of mid-gap energy states. Effective radiative transitions between the valence and conduction bands are impeded by these structural characteristics, reducing both their photo stability and photoluminescence quantum yield. In addition, QDs have also played a revolutionary role in photovoltaics. Quantum Dot Solar Cells (QDSCs) represent a ground-breaking advancement in photovoltaic technology, leveraging the unique optoelectronic characteristics of QDs to enhance solar energy conversion efficiency. QDs are semiconductor nanocrystals typically measuring between 2 and 10 nm in diameter. Due to their strong quantum confinement effects, QDs exhibit size-dependent electronic and optical properties. This property tunability enables precise control over absorption spectra and energy levels, offering the potential to exceed the efficiency limitations of conventional silicon-based solar cells and paving the way for cost-effective and adaptable photovoltaic solutions (Sharma et al. 2025).

8.3.2.3 Perovskite Quantum Dots

One kind of QD made from perovskite materials is called a perovskite quantum dot (PQD). Despite being a relatively new technology, PQDs have shown characteristics that are on par with or even better than those of conventional metal chalcogenide quantum dots. They have outstanding color purity, excellent photoluminescence quantum yields, and strong defect tolerance. PQDs are ideal for application in electrical and optoelectronic devices because of these exceptional properties. They therefore have a lot of potential for use in the real world, some of which are currently being implemented in fields like quantum dot solar cells. Usually, reprecipitation is used to create hybrid organic–inorganic perovskite (HOIP) QDs at room temperature. The hot-injection approach, room temperature method, microfluidic reaction system, and micelle method can all be used to create any inorganic perovskite. H. Wang et al. were the first to produce $CH_3NH_3PbBr_3$ PQDs using a ligand-assisted reprecipitation (LARP) technique, which involves the mixing of solvents (H. Wang et al. 2018). Perovskite QDs with strong photoluminescence can be synthesized through a relatively straightforward process, even under ambient conditions. It is especially helpful for researching how the size and chemical makeup of quantum dots affect the optical characteristics of nanomaterials since it allows them to modify their emission wavelength. Multidentate polymer micelles are used to create colloidal perovskite nanocrystals. It explains an economical technique that greatly improves the stability of the nanocrystals, especially in polar solvents, by using a copolymer to function as a nanoreactor and form a protective outer layer. This polymer shell also facilitates further surface modifications, opening possibilities for functional nanodevices constructed through self-assembly or lithography (Hou et al. 2017). Young Hoon Kim et al explores a method to enhance the performance of perovskite nanoparticles (NPs) used in light-emitting diodes (LEDs). The authors successfully synthesize formamidinium lead bromide ($FAPbBr_3$) NPs with modified surface ligands at room temperature. By controlling the length of these ligands, they significantly improve both the internal luminescence efficiency of the NPs and the ability of charges to move through NP films (Kim et al. 2017). Tom C. Jellicoe tuned the optical properties of these nanocrystals by adjusting their composition and exploiting quantum confinement effects and investigated the luminescence decay processes in these tin-based materials, identifying both fast band-to-band emission and slower emission from shallow defect sites and explored the stability of these nanocrystals, noting a decrease in luminescence quantum efficiency over time due to the formation of nonradiative defect states (Jellicoe et al. 2016).

Recent advancements include:

- **Compositional Engineering**: Stability and optoelectronic characteristics of perovskite QDs have been improved when their chemical composition is altered, for example, by adding bigger organic cations or alloying with other elements. These adjustments enhance resistance to environmental degradation and allow precise controlling of their optical characteristics (Surana et al. 2018).

- **Encapsulation Strategies**: PQDs have been successfully protected from environmental stress by advanced encapsulation techniques, such as embedding them in polymer matrices, covering them with inorganic layers like silica, or adding 2D materials like graphene. These techniques have greatly enhanced PQDs' overall performance and operational stability (Cao et al. 2016).
- The combination of PQDs with microreactor technology marks a notable advancement in areas such as photocatalysis, optoelectronics, and sustainable chemistry. Microreactors provide precise control over reaction environments, superior light–material interaction, and excellent scalability, making them highly suitable for harnessing the outstanding optical and electronic characteristics of PQDs. This synergy enables continuous, efficient, and eco-friendly processing for applications like pollutant removal, photochemical synthesis, and energy conversion. Despite ongoing challenges related to the stability and toxicity of PQDs, advancements in encapsulation methods and the development of lead-free alternatives are steadily addressing these concerns. Consequently, PQD-integrated microreactors are emerging as powerful tools in the creation of next-generation, sustainable technologies.

8.4 APPLICATIONS

QDs based microreactors find use in bioimaging, sensing, and optoelectronic device manufacturing (see Figure 8.4). They improve performance and repeatability in advanced material applications by offering precise control for consistent and continuous quantum dot production.

8.4.1 Application of Bio-Sensing

Functionalized QDs have been used to monitor the concentration and spatial distribution of biomarkers, including ions, macromolecules, and small molecules, within live organisms. Such measurements offer valuable insights into the physiological state of the subject. The fluorescence properties of the QDs, such as intensity and emission wavelength, respond sensitively to changes in biomarker concentration. The distribution of biomarkers can be efficiently mapped by researchers by keeping an eye on these fluorescence variations. Additionally, QDs surface has a large number of reactive groups that allow for selective binding to target molecules, improving the detection process accuracy and stability (Cheng et al. 2021).

Both the pH profile and metal cation gradients play essential roles in maintaining cellular homeostasis and supporting normal physiological functions (Liu et al. 2007). Mercaptoacetic acid (MAA)-capped CdSe/ZnSe/ZnS semiconductor QDs were synthesized and employed as pH sensors in living SKOV-3 human ovarian cancer cells. A tenfold increase in luminescence intensity was observed as the intracellular pH shifted from 4 to 10 (Shuhong et al. 2011). The photoluminescence (PL) of thiol-capped CdTe QDs (QDs) exhibits sensitivity to pH changes due to modifications in surface chemistry and electrostatic interactions. At higher pH values (around 10), the carboxylic acid terminal groups on the thiol ligands exist predominantly in their deprotonated, negatively charged form ($-COO^-$), resulting in strong electrostatic repulsion between QDs. As the pH decreases toward the pKa of the carboxyl group (approximately 4), these groups are gradually converted to their neutral, protonated form (–COOH). This reduction in surface charge diminishes interparticle repulsion, allowing more free ligands to approach and interact with the QD surfaces. Enhanced ligand binding improves surface passivation either by eliminating trap states or by forming a denser organic shell, both of which contribute to increased PL intensity. Due to this pH-responsive behavior, QDs are valuable tools for monitoring processes that involve pH fluctuations or for detecting pathogens that alter the pH of their environment (Yu et al. 2007). Several capping pollutants can be used to adjust the selectivity and sensitivity of CdTe QDs, which have been widely used for heavy metal ion monitoring (Labeb et al. 2018). For instance, CdTe QDs capped with thioglycolic acid (TGA) have been utilized for the sensitive detection of Hg^{2+} ions in the nanomolar range, specifically from 1.25×10^{-9} M to 1×10^{-8} M, achieving a detection limit (LOD) as low as 3.5×10^{-10} M. Additionally, sulfur-doped carbon dots (CDs) have been extensively used for the detection of Fe^{3+} ions (Q. Xu et al. 2015). In µM Fe^{3+}, whereas other metal ions – such as Zn^{2+}, Ni^{2+}, K^+, Mn^{2+}, Mg^{2+}, Cd^{2+}, Ca^{2+}, Cu^{2+}, and Ba^{2+} – exhibited negligible influence (Shao et al. 2019).

8.4.2 Applications in Bio-imaging

QDs have been widely employed for cell labelling and other in vitro applications due to their exceptional fluorescent qualities. Because QD-based cell defining has already been proven to be feasible.(Chan and Nie 1998), more focus has been placed on creating more precise and effective labelling techniques for specific cells or organelles, especially for the correct identification of certain tumor locations at the cellular and subcellular levels (MatthewáLeevy 2008). Streptavidin-coated QDs and biotinylated zinc(II) dipicolylamine (Zn–DPA) have been combined to create incredibly bright fluorescent imaging probes. Improved bacterial cell surface recognition and the capacity to differentiate between distinct mutations of the same bacterial species were demonstrated by these probes. Additionally, since the epidermal growth factor receptor (EGFR) is overexpressed in tumor cells, QDs conjugated with anti-EGFR antibodies can be utilized alongside optical imaging techniques to monitor molecular changes associated with the progression of cervical cancer (Nida et al. 2005). To detect cancer cells that express the folate receptor, folate-decorated nitrogen-doped GQDs were created.

When kept at 4°C, the resultant folic acid (FA)-functionalized GQDs demonstrated outstanding photostability, little cytotoxicity, and stability for at least two months (X. Li et al. 2020). Aggregation-induced emission (AIE) QDs with NIR-II fluorescence were created using a microfluidic device. When compared to traditional AIE dots, the resultant AIE QDs demonstrated better liver evasion and tumor targeting capacities.

When employing QDs as fluorescent sensors for in vivo bioimaging, emission in the far-red to near-infrared region (650–900 nm) is used since it lies outside of the primary absorption peaks of blood and water (Ntziachristos et al. 2003). Mercaptopropionic acid-coated ultra-small, near-infrared InAs/InP/ZnSe QDs were created and effectively used in vivo imaging applications. These QDs can travel through healthy blood arteries, extravasate when they reach angiogenic tumor capillaries, and preferentially concentrate at tumor locations through the increased permeability and retention (EPR) effect because of the leaky vasculature commonly present in malignancies. The QDs were doped with transition metal ions to improve their optical characteristics, which led to a greater Stokes shift. This doping technique prolongs the excited-state lifetimes, reduces self-absorption and energy transfer, and improves the QDs overall stability (T.-T. Sun et al. 2015).

8.4.3 Applications in Photovoltaics

High-quality, monodisperse QD production has historically presented several difficulties because traditional batch synthesis techniques frequently have variable particle size distribution, poor reproducibility, and restricted scalability. QD production has changed since the advent of microreactor-based synthesis, a continuous flow chemical technology that provides accurate control over reaction conditions, quick and even mixing, and improved heat and mass transfer efficiency (Kubendhiran et al. 2019). Optimizing QDs for solar applications requires the exact control that microreactor synthesis offers. Reactants move via channels that are normally tens to hundreds of micrometers wide in a microreactor. These small dimensions encourage laminar flow, which effectively removes turbulence and allows for incredibly quick and even molecular-scale precursor mixing. The final particle size distribution is largely determined by the initial nucleation burst, which must be controlled by this uniform mixing. Additionally, extremely effective heat dissipation and transfer are made possible by the high surface-area-to-volume ratio of microfluidic channels. Because it avoids localized overheating, which can lead to unchecked particle growth, aggregation, and increased polydispersity, this is especially crucial for managing the exothermic reactions typical of quantum dot synthesis.

8.5 CONCLUSION AND FUTURE OUTLOOK

In summary, the source material highlights the revolutionary potential of microreactor technology while clarifying the drawbacks of conventional batch methods for creating QDs. With improved heat and mass transmission, better mixing, accurate temperature and residence time control, and the capacity to create multizone temperature profiles, microreactors provide unmatched control over reaction conditions. These benefits make it easier to produce high-quality QDs with customized features in a repeatable manner. Realizing the potential of different types of quantum dots, such as noble metal, carbon, and semiconductor QDs (like CdSe, its non-cadmium alternatives, and perovskite QDs), is essential for their successful application in a variety of fields, including photovoltaics, biosensing, and bioimaging. Notwithstanding persistent obstacles, such as the toxicity of some quantum dot compositions and stability issues with others, developments in synthesis made possible by microreactors keep opening the door for the creation of sustainable, next-generation technologies that make use of QDs. The production of QDs in microreactors is gradually progressing toward material improvement and industrial use. The next goal is to achieve scales of kilogram per day throughput with many parallel microreactors because of the platform's unique benefits, which include an emphasis on control over parameter-specific reactions and the effectiveness of mass and heat transmission. More complicated microfluidic designs are part of these new paradigms, which improve the capacity to synthesize complex structural nanomaterials like QDs, including doped and core-shell varieties. High throughput with reliable reproducibility will be made possible via automation and real-time monitoring. Their demand for use in LEDs, display technologies, and multiplexed bioimaging is expected to rise, which will raise expectations for employing microreactors to produce high-performance, application-specific QDs.

REFERENCES

Amirjani, Amirmostafa, and Davoud Fatmehsari Haghshenas. 2018. "Modified Finke–Watzky Mechanisms for the Two-Step Nucleation and Growth of Silver Nanoparticles." *Nanotechnology* 29 (50). IOP Publishing: 505602.

Amreen, Khairunnisa, and Sanket Goel. 2021. "Miniaturized and Microfluidic Devices for Automated Nanoparticle Synthesis." *ECS Journal of Solid State Science and Technology* 10 (1). IOP Publishing: 017002.

Bahnemann, Detlef W., Claudius Kormann, and Michael R. Hoffmann. 1987. "Preparation and Characterization of Quantum Size Zinc Oxide: A Detailed Spectroscopic Study." *The Journal of Physical Chemistry* 91 (14): 3789–98.

Berenguel-Alonso, M., I. Ortiz-Gómez, B. Fernández, P. Couceiro, J. Alonso-Chamarro, L.F. Capitán-Vallvey, A. Salinas-Castillo, and M. Puyol. 2019. "An LTCC Monolithic Microreactor for the Synthesis of Carbon Dots with Photoluminescence Imaging of the Reaction Progress." *Sensors and Actuators B: Chemical* 296 (October): 126613.

Bose, Sandeep, Amit Chatterjee, Shantha Kumar Jenifer, Biswajit Mondal, Pillalamarri Srikrishnarka, Debasmita Ghosh, Angshuman Ray Chowdhuri, M. P. Kannan, Sailaja V. Elchuri, and Thalappil Pradeep. 2021. "Molecular Materials through Microdroplets: Synthesis of Protein-Protected Luminescent Clusters of Noble Metals." *ACS Sustainable Chemistry & Engineering* 9 (12): 4554–63.

Cao, Yu, Yiming Liu, Jing Zhou, Yijun Wang, Jian Ni, and Jianjun Zhang. 2016. "Non-Uniform Distribution in Mc-Si1-xGex: H and Its Influence on Thin Film and Device Performance." *Solar Energy Materials and Solar Cells* 151: 1–6. Elsevier.

Carlander, Ulrika, Klara Midander, Yolanda S. Hedberg, Gunnar Johanson, Matteo Bottai, and Hanna L. Karlsson. 2019. "Macrophage-Assisted Dissolution of Gold Nanoparticles." *ACS Applied Bio Materials* 2 (3): 1006–16.

Chan, Warren C. W., and Shuming Nie. 1998. "Quantum Dot Bioconjugates for Ultrasensitive Nonisotopic Detection." *Science* 281 (5385): 2016–18.

Chen, G. G., G. S. Luo, J. H. Xu, and J. D. Wang. 2004. "Membrane Dispersion Precipitation Method to Prepare Nanopartials." *Powder Technology* 139 (2): 180–85. Elsevier.

Chen, Guiguang, Guangsheng Luo, Xuerui Yang, Yiwen Sun, and Jiading Wang. 2004. "Anatase-TiO2 Nano-Particle Preparation with a Micro-Mixing Technique and Its Photocatalytic Performance." *Materials Science and Engineering: A* 380 (1–2): 320–25. Elsevier.

Cheng, Yu, Si Da Ling, Yuhao Geng, Yundong Wang, and Jianhong Xu. 2021. "Microfluidic Synthesis of Quantum Dots and Their Applications in Bio-Sensing and Bio-Imaging." *Nanoscale Advances* 3 (8): 2180–95. Royal Society of Chemistry.

Csapó, Edit, Ditta Ungor, Zoltán Kele, Péter Baranyai, Andrea Deák, Ádám Juhász, László Janovák, and Imre Dékány. 2017. "Influence of pH and Aurate/Amino Acid Ratios on the Tuneable Optical Features of Gold Nanoparticles and Nanoclusters." *Colloids and Surfaces A: Physicochemical and Engineering Aspects* 532: 601–8. Elsevier.

Czyżowska, Agnieszka, Anna Barbasz, Lilianna Szyk-Warszyńska, Magdalena Oćwieja, Edit Csapó, and Ditta Ungor. 2021. "The Surface-Dependent Biological Effect of Protein-Gold Nanoclusters on Human Immune System Mimetic Cells." *Colloids and Surfaces A: Physicochemical and Engineering Aspects* 620: 126569. Elsevier.

Efros, Alexander L., and Louis E. Brus. 2021. "Nanocrystal Quantum Dots: From Discovery to Modern Development." *ACS Nano* 15 (4): 6192–6210.

Elugoke, Saheed E., Gloria E. Uwaya, Taiwo W. Quadri, and Eno E. Ebenso. 2024. "Carbon Quantum Dots: Basics, Properties, and Fundamentals." In *ACS Symposium Series*, edited by Elyor Berdimurodov, Dakeshwar Kumar Verma, and Lei Guo, 1465: 3–42. Washington, DC: American Chemical Society.

Finney, Eric E., and Richard G. Finke. 2008. "Nanocluster Nucleation and Growth Kinetic and Mechanistic Studies: A Review Emphasizing Transition-Metal Nanoclusters." *Journal of Colloid and Interface Science* 317 (2): 351–74. Elsevier.

Goswami, Nirmal, Anupam Giri, M. S. Bootharaju, Paulrajpillai Lourdu Xavier, Thalappil Pradeep, and Samir Kumar Pal. 2011. "Copper Quantum Clusters in Protein Matrix: Potential Sensor of Pb^{2+} Ion." *Analytical Chemistry* 83 (24): 9676–80.

Haesuwannakij, Setsiri, Weeranuch Karuehanon, Vijay Laksami Mishra, Hiroaki Kitahara, Hidehiro Sakurai, Shokyoku Kanaoka, and Sadahito Aoshima. 2014. "Size-Controlled Preparation of Gold Nanoclusters Stabilized by High-Viscosity Hydrophilic Polymers Using a Microflow Reactor." *Monatshefte Für Chemie - Chemical Monthly* 145 (1): 23–28.

Hao, Nanjing, Yuan Nie, and John X. J. Zhang. 2018. "Microfluidic Synthesis of Functional Inorganic Micro-/Nanoparticles and Applications in Biomedical Engineering." *International Materials Reviews* 63 (8): 461–87.

Heng, Zeng Wei, Woon Chan Chong, Yean Ling Pang, and Chai Hoon Koo. 2021. "An Overview of the Recent Advances of Carbon Quantum Dots/Metal Oxides in the Application of Heterogeneous Photocatalysis in Photodegradation of Pollutants towards Visible-Light and Solar Energy Exploitation." *Journal of Environmental Chemical Engineering* 9 (3): 105199. Elsevier.

Hou, Shaocong, Yuzheng Guo, Yuguo Tang, and Qimin Quan. 2017. "Synthesis and Stabilization of Colloidal Perovskite Nanocrystals by Multidentate Polymer Micelles." *ACS Applied Materials & Interfaces* 9 (22): 18417–22.

Jähnisch, Klaus, Volker Hessel, Holger Löwe, and Manfred Baerns. 2004. "Chemistry in Microstructured Reactors." *Angewandte Chemie International Edition* 43 (4): 406–46.

Jellicoe, Tom C., Johannes M. Richter, Hugh F. J. Glass, Maxim Tabachnyk, Ryan Brady, Siân E. Dutton, Akshay Rao, et al. 2016. "Synthesis and Optical Properties of Lead-Free Cesium Tin Halide Perovskite Nanocrystals." *Journal of the American Chemical Society* 138 (9): 2941–44.

Jeong, Da-Woon, Basudev Swain, Tae-Yeon Seong, Kyoung-Tae Park, Chan Gi Lee, and Bum Sung Kim. 2016. "Effect of Amine Additive for the Synthesis of Cadmium Selenide Quantum Dots in a Microreactor." *International Journal of Applied Ceramic Technology* 13 (2): 223–27.

Kawasaki, Shin-Ichiro, Kiwamu Sue, Ryuto Ookawara, Yuichiro Wakashima, Akira Suzuki, Yukiya Hakuta, and Kunio Arai. 2010. "Engineering Study of Continuous Supercritical Hydrothermal Method Using a T-Shaped Mixer: Experimental Synthesis of NiO Nanoparticles and CFD Simulation." *The Journal of Supercritical Fluids* 54 (1): 96–102. Elsevier.

Kim, Young-Hoon, Geon-Hui Lee, Young-Tae Kim, Christoph Wolf, Hyung Joong Yun, Woosung Kwon, Chan Gyung Park, and T.-W. Lee. 2017. "High Efficiency Perovskite Light-Emitting Diodes of Ligand-Engineered Colloidal Formamidinium Lead Bromide Nanoparticles." *Nano Energy* 38: 51–58. Elsevier.

Kubendhiran, Subbiramaniyan, Zhen Bao, Kashyap Dave, and Ru-Shi Liu. 2019. "Microfluidic Synthesis of Semiconducting Colloidal Quantum Dots and Their Applications." *ACS Applied Nano Materials* 2 (4): 1773–90.

Kulkarni, Madhusudan B, and Sanket Goel. 2020. "Microfluidic Devices for Synthesizing Nanomaterials—a Review." *Nano Express* 1 (3): 032004.

Kurian, Manju, and Anju Paul. 2021. "Recent Trends in the Use of Green Sources for Carbon Dot Synthesis–A Short Review." *Carbon Trends* 3: 100032. Elsevier.

Labeb, Mohmed, Abdel-Hamed Sakr, Moataz Soliman, Tarek M. Abdel-Fattah, and Shaker Ebrahim. 2018. "Effect of Capping Agent on Selectivity and Sensitivity of CdTe Quantum Dots Optical Sensor for Detection of Mercury Ions." *Optical Materials* 79: 331–35. Elsevier.

Li, Jianbo, Jie Lv, Luoping Fu, Mingjin Tang, and Xiaodan Wu. 2020. "New Ecofriendly Nitrogen-Doped Carbon Quantum Dots as Effective Corrosion Inhibitor for Saturated CO2 3% NaCl Solution." *Russian Journal of Applied Chemistry* 93 (3). Springer: 380–92.

Li, Xuanyu, Menglei Zha, Yaxi Li, Jen-Shyang Ni, Tianliang Min, Tianyi Kang, Guang Yang, Hao Tang, Kai Li, and Xingyu Jiang. 2020. "Sub-10 Nm Aggregation-Induced Emission Quantum Dots Assembled by Microfluidics for Enhanced Tumor Targeting and Reduced Retention in the Liver." *Angewandte Chemie International Edition* 59 (49): 21899–903.

Liao, MC Harris, Y. H. Change, C. C. Tsai, M. H. Chieng, and Y. F. Chen. 1999. "Growth and Photoluminescence Study of ZnTe Quantum Dots." *Journal of Applied Physics* 86 (8): 4694–96. American Institute of Physics.

Lim, Shi Ying, Wei Shen, and Zhiqiang Gao. 2015. "Carbon Quantum Dots and Their Applications." *Chemical Society Reviews* 44 (1): 362–81. Royal Society of Chemistry.

Lin, Liangliang, Yijian Yin, Sergey A. Starostin, Hujun Xu, Chengdong Li, Kejun Wu, Chaohong He, and Volker Hessel. 2021. "Microfluidic Fabrication of Fluorescent Nanomaterials: A Review." *Chemical Engineering Journal* 425 (December): 131511.

Liu, Yu-San, Yinghua Sun, P. Thomas Vernier, Chi-Hui Liang, Suet Ying Christin Chong, and Martin A. Gundersen. 2007. "pH-Sensitive Photoluminescence of CdSe/ZnSe/ZnS Quantum Dots in Human Ovarian Cancer Cells." *The Journal of Physical Chemistry C* 111 (7): 2872–78.

Lu, Yue, Ling Zhang, and Hengwei Lin. 2014. "The Use of a Microreactor for Rapid Screening of the Reaction Conditions and Investigation of the Photoluminescence Mechanism of Carbon Dots." *Chemistry–A European Journal* 20 (15): 4246–50. Wiley Online Library.

Luo, Guangsheng, Le Du, Yujun Wang, Yangcheng Lu, and Jianhong Xu. 2011. "Controllable Preparation of Particles with Microfluidics." *Particuology* 9 (6): 545–58. Elsevier.

Luty-Błocho, M., M. Wojnicki, and K. Fitzner. 2017. "Gold Nanoparticles Formation via Au(III) Complex Ions Reduction with L-Ascorbic Acid: Gold Nanoparticles Formation via Au(Iii) Complex Ions Reduction." *International Journal of Chemical Kinetics* 49 (11): 789–97.

Luty-Błocho, Magdalena, Krzysztof Pacławski, Marek Wojnicki, and Krzysztof Fitzner. 2013. "The Kinetics of Redox Reaction of Gold (III) Chloride Complex Ions with l-Ascorbic Acid." *Inorganica Chimica Acta* 395: 189–96. Elsevier.

Mahamuni, Shailaja, Ali Azam Khosravi, Manisha Kundu, Anjali Kshirsagar, Anjali Bedekar, D. B. Avasare, Prabhat Singh, and S. K. Kulkarni. 1993. "Thiophenol-Capped ZnS Quantum Dots." *Journal of Applied Physics* 73 (10): 5237–40. American Institute of Physics.

Marre, Samuel, Jinyoung Baek, Jongnam Park, Moungi G. Bawendi, and Klavs F. Jensen. 2009. "High-Pressure/High-Temperature Microreactors for Nanostructure Synthesis." *JALA: Journal of the Association for Laboratory Automation* 14 (6): 367–73.

Marre, Samuel, Jongnam Park, Jane Rempel, Juan Guan, Moungi G. Bawendi, and Klavs F. Jensen. 2008. "Supercritical Continuous-Microflow Synthesis of Narrow Size Distribution Quantum Dots." *Advanced Materials* 20 (24): 4830–34. WILEY-VCH Verlag Weinheim.

Martínez-Cisneros, Cynthia S., Sara Gómez-de Pedro, Mar Puyol, Joan García-García, and Julián Alonso-Chamarro. 2012. "Design, Fabrication and Characterization of Microreactors for High Temperature Syntheses." *Chemical Engineering Journal* 211–212 (November): 432–41.

MatthewáLeevy, W. 2008. "Quantum Dot Probes for Bacteria Distinguish Escherichia Coli Mutants and Permit in Vivo Imaging." *Chemical Communications*, no. 20: 2331–33. Royal Society of Chemistry.

Mirhosseini Moghaddam, Mojtaba, Mostafa Baghbanzadeh, Amin Sadeghpour, Otto Glatter, and C. Oliver Kappe. 2013. "Continuous-Flow Synthesis of CdSe Quantum Dots: A Size-Tunable and Scalable Approach." *Chemistry – A European Journal* 19 (35): 11629–36.

Molaei, Mohammad Jafar. 2019. "Carbon Quantum Dots and Their Biomedical and Therapeutic Applications: A Review." *RSC Advances* 9 (12): 6460–81. Royal Society of Chemistry.

Murray, C. B., D. J. Norris, and M. G. Bawendi. 1993. "Synthesis and Characterization of Nearly Monodisperse CdE (E = Sulfur, Selenium, Tellurium) Semiconductor Nanocrystallites." *Journal of the American Chemical Society* 115 (19): 8706–15.

Nida, Dawn L., Mohammed S. Rahman, Kristen D. Carlson, Rebecca Richards-Kortum, and Michelle Follen. 2005. "Fluorescent Nanocrystals for Use in Early Cervical Cancer Detection." *Gynecologic Oncology* 99 (3): S89–94. Elsevier.

Nightingale, Adrian M., and John. C. De Mello. 2009. "Controlled Synthesis of III–V Quantum Dots in Microfluidic Reactors." *ChemPhysChem* 10 (15): 2612–14.

Ntziachristos, Vasilis, Christoph Bremer, and Ralph Weissleder. 2003. "Fluorescence Imaging with Near-Infrared Light: New Technological Advances That Enable in Vivo Molecular Imaging." *European Radiology* 13 (1): 195–208.

Pandey, S., D. Mukherjee, P. Kshirsagar, C. Patra, and D. Bodas. 2021. "Multiplexed Bio-Imaging Using Cadmium Telluride Quantum Dots Synthesized by Mathematically Derived Process Parameters in a Continuous Flow Active Microreactor." *Materials Today Bio* 11 (June): 100123.

Pandey, Sulaxna, and Dhananjay Bodas. 2022. "Effect of Micro-Impeller Geometries on Mixing in a Continuous Flow Active Microreactor." *Materials Science and Engineering: B* 283 (September): 115843.

Park, Ji Young, Da-Woon Jeong, Won Ju, Han Wook Seo, Yong-Ho Choa, and Bum Sung Kim. 2016. "Optical Characteristics of Cdse/Zns Quantum Dot with Precursor Flow Rate Synthesized by Using Microreactor." *Journal of Powder Materials* 23 (2): 91–94. The Korean Powder Metallurgy Institute.

Peyser, Lynn A., Tae-Hee Lee, and Robert M. Dickson. 2002. "Mechanism of Ag_n Nanocluster Photoproduction from Silver Oxide Films." *The Journal of Physical Chemistry B* 106 (32): 7725–28.

Rao, Longshi, Yong Tang, Zongtao Li, Xinrui Ding, Guanwei Liang, Hanguang Lu, Caiman Yan, Kairui Tang, and Binhai Yu. 2017. "Efficient Synthesis of Highly Fluorescent Carbon Dots by Microreactor Method and Their Application in Fe3+ Ion Detection." *Materials Science and Engineering: C* 81 (December): 213–23.

Rašljić Rafajilović, Milena, Katarina Radulović, Milče M. Smiljanić, Žarko Lazić, Zoran Jakšić, Dragomir Stanisavljev, and Dana Vasiljević Radović. 2020. "Monolithically Integrated Diffused Silicon Two-Zone Heaters for Silicon-Pyrex Glass Microreactors for Production of Nanoparticles: Heat Exchange Aspects." *Micromachines* 11 (9): 818. Multidisciplinary Digital Publishing Institute.

Shao, Meng, Qin Yu, Nan Jing, Yu Cheng, Dong Wang, Yun-Dong Wang, and Jian-Hong Xu. 2019. "Continuous Synthesis of Carbon Dots with Full Spectrum Fluorescence and the Mechanism of Their Multiple Color Emission." *Lab on a Chip* 19 (23): 3974–78. Royal Society of Chemistry.

Sharma, Shailesh Narain, Pooja Semalti, Bhawna Rajpal, and A. Srinivas Rao. 2025. "Pioneering Advancements in Quantum Dot Solar Cells: Innovations in Synthesis and Cutting-Edge Applications." *Current Opinion in Colloid & Interface Science* 101905. Elsevier.

Shibu, Edakkattuparambil Sidharth, and Thalappil Pradeep. 2011. "Quantum Clusters in Cavities: Trapped Au_{15} in Cyclodextrins." *Chemistry of Materials* 23 (4): 989–99.

Shuhong, Xu, Wang Chunlei, Wang Zhuyuan, Cui Yiping, Zhang Hao, and E. Yang Bai. 2011. "pH-Sensitive Photoluminescence for Aqueous Thiol-Capped CdTe Nanocrystals." *Nanotechnology (Print)* 22. https://www.osti.gov/etdeweb/biblio/21548491.

Sonawane, Shirish H., Mahendra L. Bari, Prashant L. Suryawanshi, Jitendra S. Narkhede, Satyendra Mishra, and Bharat A. Bhanvase. 2015. "Effect of Process Parameters on Properties of Colloids in a Continuous-Flow Microreactor System." *Chemical Engineering & Technology* 38 (10): 1765–73.

Song, Yongbo, Hadi Abroshan, Jinsong Chai, Xi Kang, Hyung J. Kim, Manzhou Zhu, and Rongchao Jin. 2017. "Molecular-like Transformation from PhSe-Protected Au_{25} to Au_{23} Nanocluster and Its Application." *Chemistry of Materials* 29 (7): 3055–61.

Sun, Ting-Ting, Ming Wu, Xi-Wen He, Wen-You Li, and Xi-Zeng Feng. 2015. "The Facile One-Step Aqueous Synthesis of near-Infrared Emitting Cu+ Doped CdS Quantum Dots as Fluorescence Bioimaging Probes with High Quantum Yield and Low Cytotoxicity." *Journal of Materials Chemistry B* 3 (34): 6971–78. Royal Society of Chemistry.

Sun, Ya-Ping, Bing Zhou, Yi Lin, Wei Wang, K. A. Shiral Fernando, Pankaj Pathak, Mohammed Jaouad Meziani, et al. 2006. "Quantum-Sized Carbon Dots for Bright and Colorful Photoluminescence." *Journal of the American Chemical Society* 128 (24): 7756–57.

Surana, Karan, Nitin A. Jadhav, Pramod K. Singh, and B. Bhattacharya. 2018. "Studies on Polarization Effect of Polyethylene-Based Polymer Electrolyte in Dye and Quantum Dot Sensitized Solar Cells." *Applied Nanoscience* 8 (8): 2065–69.

Tang, Yong, Longshi Rao, Zongtao Li, Hanguang Lu, Caiman Yan, Shudong Yu, Xinrui Ding, and Binhai Yu. 2018. "Rapid Synthesis of Highly Photoluminescent Nitrogen-Doped Carbon Quantum Dots via a Microreactor with Foamy Copper for the Detection of Hg2+ Ions." *Sensors and Actuators B: Chemical* 258 (April): 637–47.

Teh, Shia-Yen, Robert Lin, Lung-Hsin Hung, and Abraham P. Lee. 2008. "Droplet Microfluidics." *Lab on a Chip* 8 (2): 198–220.

Ungor, Ditta, Edit Csapó, Barbara Kismárton, Ádám Juhász, and Imre Dékány. 2017. "Nucleotide-Directed Syntheses of Gold Nanohybrid Systems with Structure-Dependent Optical Features: Selective Fluorescence Sensing of Fe3+ Ions." *Colloids and Surfaces B: Biointerfaces* 155: 135–41. Elsevier.

Ungor, Ditta, Krisztián Horváth, Imre Dékány, and Edit Csapó. 2019. "Red-Emitting Gold Nanoclusters for Rapid Fluorescence Sensing of Tryptophan Metabolites." *Sensors and Actuators B: Chemical* 288: 728–33. Elsevier.

Ungor, Ditta, István Szilágyi, and Edit Csapó. 2021. "Yellow-Emitting Au/Ag Bimetallic Nanoclusters with High Photostability for Detection of Folic Acid." *Journal of Molecular Liquids* 338: 116695. Elsevier.

Vannoy, Charles H., Anthony J. Tavares, M. Omair Noor, Uvaraj Uddayasankar, and Ulrich J. Krull. 2011. "Biosensing with Quantum Dots: A Microfluidic Approach." *Sensors* 11 (10): 9732–63. Molecular Diversity Preservation International (MDPI).

Wallace, William T., and Robert L. Whetten. 2002. "Coadsorption of CO and O_2 on Selected Gold Clusters: Evidence for Efficient Room-Temperature CO_2 Generation." *Journal of the American Chemical Society* 124 (25): 7499–7505.

Wang, Fu, Yong-hua Chen, Chun-yan Liu, and Dong-ge Ma. 2011. "White Light-Emitting Devices Based on Carbon Dots' Electroluminescence." *Chemical Communications* 47 (12): 3502–4. Royal Society of Chemistry.

Wang, Hung-Chia, Zhen Bao, Hsin-Yu Tsai, An-Cih Tang, and Ru-Shi Liu. 2018. "Perovskite Quantum Dots and Their Application in Light-Emitting Diodes." *Small* 14 (1): 1702433.

Wang, Yuwei, Jia-Jun Yan, Shengnan Hu, David James Young, Hong-Xi Li, and Zhi-Gang Ren. 2021. "A Photoluminescent Ag_{10} Cu_6 Cluster Stablized by a PNNP Ligand and Phenylacetylides Selectively and Reversibly Senses Ammonia in Air and Water." *Chemistry – An Asian Journal* 16 (18): 2681–86.

William, W. Yu, Emmanuel Chang, Rebekah Drezek, and Vicki L. Colvin. 2006. "Water-Soluble Quantum Dots for Biomedical Applications." *Biochemical and Biophysical Research Communications* 348 (3): 781–86. Elsevier.

Wojnicki, M., and Volker Hessel. 2022. "Quantum Materials Made in Microfluidics-Critical Review and Perspective." *Chemical Engineering Journal* 438. Elsevier BV. https://wrap.warwick.ac.uk/id/eprint/165624/

Wojnicki, Marek, Krzysztof Fitzner, and Magdalena Luty-Błocho. 2016. "Kinetic Studies of Nucleation and Growth of Palladium Nanoparticles." *Journal of Colloid and Interface Science* 465: 190–99. Elsevier.

Wu, Han, Juan Qiao, Yoon-Ho Hwang, Chengnan Xu, Tian Yu, Rongyue Zhang, Huiwu Cai, Dong-Pyo Kim, and Li Qi. 2019. "Synthesis of Ficin-Protected AuNCs in a Droplet-Based Microreactor for Sensing Serum Ferric Ions." *Talanta* 200: 547–52. Elsevier.

Xavier, Paulrajpillai Lourdu, Kamalesh Chaudhari, Ananya Baksi, and Thalappil Pradeep. 2012. "Protein-Protected Luminescent Noble Metal Quantum Clusters: An Emerging Trend in Atomic Cluster Nanoscience." *Nano Reviews* 3 (1): 14767.

Xia, Chunlei, Shoujun Zhu, Tanglue Feng, Mingxi Yang, and Bai Yang. 2019. "Evolution and Synthesis of Carbon Dots: From Carbon Dots to Carbonized Polymer Dots." *Advanced Science* 6 (23): 1901316.

Xie, Jianping, Yuangang Zheng, and Jackie Y. Ying. 2010. "Highly Selective and Ultrasensitive Detection of Hg2+ Based on Fluorescence Quenching of Au Nanoclusters by Hg2+–Au+ Interactions." *Chemical Communications* 46 (6): 961–63. Royal Society of Chemistry.

Xu, Quan, Peng Pu, Jungang Zhao, Chenbo Dong, Chun Gao, Yusheng Chen, Jiarui Chen, Yao Liu, and Hongjun Zhou. 2015. "Preparation of Highly Photoluminescent Sulfur-Doped Carbon Dots for Fe (III) Detection." *Journal of Materials Chemistry A* 3 (2): 542–46. Royal Society of Chemistry.

Xu, Xiaoyou, Robert Ray, Yunlong Gu, Harry J. Ploehn, Latha Gearheart, Kyle Raker, and Walter A. Scrivens. 2004. "Electrophoretic Analysis and Purification of Fluorescent Single-Walled Carbon Nanotube Fragments." *Journal of the American Chemical Society* 126 (40): 12736–37.

Yang, Hongwei, Weiling Luan, Rui Cheng, Haijian Chu, and Shan-tung Tu. 2011. "Synthesis of Quantum Dots via Microreaction: Structure Optimization for Microreactor System." *Journal of Nanoparticle Research* 13 (8): 3335–44.

Yu, Dahai, Zhi Wang, Yan Liu, Li Jin, Yueming Cheng, Jianguang Zhou, and Shugui Cao. 2007. "Quantum Dot-Based pH Probe for Quick Study of Enzyme Reaction Kinetics." *Enzyme and Microbial Technology* 41 (1–2): 127–32. Elsevier.

9 Degradation of Organic Pollutants Using Microreactors

Arunava Sengupta, Mekhla Banerjee, Arindam Banerjee, and Sankha Chakrabortty

9.1 INTRODUCTION

Organic pollutants represent a diverse class of carbon-based chemicals that pose significant risks to environmental integrity and human health. These substances are categorized primarily into persistent organic pollutants (POPs) and volatile organic compounds (VOCs), each with distinct properties and pathways of impact (Gaur et al., 2022). POPs, such as polychlorinated biphenyls (PCBs) and organochlorine pesticides (e.g., DDT), are characterized by their high persistence in environmental media—including soil, water, and air—with half-lives exceeding decades in some cases (Gaur et al., 2022; Marathe et al., 2021). Their lipophilic nature facilitates bioaccumulation in fatty tissues of organisms and biomagnification across food chains, leading to elevated concentrations in top predators, including humans (Alharbi et al., 2018). POPs are also prone to long-range transport via atmospheric and oceanic processes, contaminating regions remote from their original sources, such as the Arctic. The Stockholm Convention (2001) identified the "Dirty Dozen" POPs, aiming to eliminate or restrict their production and use globally due to links to immunosuppression, endocrine disruption, cancer, and neurological disorders (Gaur et al., 2022). VOCs, emitted as gases from products like paints, solvents, and fuels, contribute to indoor air pollution at concentrations 2–10 times higher than outdoors. Exposure to VOCs like benzene and formaldehyde can cause acute symptoms (e.g., respiratory irritation) and chronic conditions (e.g., liver damage or cancer). Additionally, VOCs react with nitrogen oxides to form ground-level ozone, exacerbating urban smog and ecosystem damage (Wolkoff, 1995).

This chapter focuses on exploring microreactor technology as an emerging and sustainable solution for treating POPs. It begins by outlining the major environmental issues linked to POPs, including their long-term stability, toxic nature, and tendency to accumulate in living organisms. The discussion then shifts to the drawbacks of conventional treatment techniques, such as advanced oxidation processes (AOPs), which often face challenges related to efficiency, operating cost, and large-scale implementation. The chapter further explains the working principles and design characteristics of microreactors, emphasizing how their miniaturized channels, large

 DOI: 10.1201/9781003633761-9

surface-to-volume ratios, and controlled flow dynamics enhance heat and mass transfer, leading to better reaction control and degradation performance. It examines various degradation pathways that can be employed within microreactor systems—such as photocatalytic, electrochemical, Fenton-based, ozonation, and enzymatic methods—and highlights how these processes benefit from the micro-scale environment. Real-world examples and laboratory studies demonstrating the rapid breakdown of pollutants like dyes, pharmaceutical residues, and pesticides are reviewed to showcase the technology's effectiveness. In addition, the chapter discusses the combination of microreactors with renewable energy systems and compares their efficiency and sustainability with conventional reactor designs. The closing section points toward future prospects, particularly the potential of integrating microreactor systems with membrane filtration units to achieve simultaneous pollutant degradation and product separation, advancing the field toward cleaner and more energy-efficient water treatment technologies.

9.2 THE CRITICAL NEED FOR ADVANCED DEGRADATION TECHNOLOGIES FOR ORGANIC POLLUTANTS

The extensive release of industrial wastewater; the large-scale manufacturing, use, and application of chemical products; and the rising global population—all significantly contribute to the growing buildup of priority organic pollutants (POPs) in the environment (Capocelli et al., 2012; Jiang et al., 2014). This concerning trend is particularly prevalent in developing nations, where frameworks for environmental regulation and monitoring are not rigorously enforced; as a result, the constant off-site discharge of wastewater harms human health and the overall ecosystem. A broad spectrum of these organic compounds are either not yet regulated or currently being considered for regulation, are not fully understood, and are frequently found at low concentrations in natural and engineered water systems because they are only partially removed by conventional wastewater treatment methods (Bahmani et al., 2013; Chefetz et al., 2008; Fawell and Ong, 2012). These chemicals, which include complex organic dyes, antibiotics, phenols, pesticides, and chlorinated compounds, are introduced into the environment through direct and indirect pathways, such as the uncontrolled release of pesticide residues, fertilizers, and inadequate management of agricultural runoff, all of which compromise the purity of water (Bolong et al., 2009; Tijani et al., 2014).

The well-established conventional wastewater treatment processes, such as primary sedimentation, activated sludge, and chlorination, are notoriously ineffective at completely mineralizing a vast array of synthetic organic pollutants, often merely transferring them between phases or generating even more toxic transformation by-products (Grandclément et al., 2017). This critical inadequacy is exacerbated by the continuous introduction of persistent, bioaccumulative, and toxic compounds into wastewater streams, including pharmaceuticals, endocrine-disrupting chemicals, and per- and polyfluoroalkyl substances (PFAS), which evade traditional treatment and are detected in effluent-receiving waters worldwide (Richardson & Kimura, 2020). Consequently, there is an urgent and non-negotiable need for the development, optimization, and implementation of advanced degradation technologies capable of achieving complete mineralization of these recalcitrant contaminants into innocuous end products like CO_2 and H_2O. AOPs represent a particularly promising class of such technologies, operating on the principle of generating highly reactive, non-selective hydroxyl radicals (•OH) that can unselectively oxidize and dismantle complex organic molecular structures (Miklos et al., 2018). These processes, including photocatalysis, ozonation, Fenton reactions, and electrochemical oxidation, offer a powerful alternative by leveraging intense chemical, photochemical, or electrical energy to break down pollutants that are otherwise resistant to biological degradation (Ganiyu et al., 2021). The necessity for these advanced systems is further amplified by increasingly stringent environmental regulations limiting the discharge of specific micropollutants, pushing wastewater treatment plants beyond the capabilities of their existing infrastructure to meet these new legal standards (Barbosa et al., 2016). Moreover, the evolving nature of chemical pollution, with thousands of new compounds registered annually, demands treatment solutions that are not only effective against known contaminants but also broadly applicable against emerging and unknown threats, a key advantage of radical-based AOPs (Yang et al., 2023). While challenges remain regarding energy efficiency, operational costs, and potential formation of unknown transformation products during treatment, ongoing research into catalytic materials, hybrid systems, and process integration is steadily enhancing their feasibility and effectiveness (Wang & Zhuan, 2020). Therefore, investing in and advancing these destructive technologies is not merely an academic exercise but a fundamental prerequisite for safeguarding water resources, protecting ecosystem health, and ensuring public safety in the face of relentless and complex chemical pollution.

Microreactor technology offers profound advantages over conventional macroscopic reaction systems, primarily due to their dramatically high surface-area-to-volume ratios, which fundamentally enhance transport phenomena and process control (Jensen, 2017). Firstly, this characteristic enables exceptionally efficient heat transfer, allowing for near-isothermal operation even of highly exothermic or endothermic reactions; this eliminates hot spots, minimizes thermal degradation byproducts, and enables safer operation at more severe temperatures and pressures compared to batch reactors (Noël et al., 2018). Secondly, the small diffusion paths within microchannels result in vastly superior mass transfer rates, ensuring rapid mixing of reagents and enhancing the contact between catalysts, reactants, and phases, which is particularly beneficial for multiphase reactions and leads to improved reaction kinetics and higher yields (Günther & Jensen, 2006). Furthermore, microreactors provide unprecedented precise control over reaction parameters, including residence time, temperature, and pressure, with laminar flow conditions enabling highly reproducible and predictable outcomes; this facilitates the accurate study of reaction kinetics and the safe generation of potentially hazardous intermediates (Yoshida et al., 2013). Finally, the scalability of microreactor systems is addressed not by traditional scale-up, which often alters reactor geometry and dynamics, but by a numbering-up strategy where multiple identical reactor units are operated in parallel; this approach maintains the intrinsic performance benefits of the micro-scale while achieving industrially relevant production volumes, effectively decoupling production rate from reaction efficiency (Plouffe et al., 2016). These combined advantages make microreactors powerful tools for process intensification, high-throughput screening, and the sustainable manufacturing of fine chemicals and pharmaceuticals.

9.3 TYPES AND SOURCES OF ORGANIC POLLUTANTS

Organic pollutants encompass a wide range of compounds with varying structures and properties. They can be broadly categorized into four major groups: dyes, pesticides, pharmaceuticals, and industrial chemicals. Each category has distinct sources and environmental pathways, contributing uniquely to the overall pollution burden (Table 9.1).

TABLE 9.1
Persistent Organic Pollutants Investigated in This Work (in the Alphabetical Order), Their Functional Categories, and the Years Listed by the Stockholm Convention (Tkaczyk et al., 2020)

Name	Abbreviation or Short Name	Category	Listing Year
Aldrin	Aldrin	Pesticide	2001
Chlordane	Chlordane	Pesticide	2009
Chlordecone	Chlordecone	Pesticide	2001
Commercial decabromodiphenyl ether	C-decaBDE	Industrial	2017
Commercial octabromodiphenyl ether	C-octaBDE	Industrial	2009
Commercial pentabromodiphenyl ether	C-pentaBDE	Industrial	2009
Dichlorodiphenyltrichloroethane	DDT	Pesticide	2001
Dicofol	Dicofol	Pesticide	2019
Dieldrin	Dieldrin	Pesticide	2001
Endrin	Endrin	Pesticide	2001
Heptachlor	Heptachlor	Pesticide	2001
Hexabromobiphenyl	HBB	Pesticide	2009
Hexabromocyclododecane	HBCDD	Industrial	2013
Hexachlorobenzene	HCB	Pesticide	2001
Mirex	Mirex	Pesticide	2001
Pentachlorophenol and its salts and esters	PCPs	Pesticide	2015
Perfluorooctane sulfonic acid, its salt and perfluorooctane sulfonyl fluoride	PFOS	Industrial	2009
Perfluorooctanoic acid, its salts and related compounds	PFOA	Industrial	2019
Polychlorinated biphenyls	PCBs	Industrial	2001
Short-chain chlorinated paraffins	SCCPs	Industrial	2017
Technical endosulfan and its related isomers	Endosulfan	Pesticide	2011
Toxaphene	Toxaphene	Pesticide	2001
α-Hexachlorocyclohexane	α-HCH	Pesticide	2009
β-Hexachlorocyclohexane	β-HCH	Pesticide	2009
γ-Hexachlorocyclohexane	γ-HCH	Pesticide	2009

9.3.1 Toxicity and Environmental Persistence

Organic pollutants present substantial risks to environmental and human health due to their inherent toxicity and remarkable persistence in various environmental compartments. The stable chemical structures of these compounds, often featuring halogenated aromatic rings, render them resistant to natural degradation processes including chemical, biological, and photolytic breakdown. This persistence leads to prolonged environmental contamination and continuous exposure risks.

9.3.1.1 Health Impacts on Humans and Wildlife

The toxicological profiles of organic pollutants vary considerably but share common concerns regarding their biological effects. POPs have been linked to reproductive impairment, developmental abnormalities, immune system dysfunction, and carcinogenicity in both humans and wildlife (Marathe et al., 2021). For instance, DDT and its metabolites are known to cause eggshell thinning in birds of prey, leading to population declines (Yan et al., 2024). PCBs exhibit neurotoxic effects and have been associated with developmental delays and behavioral problems in children exposed in utero. The endocrine-disrupting properties of many organic pollutants can interfere with hormonal systems at very low concentrations, potentially affecting reproduction and development. The bioaccumulative nature of these compounds results in their concentration in fatty tissues, leading to biomagnification through food chains. Top predators, including humans, consequently face the highest exposure levels. Studies of populations with high fish consumption from contaminated waters have demonstrated increased body burdens of POPs and associated health effects. Additionally, many organic pollutants can transfer from mother to offspring through the placenta and breast milk, exposing developing organisms during critical stages of development.

9.3.1.2 Mechanisms of Toxicity

The toxicity of organic pollutants operates through multiple mechanisms. Some compounds directly interact with cellular components, while others undergo metabolic activation to more toxic intermediates. For instance, certain PCBs and PBDEs can be enzymatically oxidized to quinone-type derivatives that act as Michael acceptors, reacting with cellular nucleophiles including glutathione, proteins, and DNA. This reactivity can cause oxidative stress, DNA damage, and disruption of essential cellular functions.

Theoretical studies using density functional theory have revealed that the electron-accepting capacity of PCBs and PBDEs increases with the degree of halogenation, enhancing their potential to generate oxidative stress through electron transfer processes. These compounds can also interact directly with DNA nucleobases through hydrogen bonding, potentially leading to mutagenic and carcinogenic effects. The combination of multiple pollutants in the environment may produce additive or synergistic effects that complicate risk assessment and mitigation efforts (Martinez, 2024).

9.3.1.3 Environmental Distribution and Persistence

The persistence of organic pollutants in the environment is quantified by their half-lives, which can range from years to decades in soil and sediment (Yan et al., 2024). For example, dieldrin has a half-life of approximately five years in soil, while mirex can persist for up to ten years. This longevity is attributed to the chemical stability of these compounds, particularly the strength of carbon-halogen bonds that resist hydrolysis and photolytic degradation.

The global distribution of POPs is facilitated by their semi-volatile nature, which enables them to undergo long-range atmospheric transport. They evaporate from warm regions and condense in cooler climates, leading to contamination in polar regions far from their original sources. This phenomenon has resulted in the accumulation of POPs in Arctic and Antarctic ecosystems, where they threaten wildlife populations including penguins and marine mammals. The combination of persistence, bioaccumulation potential, and toxicity makes organic pollutants a particularly challenging environmental issue that requires global cooperation for effective management.

9.3.2 Limitations of Traditional Degradation Methods

Various approaches have been developed to address organic pollutants, including physical, chemical, and biological methods. However, traditional degradation techniques face significant limitations in efficiency, cost, and applicability to diverse pollutant types. AOPs have emerged as promising technologies for organic pollutant degradation but encounter specific challenges that limit their widespread implementation (Table 9.2).

TABLE 9.2
Comparative Analysis of Traditional Degradation Methods for Organic Pollutants

Degradation Methods	Key Advantages	Major Limitations	Optimal Applications
Photocatalysis	Utilizes solar energy, no sludge production, operates at ambient conditions	High electron-hole recombination, limited visible light utilization, catalyst recovery challenges	Treatment of dilute wastewater streams with appropriate light availability
Fenton Process	Simple setup, high reaction rates, effective for many organics	Narrow pH range, high sludge production, reagent consumption	Industrial wastewater pretreatment, especially for high-strength waste
Photo-Fenton	Enhanced degradation rates, reduced sludge compared to Fenton	Higher energy requirements, complex system design, still pH-dependent	Treatment of specific recalcitrant compounds when light is available
Ozonation	Powerful oxidation, no secondary sludge, disinfection benefits	Energy-intensive, formation of toxic byproducts, selective reactivity	Polishing treatment for municipal wastewater, color removal
Sonolysis	Produces reactive radicals, no chemical additives, effective with volatile pollutants	High energy consumption, limited penetration in large volumes, equipment costs	Combination with other AOPs, treatment of volatile compounds in closed systems

9.3.3 General Challenges Across Traditional Methods

Beyond the specific limitations of individual technologies, traditional degradation methods face several common challenges:

- High operational costs associated with energy input and chemical consumption
- Formation of transformation products that may be more toxic than parent compounds
- Matrix effects from complex real-world samples that reduce treatment efficiency
- Difficulties in scaling up from laboratory to field applications
- Limited effectiveness against mixed pollutants in actual environmental sample

9.4 FUNDAMENTALS OF MICROREACTOR TECHNOLOGY

Microreactor technology has emerged as a transformative approach in chemical engineering and process technology over recent decades. These devices, characterized by their sub-millimeter confinement dimensions, represent a significant departure from traditional chemical reaction vessels and have enabled unprecedented control over chemical processes. The development of microreactors began with gas-phase systems but expanded significantly to liquid-phase applications in the late 1990s, driven by advances in microfabrication techniques and growing interest in process intensification. The fundamental concept underlying microreactor technology is the substantial reduction in physical dimensions of reaction spaces, typically to microchannels with lateral dimensions below 1 mm, which creates unique physical and chemical environments that differ dramatically from conventional macro-scale reactors (Capocelli et al., 2012).

The historical development of microreactor technology is closely tied to advances in microfluidics and micro-electro-mechanical systems (MEMS). Early innovations included static micro mixers with split-and-recombine structures and chip-based flow thermocyclers for polymerase chain reaction (PCR) applications. These developments demonstrated the potential for miniaturized systems to achieve superior performance in chemical processing, particularly for reactions requiring precise temperature control or handling of unstable intermediates. The technology has since evolved to encompass increasingly sophisticated designs incorporating integrated heat exchangers, multiple reagent inputs, and real-time monitoring capabilities.

9.4.1 Definition and Basic Concepts

Microreactors, also known as microstructure reactors or microchannel reactors, are devices in which chemical reactions take place within confined spaces with typical lateral dimensions below 1 millimeter. The most common configuration involves fluid flow through microchannels, though various other geometries exist depending on specific applications. Unlike conventional batch reactors, microreactors typically operate as continuous flow systems, allowing for uninterrupted processing of reactants and products. This continuous operation eliminates the delays associated with batch workup and enables immediate processing of unstable intermediates, often resulting in improved selectivity and yield.

The defining characteristic of microreactors is their miniaturized internal structure, which creates fundamentally different physical conditions compared to macroscopic reactors. The small dimensions result in high surface-to-volume ratios—typically orders of magnitude greater than conventional reactors—which profoundly influence heat and mass transfer efficiencies. This structural feature enables precise control over reaction parameters including temperature, concentration, and residence time, leading to reproducible composition, structure, and physicochemical properties of products. Microreactors can be designed for a wide range of chemical processes, including single-phase and multiphase reactions, catalytic transformations, and biochemical reactions.

9.4.2 Design Principles

The design of microreactors is guided by several fundamental principles that leverage the advantages of miniaturization. Fluid dynamics at micro-scales is characterized by low Reynolds numbers, typically resulting in laminar flow conditions. This flow behavior allows precise prediction and control of fluid elements within the reactor, enabling accurate modeling and simulation of reactor performance. Designers must carefully consider channel geometry, aspect ratios, and surface properties to optimize reactant mixing and heat transfer while minimizing pressure drops and clogging potential (Figure 9.1).

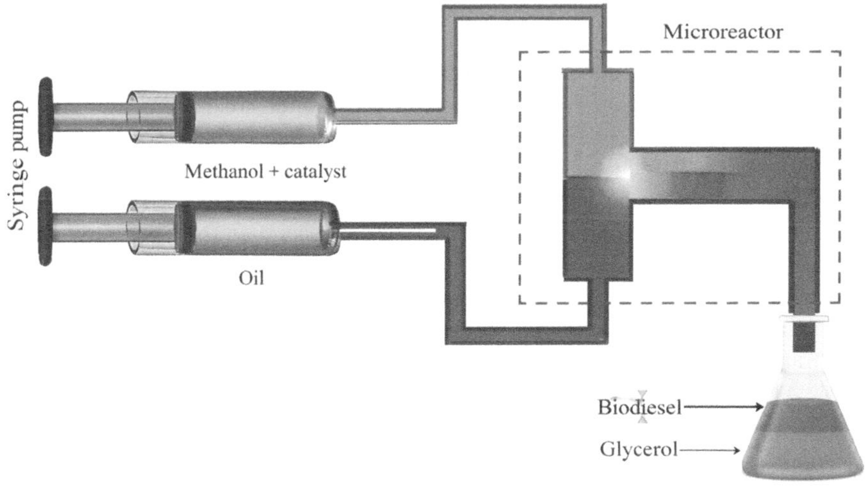

FIGURE 9.1 Design and application of microreactor in energy synthesis sector.

9.4.3 Key Features of Microreactors

9.4.3.1 High Surface-to-Volume Ratio

The most significant characteristic of microreactors is their exceptionally high surface-to-volume ratio, which can exceed 10,000 m^2/m^3 compared to typically less than 100 m^2/m^3 in conventional reactors. This geometric feature has profound implications for transport phenomena within the reactor. Heat transfer coefficients in microreactors can reach 1–500 $MW/m^3/K$, dramatically higher than the few $kW/m^3/K$ achievable in traditional glassware. This exceptional heat transfer capability allows efficient temperature control even for highly exothermic reactions, enabling safe operation at higher temperatures and preventing thermal runaway incidents (Hessel et al., 2005; Yoshida et al., 2008).

The large surface area relative to volume also enhances mass transfer rates between reactants and catalysts or between immiscible phases in multiphase reactions. This characteristic is particularly beneficial for reactions limited by mass transfer rates, as the reduced diffusion paths significantly shorten the time required for reactants to come into contact. The enhanced mass transfer enables faster reaction kinetics and often improves selectivity by reducing side reactions that might occur in concentration gradients present in larger reactors. Additionally, the large surface area facilitates efficient interfacial reactions and enhances the effectiveness of heterogeneous catalysts immobilized on channel surfaces.

9.4.3.2 Laminar Flow Regime

Fluid flow in microreactors is characterized by low Reynolds numbers (typically <100), resulting in laminar rather than turbulent flow. In laminar flow, fluids move in parallel layers with minimal mixing between layers except by molecular diffusion. This flow behavior allows precise prediction and control of fluid elements within the reactor, enabling accurate modeling and simulation of reactor performance. The absence of turbulence eliminates unpredictable eddies and vortices, resulting in highly uniform residence time distributions compared to conventional reactors.

The laminar flow regime enables several unique applications that leverage the predictable fluid behavior. For example, multiple parallel streams can flow side-by-side without turbulent mixing, allowing controlled interfacial reactions between adjacent phases. This characteristic has been exploited for liquid–liquid extractions, nanoparticle synthesis, and studies of reaction kinetics at interfaces. The predictable flow also facilitates the implementation of gradient generation for screening reaction conditions and the creation of well-defined microenvironment for cell culture and biological studies. Although laminar flow lacks the chaotic mixing of turbulent flow, microreactor designs incorporate various strategies to enhance mixing within the constraints of laminar conditions, as discussed in the following section.

9.4.3.3 Rapid Mixing Capabilities

Efficient mixing is critical for chemical reactions where mass transfer limitations can influence reaction rates, selectivity, and yield. In microreactors, the small dimensions and laminar flow regime make turbulent mixing impossible, necessitating alternative approaches to achieve rapid homogenization of reactants. Microreactors address this challenge through innovative micromixer designs that enhance mixing within the laminar flow constraint. These mixers can be categorized as passive or active systems.

9.4.4 Materials and Fabrication Techniques

9.4.4.1 Materials for Microreactor Construction

The selection of appropriate materials is critical for microreactor performance and application suitability. Materials must exhibit compatibility with reactants and products, fabrication ease, and appropriate thermal and mechanical properties. The most common materials can be categorized into three classes: inorganic materials, elastomers and plastics, and hybrid/composite materials.

Silicon was among the first materials used for microreactor fabrication due to its well-established processing techniques derived from microelectronics manufacturing. Silicon offers excellent thermal conductivity, high mechanical strength, and good resistance to organic solvents. Its crystalline structure allows precise etching of microchannels with high aspect ratios. However, silicon is fragile, opaque in visible light (limiting visual monitoring), and requires expensive fabrication techniques. Glass is widely used for microreactors due to its optical transparency (enabling visual observation and spectroscopic monitoring), good chemical resistance to many solvents, and well-developed etching and bonding techniques. Various glass types (borofloat, fused silica, quartz) offer different thermal and optical properties suitable for specific applications. Glass microreactors are particularly valuable for photochemical reactions and applications requiring corrosion resistance. The main limitations include difficulty in creating high-aspect-ratio structures and relatively lower thermal conductivity compared to silicon. Polymers have gained popularity due to their low cost, ease of fabrication, and wide range of material properties. Polydimethylsiloxane (PDMS) is commonly used for rapid prototyping and biological applications due to its optical transparency, gas permeability, and flexibility. However, PDMS swells in organic solvents, limiting its chemical compatibility. Polymethyl methacrylate (PMMA) offers better solvent resistance but lower optical transmission and thermal stability. Other polymers like polycarbonate (PC), polyvinyl chloride (PVC), and polylactic acid (PLA) are also used depending on application requirements.

9.4.4.2 Fabrication Techniques

Microreactor fabrication employs diverse techniques adapted from microelectronics, MEMS, and microfluidics manufacturing. The choice of method depends on the material, feature size requirements, production scale, and cost considerations.

Photolithography and etching represent the most precise fabrication methods, particularly for silicon and glass microreactors. The process begins with applying a UV-sensitive photoresist coating on a substrate (silicon or glass wafer). A photomask containing the reactor design is then used to selectively expose the photoresist to UV light. Subsequent development removes the exposed (or unexposed, depending on photoresist type) regions, creating a pattern that serves as an etch mask. Wet chemical etching (using solutions like potassium hydroxide for silicon) or dry etching (plasma etching) transfers the pattern into the substrate. Finally, the patterned substrate is bonded to another substrate to enclose the channels. This approach enables feature sizes down to micrometers but involves high equipment costs and complex processes.

Soft lithography has revolutionized polymer microreactor fabrication, particularly for PDMS devices. This technique involves creating a master mold (typically in silicon or SU-8 photoresist) using photolithography. PDMS prepolymer is then cast against the master and cured to create a replica containing the channel patterns. The PDMS replica is finally bonded to another PDMS layer or glass substrate using oxygen plasma treatment. Soft lithography offers lower cost and rapid prototyping capabilities compared to silicon fabrication, making it popular for research applications. However, the elastomeric nature of PDMS can limit pressure handling capability and long-term stability.

Micromachining techniques including laser ablation, mechanical milling, and hot embossing are used for various materials, particularly thermoplastics. Laser ablation uses focused laser beams to selectively remove material and create channel patterns. Mechanical milling employs miniature cutting tools to physically remove material. Hot embossing involves pressing a patterned master into a heated thermoplastic sheet to create impressions. These methods offer flexibility in material selection and moderate feature sizes (typically tens of micrometers) with lower equipment costs than cleanroom-based techniques.

Recent advances in additive manufacturing (3D printing) have opened new possibilities for microreactor fabrication with increasingly fine resolution. Stereolithography (SLA) and digital light processing (DLP) can now achieve feature sizes below 50 μm, enabling direct fabrication of complex microchannel networks without master molds. These techniques allow creation of truly three-dimensional channel geometries that would be impossible with conventional planar lithography, offering potential for further performance enhancements through innovative reactor designs

9.5 DEGRADATION MECHANISMS IN MICROREACTORS

Microreactor technology has revolutionized the field of contaminant degradation by providing enhanced efficiency, precision, and control over traditional macroscopic systems. These miniaturized devices typically feature channel dimensions ranging from tens to hundreds of micrometers, creating unique environmental conditions that significantly improve degradation kinetics and selectivity. The fundamental advantage of microreactors lies in their exceptionally high surface-to-volume ratios (often exceeding 10,000 m^2/m^3), which dramatically enhance mass and heat transfer rates while reducing diffusion path lengths to micrometer scales. This characteristic is particularly valuable for degradation processes where reaction rates are often limited by mass transfer constraints rather than intrinsic kinetics. The confinement effects inherent to microreactors lead to laminar flow conditions (Reynolds numbers typically <100) that enable precise fluid control and predictable residence time distributions. Additionally, the small volumes minimize the consumption of expensive catalysts and reagents while improving safety when handling toxic compounds. These systems also facilitate excellent spatial and temporal control over reaction parameters, allowing for optimized degradation pathways that are difficult to achieve in conventional reactors (Rashmi Pradhan et al., 2019).

9.5.1 Photocatalytic Degradation

Photocatalytic degradation in microreactors represents one of the most extensively studied applications of microreactor technology for environmental remediation. This process utilizes semiconductor catalysts such as titanium dioxide (TiO_2) and zinc oxide (ZnO) which, upon irradiation with appropriate light energy, generate electron-hole pairs that subsequently produce highly reactive oxygen species (ROS) including hydroxyl radicals (•OH) and superoxide anions ($•O_2^-$). These ROS non-selectively oxidize organic pollutants, ultimately mineralizing them to CO_2, water, and inorganic ions.

The exceptional performance of photocatalytic microreactors stems from several synergistic factors. Their high surface-to-volume ratios allow for efficient immobilization of catalyst materials as thin films or nanoparticles, significantly increasing the available active sites for reactions while minimizing mass transfer limitations. Additionally, the narrow channel dimensions ensure uniform light penetration throughout the reactor volume, addressing a critical limitation of conventional photoreactors where light intensity decreases exponentially from the irradiation source. Computational fluid dynamics (CFD) studies have revealed that microreactors exhibit significantly improved photon utilization efficiency, with nearly 100% of the catalyst surface receiving uniform illumination in optimized designs (Figure 9.2).

9.5.2 Electrochemical Oxidation/Reduction

Electrochemical degradation in microreactors utilizes electrical energy to drive oxidation and reduction reactions that break down pollutants directly at electrode surfaces or via electrogenerated mediators. This approach offers precise control over reaction parameters through adjustment of applied potential or current, enabling selective degradation pathways and minimizing unwanted side reactions. The confined geometry of microreactors significantly enhances electrochemical efficiency by minimizing the interelectrode distance (often <1 mm), which reduces ohmic losses and decreases the required supporting electrolyte concentration. The fundamental mechanisms involve direct electron transfer at the electrode-solution interface or indirect oxidation through electrogenerated mediators such as hypochlorite, ozone, or peroxydisulfate. In direct oxidation, pollutants adsorb onto the electrode surface and undergo electron transfer reactions, while indirect oxidation utilizes electrogenerated species that diffuse into the solution to oxidize target compounds. The choice of electrode material critically influences the reaction mechanism and efficiency, with boron-doped diamond (BDD), mixed metal oxides (MMO), and platinum representing the most common anode materials for oxidative applications. Microreactors enhance electrochemical processes through several

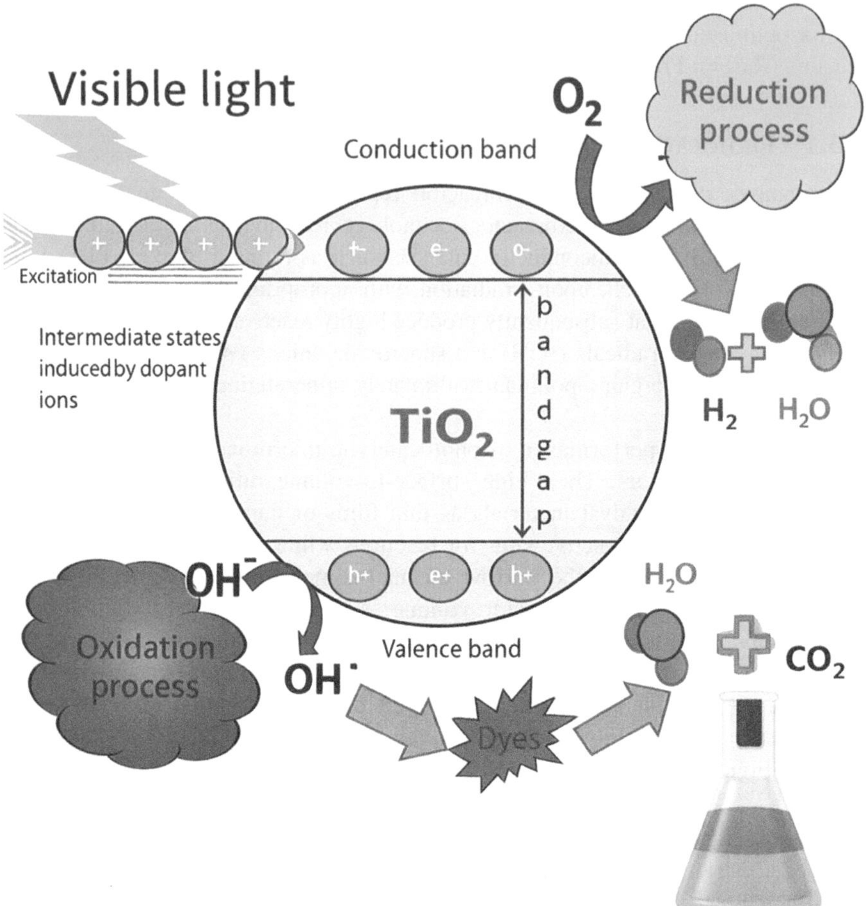

FIGURE 9.2 Mechanisms of photocatalytic degradation in TiO_2-coated microreactors.

unique advantages. The high surface-to-volume ratio increases the effective electrode area per unit volume, leading to higher reaction rates and improved current efficiency. Laminar flow conditions enable precise prediction and control of mass transfer rates, allowing for optimization of residence time distributions and reaction kinetics. Additionally, the efficient heat dissipation in microreactors prevents thermal runaway reactions that can occur in macroscopic electrochemical cells operating at high current densities (Atobe et al., 2018).

9.5.3 Advanced Oxidation Processes

AOPs encompass a range of techniques that generate highly reactive species, primarily hydroxyl radicals (•OH), for the destructive oxidation of organic pollutants.

Microreactors significantly enhance AOP efficiency by providing intensified mixing, improved mass transfer, and better control over reaction parameters compared to conventional systems. The most common AOPs implemented in microreactors include Fenton and photo-Fenton processes, ozonation, and persulfate-based oxidation, each leveraging the unique features of micro-scale operation. The Fenton process utilizes a mixture of ferrous iron (Fe^{2+}) and hydrogen peroxide (H_2O_2) to generate hydroxyl radicals through the reaction: $Fe^{2+} + H_2O_2 \rightarrow Fe^{3+} + \bullet OH + OH^-$. In microreactors, the precise control over stoichiometric ratios and residence time prevents catalyst deactivation and optimizes radical production. The photo-Fenton variant incorporates UV or visible light to enhance the process by photo reducing Fe^{3+} back to Fe^{2+} and generating additional radicals, thereby improving efficiency and reducing sludge production. Microreactors significantly enhance light penetration and distribution in photo-Fenton systems, addressing a major limitation of conventional photoreactors. Ozonation microreactors exploit the powerful oxidizing capability of ozone (O_3) to degrade refractory organic compounds. The confined dimensions of microreactors enable efficient gas-liquid mass transfer, a critical factor often limiting ozonation efficiency in conventional contactors. Microreactor designs incorporating porous membranes or microchannels with tailored surface properties create large interfacial areas that promote ozone transfer from the gas to the liquid phase. Additionally, the precise control over gas-liquid flow ratios prevents channeling and ensures uniform ozone distribution throughout the reactor volume.

9.5.4 Enzymatic Degradation

Enzymatic degradation in microreactors utilizes biological catalysts, primarily oxidoreductases and hydrolases, for the specific breakdown of pollutants under mild environmental conditions. This approach offers exceptional selectivity, operation at ambient temperature and pressure, and minimal energy requirements compared to conventional physicochemical methods. Microreactors significantly enhance enzymatic processes by providing efficient mass transfer to immobilized enzymes, precise control over reaction conditions, and reduced enzyme leaching and deactivation. The key mechanism involves the specific binding of pollutant molecules to the enzyme's active site, followed by catalytic transformation into less harmful products. Peroxidases, laccases, and cutinases represent the most studied enzyme classes for environmental applications. Peroxidases (e.g., soybean peroxidase) utilize hydrogen peroxide to oxidize a wide range of phenolic compounds, including chlorophenols and dyes. Laccases employ molecular oxygen to catalyze similar reactions, while cutinases specifically target ester bonds in plastics like polyethylene terephthalate (PET). The confined geometry of microreactors enhances enzymatic efficiency by reducing diffusion limitations and maintaining optimal enzyme-to-substrate ratios throughout the reaction volume. Enzyme immobilization strategies critically influence microreactor performance, with common approaches including physical adsorption, covalent binding, encapsulation, and cross-linking. Recent advancements have focused on novel support materials such as Janus microspheres with asymmetric porosity that provide high surface area while minimizing mass transfer resistance. For instance, cutinase immobilized on functionalized Janus microspheres retained over 78% of its initial activity after eight reuse cycles in a microreactor system for nanoplastic degradation. Similarly, soybean peroxidase exhibited enhanced stability and reusability when covalently immobilized in microchannel reactors compared to free enzyme formulations. The microenvironment optimization within enzymatic microreactors significantly influences catalytic efficiency. Factors including pH, temperature, ionic strength, and cofactor concentration can be precisely controlled to maintain enzyme activity and stability. Microreactors also facilitate efficient product removal, preventing inhibition effects that often plague batch enzymatic processes. Computational fluid dynamics (CFD) simulations have revealed that microreactors can achieve near-ideal mixing conditions under laminar flow, ensuring uniform substrate distribution to immobilized enzymes. Enzymatic microreactors have demonstrated impressive performance for various environmental applications. For example, a microreactor with immobilized soybean peroxidase achieved over 95% degradation of 2,4,6-trichlorophenol at residence times of just a few minutes. Similarly, cutinase-based microreactors effectively degraded PET nanoplastics, maintaining over 80% degradation efficiency after multiple operational cycles. These systems are particularly valuable for treating wastewater containing recalcitrant compounds that resist conventional biological treatment, such as endocrine-disrupting chemicals, pharmaceuticals, and microplastics (Mohanan et al., 2020).

9.6 OTHER ENVIRONMENTAL REMEDIATION APPLICATIONS

Microreactor technology has transitioned from laboratory curiosity to practical applications across various environmental domains, particularly in the degradation of POPs. The unique advantages of microreactors—including enhanced mass transfer, precise process control, and exceptional thermal management—make them particularly suitable for addressing complex environmental challenges that conventional systems struggle to handle efficiently. The applications span diverse areas including textile wastewater treatment, pharmaceutical removal, pesticide degradation, and integration with renewable energy systems, demonstrating the versatility and effectiveness of this technology. The fundamental principles underlying microreactor performance derive from their miniature dimensions, which typically range from micrometers to millimeters. These compact dimensions create extremely high surface-to-volume ratios (often exceeding 10,000 m^2/m^3) that dramatically enhance heat and mass transfer rates compared to conventional macroscopic reactors. Additionally, the laminar flow conditions (typically Reynolds numbers <100) enable precise control over residence time distributions and reaction parameters, resulting in more efficient and predictable degradation processes. These characteristics collectively contribute to the superior performance of microreactors in environmental applications. This comprehensive review examines the practical applications and case studies of microreactor technology in environmental remediation, with particular emphasis on four key areas: degradation of dyes in textile wastewater, removal of pharmaceuticals and pesticides, integration with renewable energy sources, and comparative performance against conventional batch reactors. Each application area demonstrates how microreactor technology addresses specific limitations of traditional treatment methods while offering additional advantages in efficiency, sustainability, and operational control.

9.6.1 Dye Degradation in Textile Wastewater

The textile industry represents one of the most significant sources of industrial water pollution globally, producing approximately 100 tons of dye effluent annually. These wastewaters contain various synthetic dyes characterized by complex aromatic structures that impart high chemical stability and resistance to biological degradation. Among these, azo dyes account for over 60% of commercial dyes due to their cost-effectiveness and structural versatility, but they also present significant environmental challenges. The persistence of these dyes in aquatic ecosystems reduces light penetration, adversely affecting photosynthetic activity and oxygen transfer, while many dyes exhibit toxic, carcinogenic, or mutagenic properties even at low concentrations. Conventional wastewater treatment methods often prove inadequate for complete dye removal. Physical methods such as adsorption, membrane filtration, coagulation, and precipitation primarily transfer pollutants to another phase rather than degrading them, generating secondary waste streams that require further treatment. Biological treatments typically show limited effectiveness against synthetic dyes due to their resistance to microbial degradation. These limitations have motivated the development of alternative approaches, particularly AOPs, which utilize highly reactive species to mineralize organic pollutants into harmless inorganic compounds (Ramamurthy et al., 2024).

9.6.2 Microreactor Applications for Dye Degradation

Microreactors have demonstrated exceptional performance in dye degradation through various mechanisms, including photocatalytic, electrochemical, and AOPs. The confined geometry of microreactors enhances mass transfer rates and provides uniform irradiation conditions for photocatalytic reactions, addressing key limitations of conventional photoreactors. For instance, TiO_2-coated microreactors have achieved complete degradation of ethylene in gas-phase reactions, while ZnO-based microreactors have shown exceptional performance in degrading azo dyes with efficiencies exceeding 95% within residence times of just a few minutes.

9.6.3 Pharmaceutical and Pesticide Removal

Pharmaceuticals and pesticides represent particularly challenging classes of aquatic pollutants due to their biological activity, persistence, and potential to cause adverse ecological effects even at extremely low concentrations (nanograms to micrograms per liter). These compounds enter water bodies primarily through municipal wastewater discharges, agricultural runoff, and industrial effluents. Conventional wastewater treatment plants are often ineffective at completely removing these micropollutants, especially those with high polarity or resistance to biodegradation. The environmental concerns associated with pharmaceuticals include endocrine disruption, development of antibiotic resistance, and toxic effects on aquatic organisms. For instance, compounds like ethinyl oestradiol, diclofenac, and paracetamol are known to cause endocrine disruption in wildlife, with effects reported at concentrations as low as nanograms per liter. Similarly, pesticides can persist in the environment and bioaccumulate through food chains, posing risks to ecosystem health and human safety. The European Union Strategic Approach to Pharmaceuticals in the Environment reflects growing recognition of this problem and calls for more advanced wastewater treatment technologies.

9.6.3.1 Microreactor Applications for Micropollutant Removal

Microreactors have demonstrated significant advantages for pharmaceutical and pesticide removal through various mechanisms including advanced oxidation, enzymatic degradation, and enhanced biological treatment. The high surface-to-volume ratios and excellent mass transfer characteristics of microreactors significantly improve the efficiency of these processes compared to conventional systems.

Membrane bioreactors (MBRs) integrated with microreactor principles have shown particular promise for pharmaceutical removal. MBR technology combines biological treatment with membrane filtration, operating at high biomass concentrations and relatively long sludge ages that enhance the biodegradation of refractory compounds. Studies comparing MBR and conventional activated sludge (CAS) systems have demonstrated superior removal of pharmaceuticals in MBRs, particularly for compounds with complex structures that resist degradation in conventional systems. AOPs in microreactors have shown remarkable efficiency for pharmaceutical degradation. The precise control over reaction parameters in microreactors optimizes the generation of reactive species while minimizing the formation of harmful byproducts. For instance, the Fe (II)/periodate system mentioned previously for dye degradation has also demonstrated potential for pharmaceutical removal, with over 80% degradation of compounds like sulfamethoxazole, carbamazepine, ciprofloxacin, acetaminophen, and ibuprofen achieved within just 2 minutes in acidic media (pH ≈ 3). The formation of high-valent iron-oxo species (Fe (IV)=O) as the primary oxidant, with hydroxyl radicals playing a secondary role, provides a highly efficient degradation pathway with benign byproducts (primarily iodate).

Biological approaches using microalgae in microreactor configurations have also shown promise for pharmaceutical removal. A study assessing the performance of free and immobilized cells of microalgae Nannochloropsis sp. demonstrated effective removal of pharmaceuticals, including paracetamol, ibuprofen, olanzapine, and simvastatin. The removal efficiency varied between free and immobilized cells, with immobilized cells showing significantly higher removal of paracetamol and ibuprofen, while free cells were more effective for olanzapine removal. These findings suggest that tailored approaches based on target pollutants could optimize microreactor systems for specific pharmaceutical removal applications.

9.6.4 Integration with Renewable Energy

9.6.4.1 Solar-Driven Microreactor Systems

The integration of renewable energy with microreactor technology represents a promising direction for sustainable environmental remediation. Solar-driven microreactors have attracted particular interest due to the abundance of solar energy and the potential for autonomous operation in remote locations. These systems typically utilize photocatalytic processes where semiconductor materials (e.g., TiO_2, ZnO) generate electron-hole pairs upon solar irradiation, initiating oxidative degradation of organic pollutants. The design advantages of microreactors significantly enhance the efficiency of solar-driven processes. The small channel dimensions ensure uniform light penetration throughout the reactor volume, addressing a critical limitation of conventional photoreactors where light intensity decreases exponentially from the irradiation source. Additionally, the high surface-to-volume ratios allow for efficient immobilization of catalyst materials, increasing the available active sites while minimizing mass transfer limitations. Computational fluid dynamics (CFD) studies have revealed that microreactors exhibit significantly improved photon utilization efficiency, with nearly 100% of the catalyst surface receiving uniform illumination in optimized designs.

Material innovations have further advanced solar-driven microreactor technology. Recent developments include novel catalyst formulations such as doped TiO_2, composite semiconductors, and plasmonic catalysts that extend light absorption into the visible spectrum and reduce charge carrier recombination. For instance, brookite and anatase TiO_2 phases with enhanced photocatalytic activity have demonstrated exceptional performance in degrading indigo carmine and other persistent dyes. Similarly, hierarchical nanostructures and hybrid materials have shown improved catalytic activity and stability under solar irradiation.

9.6.4.2 Hybrid Energy Systems

Beyond purely solar-driven systems, microreactors have been successfully integrated with various renewable energy sources including ultrasound, electrical energy from photovoltaic cells, and temperature gradients from solar thermal collectors. These hybrid approaches leverage multiple energy inputs to enhance degradation efficiency and overcome limitations of individual energy sources. Ultrasound integration has demonstrated particular synergy with microreactor technology. The combination of ultrasonic irradiation (typically 40 kHz) with chemical oxidation processes in microreactors enhances mixing and mass transfer while potentially generating additional reactive species through cavitation effects. A study on the tubular microreactor-based Fe(II)/periodate system showed that ultrasonic irradiation improved degradation efficiency by maintaining plug-flow conditions and enhancing reactant mixing. The ultrasound assistance also helped mitigate potential clogging issues and improved the overall robustness of the system. Energy-efficient designs represent another important aspect of renewable energy integration. Microreactors inherently require less energy input than conventional systems due to their small volume and enhanced mass transfer characteristics. The laminar flow conditions reduce pumping energy requirements, while efficient heat transfer minimizes energy needs for temperature control. These characteristics make microreactors particularly suitable for operation with limited energy resources, such as in remote locations or with intermittent renewable energy sources.

9.6.5 Comparative Performance Analysis: Microreactors vs. Batch Reactors

9.6.5.1 Technical Performance Comparison

The comparative performance between microreactors and conventional batch reactors has been extensively studied across various applications, consistently demonstrating superior efficiency and control in microreactor systems. The key advantages derive from the fundamental differences in transport phenomena and reaction environments between these reactor types. Mass transfer characteristics represent perhaps the most significant difference between microreactors and batch reactors. Studies comparing microreactors and semi-batch reactors for catalytic hydrogenation reactions have found that mass transfer coefficients in microreactors are typically two orders of magnitude higher than in batch reactors. This dramatic enhancement enables microreactors to achieve reaction rates much closer to intrinsic kinetics, particularly for fast reactions where mass transfer limitations dominate in conventional systems. For hydrogenation reactions with half-lives in the order of 100–102 seconds, microreactors have demonstrated clear advantages in obtaining intrinsic kinetic data and maintaining optimal reaction conditions. Residence time distribution is another critical factor differentiating microreactor performance. The laminar flow conditions in microreactors result in narrow residence time distributions, ensuring that all fluid elements experience similar reaction conditions. This characteristic is particularly valuable for consecutive reactions where intermediate products are desired, as it minimizes over-reaction and byproduct formation. In contrast, batch reactors often exhibit broad residence time distributions due to mixing limitations, especially as scale increases.

9.6.5.2 Economic and Operational Considerations

Beyond technical performance, economic considerations play a crucial role in determining the suitability of microreactors versus batch reactors for specific applications. While microreactors typically involve higher capital costs per unit volume, their superior performance often results in better overall economic efficiency when considering total lifecycle costs. Operational flexibility represents an important advantage of microreactor systems. The continuous flow operation allows for steady-state processing with consistent product quality, while rapid parameter changes can be implemented without the downtime associated with batch reactor cleaning and preparation. This characteristic is particularly valuable for wastewater treatment applications where feed composition may vary, as microreactors can adjust residence time, reactant ratios, and other parameters dynamically to maintain optimal performance.

Scale-up considerations differ significantly between microreactors and batch reactors. Conventional scale-up of batch reactors often faces challenges maintaining similar transport characteristics as size increases, potentially leading to reduced performance at larger scales. In contrast, microreactors are typically scaled through numbering-up (adding parallel units), which maintains identical transport phenomena and reaction conditions regardless of total capacity. This approach provides more predictable performance during scale-up but requires careful design of distribution systems to ensure equal flow through all parallel units. Despite these advantages, microreactors face certain limitations and challenges that must be considered for specific applications. Potential clogging from particulates or precipitates represents a significant concern, particularly when treating complex wastewater matrices with suspended solids or macromolecular organic compounds. Material compatibility issues may also arise with aggressive chemicals or extreme pH conditions, necessitating careful selection of construction materials. Additionally, the initial capital investment for microreactor systems can be higher than conventional batch reactors, though this is often offset by operational savings and superior performance.

9.7 FUTURE PERSPECTIVES: HYBRID MICROREACTOR-MEMBRANE FILTRATION SYSTEMS

The evolution of microreactor technology continues to advance toward increasingly integrated systems that combine multiple unit operations within compact, efficient platforms. Among the most promising future directions is the development of hybrid systems that couple microreactors with membrane filtration processes. These integrated systems offer the potential to achieve simultaneous reaction and separation, addressing one of the fundamental challenges in continuous flow processes while enhancing overall efficiency and sustainability. The combination of microreactors' superior mass/heat transfer capabilities with membranes' selective separation properties creates synergistic effects that surpass what either technology can achieve independently. The rationale for hybridization stems from the complementary strengths of both technologies. Microreactors provide precisely controlled reaction environments with exceptional transport properties, while membranes enable selective separation based on size, charge, or affinity characteristics. This combination is particularly valuable for processes requiring continuous product removal to drive equilibrium-limited reactions, catalyst retention and recycling, or purification of products from complex reaction mixtures. The integration of these technologies represents a significant step toward the realization of truly continuous, efficient, and sustainable chemical processes for environmental remediation and beyond.

9.7.1 Synergistic Integration of Microreactors and Membrane Filtration

The integration of microreactors with membrane filtration creates systems with enhanced functionality that leverage the advantages of both technologies. Several configuration approaches have emerged, each offering distinct benefits for specific applications. The most common designs include membrane-integrated microreactors, where the membrane serves as both reactor and separator; recycle systems where microreactor effluent passes through a membrane unit with retentate returned to the reactor; and staged systems with multiple microreactors and membranes in series for complex transformations.

The MBR concept represents one of the most successful examples of this hybrid approach, particularly for wastewater treatment applications. MBRs combine biological degradation in a reactor with membrane filtration, operating at high biomass concentrations and relatively long sludge ages that enhance the biodegradation of refractory compounds. Studies comparing MBR and conventional activated sludge (CAS) systems have demonstrated superior removal of pharmaceuticals in MBRs, particularly for compounds with complex structures that resist degradation in conventional systems. The high sludge concentration and extended sludge age in MBRs make the presence of specialized microorganisms capable of degrading specific micropollutants more likely.

9.7.2 Enhanced Performance and Applications

Hybrid microreactor-membrane systems demonstrate superior performance across various environmental applications, particularly in wastewater treatment and resource recovery. The continuous removal of reaction products through membranes can drive equilibrium-limited reactions toward completion, enhance reaction rates by eliminating product inhibition, and improve overall process efficiency. These advantages are particularly valuable for AOPs, enzymatic transformations, and catalytic reactions where product accumulation negatively impacts reaction kinetics. In wastewater treatment applications, hybrid systems have shown remarkable effectiveness for removing POPs that resist conventional treatment methods. The combination of AOPs in microreactors with membrane filtration enables complete degradation of recalcitrant compounds while retaining catalysts and preventing secondary pollution. For pharmaceutical removal, MBR systems have demonstrated significantly better performance than conventional activated sludge processes, particularly for compounds like carbamazepine, diclofenac, and various antibiotics that are poorly removed in traditional systems. The extended sludge age and high biomass concentration in MBRs promote the development of specialized microbial communities capable of degrading these persistent compounds. The resource recovery potential of hybrid systems represents another promising application area. Microreactors can facilitate the conversion of waste components into valuable products, while membranes enable simultaneous concentration and purification of these products. For instance, nitrogen and phosphorus recovery from wastewater can be enhanced through hybrid systems that combine biological or chemical transformation with membrane separation. Similarly, precious metal recovery from industrial effluents can be achieved through integrated reduction and filtration processes. This approach aligns with circular economy principles, transforming waste treatment from a disposal problem to a resource opportunity. Despite their significant potential, hybrid microreactor-membrane systems face several technical challenges that require further research and development. Membrane fouling remains a primary concern, particularly when treating complex wastewater matrices containing organic macromolecules, colloidal materials, and suspended solids. Fouling reduces membrane permeability, increases energy consumption, and necessitates frequent cleaning or replacement, undermining the economic viability of these systems. Developing effective anti-fouling strategies through membrane surface modification, optimized hydrodynamic conditions, and advanced cleaning protocols represents a critical research direction. System integration and control present additional challenges for hybrid microreactor-membrane systems. The dynamic interaction between reaction and separation processes requires sophisticated control strategies to maintain optimal performance under varying feed conditions. Future research should focus on developing advanced process control algorithms that can respond to changes in feed composition, flow rates, and reaction kinetics while minimizing energy consumption and maximizing resource recovery. The integration of real-time monitoring and adaptive control systems will be essential for realizing the full potential of these hybrid systems.

9.8 CONCLUSION

This comprehensive analysis of microreactor technology for environmental applications has revealed several fundamental advantages over conventional treatment systems. The exceptional surface-to-volume ratios (exceeding 10,000 m^2/m^3) characteristic of microreactors dramatically enhance mass and heat transfer rates, enabling reaction efficiencies that often surpass traditional systems by orders of magnitude. The laminar flow conditions (Reynolds numbers typically <100) provide precise control over residence time distributions and reaction parameters, resulting in more predictable and efficient degradation processes for POPs. These intrinsic properties make microreactors particularly suitable for addressing complex environmental challenges that conventional systems struggle to handle effectively. The research demonstrated remarkable effectiveness across various degradation mechanisms, including photocatalytic, electrochemical, advanced oxidation, and enzymatic processes. Photocatalytic microreactors utilizing TiO_2 and ZnO-based catalysts achieved degradation efficiencies exceeding 95% for azo dyes within minutes, leveraging uniform illumination and reduced electron-hole recombination. Electrochemical systems demonstrated precise control over reaction pathways with minimized ohmic losses due to reduced interelectrode distances. AOPs, particularly Fenton and photo-Fenton systems, benefited from optimal radical production and utilization through precise stoichiometric control. Enzymatic degradation exhibited enhanced stability and reusability when enzymes were immobilized in microreactor configurations, maintaining over 80% efficiency after multiple operational cycles. The transition from laboratory to industrial implementation of microreactor technology presents both significant opportunities and challenges. The demonstrated technical advantages position microreactors as promising solutions for various industrial applications, particularly in wastewater treatment, pharmaceutical manufacturing, and specialty chemical production. The numbering-up approach to scale-up, where multiple identical units operate in parallel, provides a more predictable pathway to industrial capacity than conventional scale-up methods that often face challenges maintaining similar transport characteristics as size increases. Several industry sectors show particular promise for early adoption. The textile industry, facing increasing regulatory pressure and public scrutiny regarding wastewater discharge, could benefit significantly from microreactor-based treatment systems that achieve superior degradation of complex dyes without generating secondary waste streams. Pharmaceutical manufacturing, with its need for precise process control and efficient handling of active ingredients, represents another promising application area where microreactors' advantages in heat and mass transfer could improve both environmental performance and production economics. The widespread adoption of microreactor technology has the potential to generate substantial environmental benefits across multiple dimensions. The significantly enhanced efficiency in degrading POPs could dramatically reduce the release of these harmful compounds into aquatic ecosystems, addressing one of the most challenging aspects of water pollution control. The ability to achieve complete mineralization of complex pollutants, rather than simply transferring them to another phase, represents a fundamental improvement over many conventional treatment approaches that generate secondary waste streams requiring further treatment or disposal. The resource efficiency of microreactor systems contributes significantly to their environmental advantages. Reduced reagent consumption, particularly for AOPs that often require substantial chemical inputs, decreases both environmental impacts and operational costs. Lower energy requirements, resulting from enhanced transport properties and more efficient processes, translate to reduced greenhouse gas emissions and fossil fuel consumption. The small physical footprints of microreactor systems minimize land use impacts and enable decentralized treatment approaches that can reduce transportation requirements.

REFERENCES

Alharbi, O. M., Khattab, R. A., & Ali, I. (2018). Health and environmental effects of persistent organic pollutants. *Journal of Molecular Liquids*, 263, 442–453.

Atobe, M., Tateno, H., & Matsumura, Y. (2018). Applications of flow microreactors in electrosynthetic processes. *Chemical Reviews*, 118(9), 4541–4572.

Bahmani, P., Maleki, A., Ghahramani, E., & Rashidi, A. (2013). Decolorization of the dye reactive black 5 using Fenton oxidation. *African Journal of Biotechnology*, 12(26), 4115–4122.

Barbosa, M. O., Moreira, N. F., Ribeiro, A. R., Pereira, M. F., & Silva, A. M. (2016). Occurrence and removal of organic micropollutants: An overview of the watch list of EU Decision 2015/495. *Water Research*, 94, 257–279.

Bolong, N., Ismail, A. F., Salim, M. R., & Matsuura, T. (2009). A review of the effects of emerging contaminants in wastewater and options for their removal. *Desalination*, 239, 229–246.

Capocelli, M., Joyce, E., Lancia, A., Mason, T. J., Musmarra, D., & Prisciandaro, M. (2012). Sonochemical degradation of estradiols: Incidence of ultrasonic frequency. *Chemical Engineering Journal*, 210, 9–17.

Chefetz, B., Mualem, T., & Ben-Ari, J. (2008). Sorption and mobility of pharmaceutical compounds in soil irrigated with reclaimed wastewater. *Chemosphere*, 73(8), 1335–1343.

Fawell, J., & Ong, C. N. (2012). Emerging contaminants and the implications for drinking water. *Water Resources Development*, 28(2), 247–263.

Ganiyu, S. O., Martinez-Huitle, C. A., & Oturan, M. A. (2021). Electrochemical advanced oxidation processes for wastewater treatment: Advances in formation and detection of reactive species and mechanisms. *Current Opinion in Electrochemistry*, 27, 100678.

Gaur, N., Dutta, D., Jaiswal, A., Dubey, R., & Kamboj, D. V. (2022). Role and effect of persistent organic pollutants to our environment and wildlife. In M. N. V. Prasad (Ed.), *Persistent Organic Pollutants (POPs)-Monitoring, Impact and Treatment*. IntechOpen.

Grandclément, C., Seyssieq, I., Piram, A., Wong-Wah-Chung, P., Vanot, G., Tiliacos, N., … & Doumenq, P. (2017). From the conventional biological wastewater treatment to hybrid processes, the evaluation of organic micropollutant removal: A review. *Water Research*, 111, 297–317.

Günther, A., & Jensen, K. F. (2006). Multiphase microfluidics: From flow characteristics to chemical and materials synthesis. *Lab on a Chip*, 6(12), 1487–1503.

Hessel, V., Löwe, H., & Schönfeld, F. (2005). Micromixers—A review on passive and active mixing principles. *Chemical Engineering Science*, 60(8–9), 2479–2501.

Jensen, K. F. (2017). Flow chemistry—Microreaction technology comes of age. *AIChE Journal*, 63(3), 858–869.

Jiang, B., Qiu, J., Wang, M., Yan, Q., & Xue, Q. (2014). Review on electrical discharge plasma technology for wastewater remediation. *Chemical Engineering Journal*, 236, 348–368.

Marathe, D., Balbudhe, S., & Kumari, K. (2021). Persistent organic pollutants: A global issue, a global response. In D. Marathe, S. Balbudhe, & K. Kumari (Eds.), *Persistent Organic Pollutants* (pp. 1–32). CRC Press.

Martínez, A. (2024). Toxicity of persistent organic pollutants: A theoretical study. *Journal of Molecular Modeling*, 30(4), 97.

Miklos, D. B., Remy, C., Jekel, M., Linden, K. G., Drewes, J. E., & Hübner, U. (2018). Evaluation of advanced oxidation processes for water and wastewater treatment–A critical review. *Water Research*, 139, 118–131.

Mohanan N, Montazer Z, Sharma PK and Levin DB. (2020). Microbial and enzymatic degradation of synthetic plastics. *Frontiers in Microbiology*, 11, 580709. doi:10.3389/fmicb.2020.580709

Noël, T., Naber, J. R., Hartman, R. L., McMullen, J. P., Jensen, K. F., & Buchwald, S. L. (2018). Palladium-catalyzed amination reactions in flow: Overcoming the challenges of clogging via acoustic irradiation. *Chemical Science*, 2(2), 287–290.

Plouffe, P., Macchi, A., & Roberge, D. M. (2016). From batch to continuous chemical synthesis: A tool-based approach. *Organic Process Research & Development*, 20(2), 262–269.

Ramamurthy K, Priya PS, Murugan R, Arockiaraj J. (2024). Hues of risk: Investigating genotoxicity and environmental impacts of azo textile dyes. *Environmental Science and Pollution Research International*, 31(23):33190–33211. doi:10.1007/s11356-024-33444-1.

Rashmi Pradhan, S., Colmenares-Quintero, R. F., & Colmenares Quintero, J. C.. (2019). Designing microflowreactors for photocatalysis using sonochemistry: A systematic review article. *Molecules*, 24(18), 3315. doi:10.3390/molecules24183315

Richardson, S. D., & Kimura, S. Y. (2020). Water analysis: Emerging contaminants and current issues. *Analytical Chemistry*, 92(1), 473–505.

Stockholm Convention. (2001). *The 12 Initial POPs Under the Stockholm Convention*. Châtelaine: Secretariat of the Stockholm Convention Clearing House.

Tijani, J. O., Fatoba, O. O., Madzivire, G., & Petrik, L. F. (2014). A review of combined advanced oxidation technologies for the removal of organic pollutants from water. *Water, Air, & Soil Pollution*, 225(9), 2102.

Tkaczyk, A., Mitrowska, K., & Posyniak, A. (2020). Synthetic organic dyes as contaminants of the aquatic environment and their implications for ecosystems: A review. *Science of the Total Environment*, *717*, 137222.

Wang, J., & Zhuan, R. (2020). Degradation of antibiotics by advanced oxidation processes: An overview. *Science of The Total Environment*, 701, 135023.

Wolkoff, P. (1995). Volatile organic compounds. *Indoor Air, Suppl*, 3, 1–73.

Yan, Z. G., Li, Z. G., & Du, J. Z. (2024). Effects of environmental organic pollutants on environment and human health: The latest updates. *Toxics*, 12(4), 231.

Yang, Y., Ok, Y. S., Kim, K. H., Kwon, E. E., & Tsang, Y. F. (2023). Occurrences and removal of pharmaceutical and personal care products from aquatic systems using advanced treatment—A review. *Environmental Research*, 115, 115–126.

Yoshida J, Nagaki A, Yamada T. (2008). Flash chemistry: Fast chemical synthesis by using microreactors. *Chemistry*, 14(25), 7450–7459. doi:10.1002/chem.200800582

Yoshida, J. I., Kim, H., & Nagaki, A. (2013). Green and sustainable chemical synthesis using flow microreactors. *ChemSusChem*, 4(3), 331–340.

10 Use of Microphotobioreactor in Algal Bioremediation

A Hypothetical Discussion for a Potential Future

Soumya Banerjee

10.1 INTRODUCTION

With the advent of neo-industrialization era, dependency of life on industrial products has increased significantly (AbdelJawad and Nagy 2023). This has been quite evident from the profuse use of industrial processed products, both for luxury and necessity. With the change in the nature of industrial products, the nature of waste being generated from industries has also changed. From processing to production till the product is delivered, huge amounts of waste are being generated in gaseous, liquid, and solid forms. In most of the cases, these products are overlooked, which is causing huge environmental nuisances. Several studies have substantiated the claim that modern-day life, although it tries to maintain its aesthetic, has been failing in keeping the environment clean. As a result, the world has been coping with serious challenges like sustainable food production, water shortage, and energy crisis followed by maintaining a healthy environmental balance. These could be connected to certain underlying topics like water-food-energy dynamics. As a result, these have created not only a global environmental imbalance, but at the same time they have affected humans and the ecosystem they are a part of (Molajou et al. 2023; Padder and Bashir 2023; Singh and Su 2022). Recently, consequences of modern-day pollution could be seen to affect air quality which has deteriorated by 66% due to the introduction of newer toxic substances (Banaduc et al. 2022; Dou et al. 2023). This could be attributed to the inefficacy of national and international regulation agencies in controlling the use of chemicals in human activities.

It has to be kept in mind that pollution is not an isolated incident, but a combination of many such incidents which have infiltrated life into various levels like ecosystem, health of humans, and their socio-economic welfare (Afshar et al. 2022; Ajibade et al. 2021). In other words, pollution is a combination of both human and natural activities resulting in chemical, physical, and biological changes among every strata of nature. Thus, it could be said that each and every component of nature, along with

DOI: 10.1201/9781003633761-10

pollutants, is dependent on each other, emphasizing the overall impact and a broader sense of socio-economic area in which pollution exists. Such an understanding, therefore, can create a holistic approach toward understanding and mitigating environmental pollutants along with sustainable environmental development.

10.2 ENVIRONMENTAL POLLUTANTS

In an ideal scenario, balanced soil and clean water have been the backbone of a healthy society; unfortunately, it remains as an underrated statement in practice. Soil, being the foundation of terrestrial ecosystem, helps in the production of food, supports food chain, and aids in maintaining vital processes of nature like photosynthesis (Munzel et al. 2025a). At the same time, soil helps in retaining water and prevents flood and water-borne diseases. It also helps in reducing climate change by retaining carbon within it. It has been reported that degradation of soil content and quality has affected the health of 3.2 billion people (Munzel et al. 2025a). As of a report published in 2021, it was highlighted that around 2 billion people were forced to live in a water-stressed condition due to the increasing human population and pollution of air, water, and soil, which resulted in climate change and global health emergency (Fuller et al. 2022). The Lancet Commission on Pollution and Health designated environmental pollution as one of the leading cause of diseases in humans and premature death. It has been reported that in 2019, 9 million cases of untimely death were listed, which is 16% of global cause of death, surpassing mortality from AIDS, malaria, and tuberculosis (Landrigan et al. 2018). Several other cases, such as disability-adjusted life years which occurs due to air and water pollution, have affected 268 million globally (Munzel et al. 2023; 2025b). Death of 8.3 million has been recorded annually caused due to air pollution, mostly among elderly individuals. Interestingly, water pollution has been found to cause a higher number of infant deaths. Around 70% of the diseases caused by environmental pollution are non-communicable and have affected life in every age group by deteriorating their quality of life (Munzel et al. 2025b).

According to a report published by World Health Organization in 2019, due to exposure to harmful pollutants, approximately 2 million lives were affected (Munzel et al. 2025a). Along with heavy metals, other chemical components like polycyclic aromatic hydrocarbons, per- and polyfluorinated substances, textile dyes, pesticides, fertilizers, organic solvents, etc., have been found to be responsible in claiming life annually. Diseases like cancer, respiratory diseases, skin irritation, and cardiovascular-related diseases could be seen (Münzel et al. 2025a).

Again, heavy metals like arsenic, cadmium, lead, zinc, chromium, etc., are also responsible for causing diseases like cancer, neurodevelopment disorder, hypertension, ischemic heart disease, and cardiovascular-related diseases (Fawell and Nieuwenhuijsen 2003; Landrigan et al. 2018; Munzel et al. 2025a). Similarly, components like persistent organic pollutants have been found to accumulate within living tissues resulting in abnormality of endocrine pathways. Such bioaccumulation has caused insulin resistance, hypertension, and diabetes (Munzel et al. 2023). Similarly, products like estrogen and its derivatives like estrone have been found to make their way from contaminated water to human body causing interruption in energy balance, obesity, and breast cancer (Saito et al. 2015).

Similarly, other components, such as heavy doses of nutrients in water bodies which reach either as industrial, domestic, and water run-off from agricultural fields, produce breeding ground for various microorganisms. These microorganisms include various pathogenic bacteria, virus, and protozoa along with fungi.

Again, components like dyes which are used to color fabrics have also been found to contaminate water bodies upon being discharged. Such dyes, upon exposure to humans, cause skin irritation, respiratory problems, neurological and endocrine disruption, photosensitivity, and cancer.

10.3 BIOREMEDIATION

Removal of environmental pollutants using microorganisms has been proven to be an effective remedial tool for its greener approach, sustainability, and for supporting circular economy (Prasad 2024). Although the science of bioremediation might be a recent topic among scientists, it has been maintaining ecological balances since their existence. The uniqueness of microbes resides in their ability to break down their surroundings via enzymatic reactions. This very feature of these organisms has been used in eliminating environmental pollutants via digestion, conversion, or accumulation. Various species of bacteria, algae, fungi, and virus are used in remediating pollutants. These organisms have been found to be effective in removing both organic and inorganic pollutants with minimal byproducts. For the last couple of decades, it has been concluded that microalgae could be used as an effective tool for bioremediation due to its cost-effectiveness and biomass growth which could be used for various other activities.

10.3.1 Microalgae

One of the advantages of using microalgae for bioremediation is their photosynthetic behavior. Microalgae can be grown in natural water, industrial, or in synthetic media overlooking the physical and chemical components. While growing in wastewater, microalgae could use light energy and atmospheric carbon dioxide for their metabolic activity (Li et al. 2019). Therefore, using microalgae for environmental remediation could produce newer scope in circular economy.

Wastewater carries both organic and inorganic components, among which amino acids, protein, carbohydrate, and fats could be found abundantly depending on the source. In case of inorganic components, both macro and micro elements are present in adequate amounts to support the growth of microalgae. This could be attributed to frequent availability of sodium, potassium, calcium, nitrogen, chlorine, phosphate and sulfur, etc. (Abdel-Raouf et al. 2012). Again, in some cases, it was observed that certain factors like temperature, water pH, and nature of pollutant can affect growth of microalgae. In other words, if chemical requirement of microalgae and availability of chemicals in wastewater do not match, then it might hinder microalgal growth (Li et al. 2019).

Although microalgae have certain constraints, they are still considered as a promising tool for elimination of environmental pollutants. This could be credited to its ability to use inorganic nutrients like phosphate and nitrogen for growth and production of oxygen as metabolic outcome. Again, their growth could be used to control coliform bacteria in wastewater due to the environment created by increased pH and oxygen levels, depleted nutrient levels, and increase in oxygen-rich species in water (Abdel-Raouf et al. 2012).

Presence of microalgae in wastewater promotes growth of bacteria, virus, and protozoa along with certain pathogens (Abdel-Raouf et al. 2012; Javed et al. 2019; Mat Aron et al. 2021). Therefore, constant monitoring has to be maintained. Microalgae, again, have the potential to propagate under a wide range of environmental factors like water pH, temperature, nutrient percentage, salinity, and concentration of carbon dioxide, making them a cheaper tool than tertiary biological methods. Therefore, they are used in the treatment of wastewater for removal of inorganic nutrients like phosphorous and nitrogen along with pollutants or by lowering of biological and chemical oxygen demands (Abdel-Raouf et al. 2012).

In several studies, it has been shown that microalgae promote symbiotic relation with bacteria, where bacteria help microalgal growth by removing toxic waste from wastewater. Again, cultivation of microalgae reduces the chance of eutrophication by reducing nutrient load in wastewater, and the biomass produced could be used for the extraction of lipid for biofuel (Gentili 2014). The biomass could also be used for the production of biogas, antioxidants, animal feeder, food supplements, enzymes, pigments, beauty products, and other bio-products (Lutzu et al. 2021).

10.3.2 Remedial Setup in Algal-Based Remediation

As compared to other bioremedial technologies, use of microalgae provides certain advantages in removal of pollutants from contaminated wastewater. This method uses the natural capability of microalgae in reducing contaminants, making it a cost-effective and eco-friendly technology. However, this method depends largely on its process parameters like pH, temperature, redox, nutrient, and sunlight. Again, removal of pollutants from wastewater depends on several other factors like nature of pollutants, algal species, and their efficiency in removing such contaminants via their metabolic activities (Samuel et al. 2021). Therefore, selection of appropriate species for removal of contaminant depends entirely on the efficacy of the process.

pH, being one of the major physicochemical factor, plays a vital role in catering the role of microalgae in removal of pollutants from wastewater. In general, propagation of microalgae requires a pH of 6.5–9 approximately. Therefore, if the pH is beyond the optimal range, then it can affect survival of microalgae, leading to insignificant removal of pollutants from wastewater. pH of water depends on the type of pollutants present (Dong et al. 2021). Some pollutants could be easily removed at neutral while others might require highly acidic or alkaline pH for better removal. Therefore, in such cases, pH is adjusted for the remedial process to happen.

Similarly, a system is incapable of working without a proper balance of acquiring and losing electrons. Thus, it can be said that oxidation-reduction balance plays a crucial role in bioavailability of pollutants and their remediation by algae. In general, microalgae require electron acceptors for its metabolic activities. As a result, they promote accumulation of electron acceptors in and around their system. This behavior of microalgae helps in remediation of highly oxidized pollutants like pharmaceutical and personal care from remediation (Kalra et al. 2021). Unfortunately, lack of oxygen could reduce the possibility of bioremediation of pollutants from wastewater. It has also been noticed that redox potential influences interaction between desired pollutants and co-contaminants which could affect the overall process of bioremediation. In a report, it was stated that *Chlorella vulgaris* was used to treat wastewater derived from rubber and latex industry. In that report, the algal species were able to reduce the nitrogen content and COD by 80% and 90% respectively. In several other instances, microalgal species were used to treat wastewater from different sources for eliminating heavy metal contaminants like cadmium, iron, arsenic, chromium, magnesium, copper, nickel, and zinc within a range from 40% to 80% (Jayaraman et al. 2024).

Growth and propagation of algal biomass largely depends on the surrounding temperature. Without proper temperature, growth of algae would stop leading to disruption of the entire bioremediation process. With abundant biomass production of algal biomass, maximum removal of contaminants would be possible. Generally, algal species thrive within a range of 15 °C to 30 °C. Interestingly, certain species of algae have been found to survive and multiply under extreme temperature condition, from as low as 5 °C to as high as 35 °C. Unfortunately, if the temperature tends to increase more, then it would lead to killing of algal cells (Shen 2014). Similarly, if the temperature reduces more, then algal cells would go into hibernation. Apart from propagation of microalgae, temperature also plays an important role in bioavailability of pollutants, maintaining water pH and redox condition. Therefore, efficacy of the process could only be achieved via proper maintenance of water temperature.

Sunlight or light of similar intensities, being an important physicochemical factor, plays a crucial role in growth of microalgae and bioremediation. Microalgae, being photosynthetic in nature, need required amount of light exposure throughout the day. With optimum exposure of light, the cell formation and photosynthesis will be maintained (Yadav et al. 2021). This would help in the assimilation of contaminants during metabolic activities of microalgae. It has been observed that with decreased intensities of sunlight, multiplication of algal cells would halt due to inadequate energy generation out of photosynthesis. Also, with an increase in light intensities, photosynthetic organisms experience a phenomenon called photo-inhibition. It is a condition characterized by generation of reactive oxygen species inside the algal cell that would increase, resulting in damage of the cell. Like other photosynthetic organisms, algae need a 12/12 hours balance of light and dark. This helps in maintaining their growth, adjustment with change in light intensities, and regulation of their metabolic processes. Therefore, this would help the microalgae to remediate pollutants from their surroundings.

10.3.3 Arrangement in a Traditional Photobioreactor

As compared to conventional bioremedial tools like bacteria and fungi, algae provide certain added advantages which uplift the spirit of contamination removal from wastewater. In this method, the natural ability of algae is used to eliminate pollutants from wastewater. Thus, this method provides an eco-friendly and cost-effective outcome. Advantages of using microalgae for remediation of contaminants from aqueous solution could be attributed to their efficacy in reducing biological oxygen demand, nitrogen, phosphorus, elimination of pathogenic bacteria like coliforms, along with removal of pollutants. Algal biomass produced out of the bioremediation process could be used for the production of methane gas, bio-fertilizer, biofuel, feed for animals, and chemical production.

Open and closed photobioreactors (PBRs) are used for cultivating algae and remediating pollutants from aqueous solution. In case of open pond system, microalgae are grown in open water containers made of plastic, concrete, and other materials. This system is comparatively cheaper since it does not require excess of energy consumption which is otherwise needed for maintaining temperature, light, and supply of carbon dioxide. As the name suggests, it is basically a water tank with open surface. Here, from one end, wastewater is supplied and from the other end, remediated water is removed and in between, density of algal biomass is maintained for better continuity of the process. Although it is an eco-friendly and cheaper technique, it has certain limitations (Fallahi et al. 2021). This could be attributed solely to environmental factors like amount of sunlight, gas, water, temperature, rainfall, etc., which directly affect algal growth and their role in removing contaminants from wastewater. On the other hand, a closed system, also known as PBR, is more predictable with defined outcomes. This could be monitored with probes that maintain growth factors and monitor growth of microalgae. Although the cost of installation and operation of a PBR is comparatively more than an open pond system, its remedial efficacy is higher (Al Sharabati et al. 2021). Lastly, there has been a third kind of algal remedial system which uses both open and closed systems. This has been done to distribute the remedial load into two sections for better outcomes. As compared to the single systems, this hybrid system has been found to be costlier, but it has been found that their role in circular economy could overcome the excess expenditure (Zhang et al. 2021).

10.4 MICROREACTOR

Recently, a newer trend of using "microreactor" has been gaining importance in scientific research and development for its minimal requirement of materials and later providing with a result which could be used for setting up production on larger scale (Velugula-Yellela et al. 2018). Previously, such an idea was restricted to only laboratory-scale setup where several trials were made for optimizing a process for industrial scale functioning. Although such setup was sufficient enough to provide information on process optimization, the process cost was more. As of now, we live in a world which is growing faster, it requires solution at a faster rate with proper accuracy. In this aspect, the use of microreactor in research and development is being incorporated. This is a nascent method which is still under development, and it concentrates on culturing of a living cell (Velugula-Yellela et al. 2018). Interestingly, so far microbioreactor has not been used in bioremedial approach. Therefore, in the following paragraph, a theoretical approach would be made by considering this idea in a microphotobioreactor (μPBR) and its potential role in the removal of environmental pollutants via microalgae. Unfortunately, no scientific papers are available to substantiate such claims. Therefore, the following hypothesis is open for scientific discussion, so that an actual μPBR could be made which could save time and resource in calibrating a bioremedial process. Thus, in the following section, a comparative view between traditional PBR and μPBR would be discussed.

10.4.1 Potential Role of μPBR in Bioremediation

In recent times, microalgae and other photosynthetic microorganisms are in demand for their significant role in producing alternative and sustainable food sources. They are also prominent in producing feed for livestock, biofuels, nutraceuticals, pigments, and bioactive compounds. Again, their robust nature, photosynthetic efficacies, and ability to fix atmospheric carbon dioxide make them suitable for bioremediation and resource recovery. However, large-scale cultivation faces operational constraints like light attenuation in dense cultures, oxygen accumulation, nutrient gradients, and temperature fluctuations, which cumulatively could affect overall microalgal biomass production.

Traditional PBRs, which include tubular, flat-panel, and stirring, require huge amount of biomass for execution of the process due to which the process gets affected by uneven light distribution and sub-optimal mass transfer. As a result, such problems could be addressed in case of μPBR. In such a microfluidic device with channel dimensions from tens to hundreds of micrometers, such functional problems like shortening of light path, optimizing CO_2, and nutrient delivery in a precise controlled environment could be achieved. Again, small culture volumes allow rapid screening of strains and conditions, while high surface-to-volume ratios maximize usage of light.

10.5 PHOTOSYNTHETIC GROWTH MECHANISM

Adequate understanding of photosynthetic energy conversion has to be prioritized for designing and optimization of μPBR. Microalgal photosynthesis involves light absorption by chlorophyll and accessory pigments, followed by charge separation in photosystems I and II. The electrons produced get transported through the thylakoid membrane. Lastly, biochemical CO_2 fixation takes place through Calvin–Benson cycle. The short optical path length in a μPBR would maintain a uniform illumination and maximize the fraction of cells operating at optimum light intensity and minimize photo-inhibition and dark zones. Thus, it would result in a uniform excitation leading to enhanced biomass productivity per unit volume.

10.6 SCREENING OF MICROORGANISMS

One of the important aspects in constructing a μPBR depends on the kind of microalgal species being selected. Selection of species depends on criteria like growth rate, light utilization, tolerance to alteration to intensity of physical growth factors and the target output profile. Among the above, growth rate could be considered as one of the most important criteria since it reciprocates process outcome. Therefore, species that could sustain under adverse conditions with optimum growth are considered. Another important aspect requires ability to absorb maximum amount of light energy from the given spectrum. This could be attributed to generation of chemical and subsequently algal biomass.

Since this process requires adequate amount of exposure to light, in certain cases, sensitivity to high-intensity light could create inhibition of microalgal growth in a reactor. As a result, species like *Ankistrodesmus* sp could not be considered due to its intolerance toward high-intensity light. Similarly, the species should be capable of withstanding fluctuation of light intensities, temperature, and sheer stress. Lastly, selection of algal species should be made based on their capability of optimum execution of desired contaminant from water. Thus, some of the commonly used microalgal species are *Chlorella vulgaris*, *Scenedesmus obliquus*, *Nannochloropsis gaditana*, and cyanobacteria like *Synechocystis* sp. and *Arthrospira platensis*

Again, certain morphological traits like cell size, motility, and colony formation have to be considered for understanding sedimentation, biofilm formation, and light absorption efficiency in μPBR channels. Also, the ability of species to grow under immobilized conditions should be considered since this technique could provide certain advantages like enhanced volumetric productivity and reduced harvesting costs, and could facilitate continuous operation. However, growth should be monitored regularly, since overgrowth could lead to light shading and clogging, resulting in inefficacy of the process.

10.7 MICRO-PHOTOBIOREACTOR CONFIGURATIONS

The configuration of a μPBR plays a decisive role in determining light utilization, gas–liquid mass transfer, and biomass productivity. Several architectures could be adapted from microreactor technology and could be explored for cultivating photosynthetic microorganisms. In such aspect, four different assumptions could be made for construction of μPBR.

10.7.1 Microcapillary Systems

In such a setup, cultures would flow through transparent capillary tubing with inner diameters within the range of 100–500 μm range. Gas exchange could occur through the medium or via gas-permeable tubing walls (e.g., fluorinated ethylene propylene or FEP). The inner surface might be functionalized for anti-fouling or, conversely, to promote biofilm growth for immobilized cultivation. Micro-capillary μPBR could be particularly useful for kinetic studies, light-response characterization, and monitoring cell morphology under defined shear conditions. Their main limitation could be potential fouling or bubble formation along the narrow bore, which could disrupt flow and light transmission.

10.7.2 Single-Microchannel Reactors

These consist of a single, shallow rectangular channel (often 100–500 μm deep) fabricated from glass, quartz, or transparent polymers. Light could be applied from one or both sides, and CO_2-enriched air could be diffused into the channel. The simplicity of this geometry would allow direct optical monitoring like chlorophyll fluorescence but it would limit throughput and scale.

10.7.3 Multi-Microchannel Arrays

An array of parallel micro-channels could increase culture volume and productivity while preserving high surface-to-volume ratio. Multi-channel designs could enable parallel testing of different light spectra, nutrient conditions, or strains in a single device. Therefore, such folding channels into serpentine patterns could help in achieving uniform residence time distribution, and exposure to illumination has been critical for photosynthetic performance.

10.7.4 Planar Microchamber Reactors

Thus, a wide but shallow culture chamber could spread the culture over a large illuminated surface area. This configuration could be optimal for maximizing photon capture and imaging, making it popular for high-content screening of algal strains or monitoring morphological changes under varying environmental conditions.

Again, configuration like coiled flow inverters or their microscale equivalents could be used, which has the ability to enhance mixing through flow inversion without introducing mechanical shear. These geometries might improve light–dark cycling, potentially boosting photosynthetic efficiency. However, ensuring uniform illumination over complex 3D paths remains a design challenge.

10.8 FABRICATION OF μPBR

Fabrication choices for μPBRs could be dictated by optical, mechanical, chemical, and biological considerations. The selected materials and methods must ensure transparency for photosynthetically active radiation (PAR), biocompatibility with the chosen microorganism, maintenance of an aseptic environment, and compatibility with integrated sensing or control systems.

Selection of materials for fabrication could be quite crucial for the process since it could influence light transmission, nutrient compatibility, and gas permeability. The choice depends on parameters responsible for optimum microbial growth, optical properties of the material, chemical resistance, and ease in system operation. In general, glass and quartz have been found to be maintaining transparency and chemical resistance. Among them quartz has been effective for near-UV or solar simulation due to low cut-off wavelength. On the other hand, glass μPBR could be ideal for integrating optical detection. Whereas, silicon substrates could be used for sensor integration, and in such case, transparent windows have been found to be effective. In fabricating a reactor, metals are used mainly for frames and or opaque structural elements and low-temperature co-fired ceramics, if used, allow embedded heating or sensing (Shukla et al. 2021).

Polymers such as polydimethylsiloxane (PDMS), polymethyl methacrylate (PMMA), polystyrene (PS), polyethylene terephthalate (PET), and fluoropolymers like fluorinated ethylene propylene (FEP) or perfluoroalkoxy alkane (PFA) could be options for the construction of a μPBR due to their low cost, transparency, and process ability (Shukla et al. 2021). PDMS has been widely used for prototyping due to high optical clarity, oxygen permeability, and ease of bonding to glass, though hydrophobicity and gas permeability could require careful designing. Again, FEP/PFA could provide significant chemical inertness and low biofouling tendencies. Although PMMA and PS are cheap and provide clarity, their lesser permeability to gas could make them a poor choice.

10.9 PROCESS PARAMETER IN μPBR

There are several process parameters which should be thoroughly optimized for better output. These include flow rate which ensures balanced residence time with nutrient and gas renewal. At the same time, it has to be kept in mind that low rates risk depletion and high rates risk washout which should be monitored properly.

10.9.1 Microchannel Dimensions

Unlike, conventional PBR, pathways and sections in a microreactor are comparatively narrower; therefore, execution of the process precision should be maintained properly. In this aspect, shallow channels should ensure uniform light but limit volume. This could be attributed to the fact that deeper channels might increase volume and, at the same time, would cause light gradients. Similarly, in case of nutrients, their supply should be maintained for proper growth, else it would induce metabolite accumulation. Again, supply of CO_2 should be made depending on its requirement in the process since excess of it in the reactor might result in fluctuation in algal growth due to alteration in pH.

Lastly, the process is solely dependent on the availability of light and its intensity. Therefore, intensity of light should be regulated properly from damaging of algal cells since light requirement varies from species to species. In this aspect, pulsed or modulated light could be used to avoid disruption in the process.

10.9.2 Hydrodynamics and Light Distribution

In a μPBR, hydrodynamics and light distribution play a critical role in optimizing growth of microalgae and in maintaining overall productivity. Hydrodynamics in a PBR controls availability of nutrients, gases (especially CO_2 and O_2), and distribution of microalgal cells inside the reactor. Again, μPBR being smaller in size, ensures mixing and uniform light exposure, which could be both a challenge and an opportunity. Thus, a proper flow regime, often achieved through microfluidic designs or mechanical mixing could help in preventing cell sedimentation and would enhance mass transfer.

Simultaneously, light distribution has to be optimized so that every cell receives adequate photon flux density without photo-inhibition or light limitation. Since light penetration could be limited by biomass concentration and reactor geometry, achieving a balance between culture density and light availability should be prioritized. Advancement in designing μPBR designs often consists of internal illumination or light-guiding structures to maximize photosynthetic efficiency of the process. As a result, interplay between hydrodynamics and light distribution should be considered as the fundamental for high-performance and scalable cultivation of microalgae in μPBR.

10.9.3 Evaluation of Parameters

In a μPBR, in-situ monitoring of parameters such as optical density (OD), fluorescence, dissolved oxygen, and pH could provide real-time insights into microalgal physiology and culture dynamics. Collection of continuous data enable accurate model fitting to extract key kinetic and photo-biological parameters like maximum specific growth rate, half-saturation constant, and optimal light intensity. OD measurement could be used to interpret concentration of algal biomass, while fluorescence would offer information on chlorophyll content and photosynthetic activity. Monitoring of oxygen evolution would provide a direct measurement of photosynthetic efficacy and pH alteration would be an indicator of metabolic activity and CO_2 uptake. These parameters being fundamental in nature, they would help in developing predictive models that would help in optimizing operation of a μPBR under varying environmental conditions. Again, integration of sensors into μPBR would help in making precise control and automation, thus, minimizing experimental variability and enhancing reproducibility. Ultimately, in-situ monitoring could transform μPBR into powerful tools for high-throughput screening, strain selection, and optimization of culture conditions in research and bioprocess development

10.9.4 Kinetic Modelling of Microalgal Growth in μPBR

The Monod (Jin et al. 2022) and Droop (Mairet and Baron 2019.) models could be considered as foundational in understanding and optimizing functioning of a μPBR, particularly in cultivation of microalgae for biomass and in its potential role in bioremediation. The Monod model could be used to describe the relationship between specific growth rate of microorganisms and concentration of a limiting substrate, along with nutrients like nitrogen or phosphorus. This model helps in predicting algal growth dynamics under various nutrient concentrations, enabling effective nutrient dosing. On the other hand, Droop model emphasizes more on considering the concept of internal nutrient quota, explaining the dependency of cell growth not only on external nutrient availability but also on intracellular accumulation of nutrients. This model could prove to be effective for μPBRs operating under fluctuating nutrient regimes which would allow for better control of nutrient loading and harvesting schedules. Thus, these models would help in designing a μPBR, enabling precise control over environmental parameters in a μPBR via supporting efficient biomass production, reducing waste, and maximizing productivity through dynamic control strategies designed for biological and physical limitations.

10.10 MICROALGAE AND μPBR

μPBR could offer a promising approach for bioremediation of wastewater through the use of microalgae by harnessing the organisms' ability to absorb pollutants and convert them into valuable biomass. Since these reactors aim to work on lesser volume than traditional PBR, it would make them suitable for laboratory settings. In wastewater treatment, microalgae could prove effective in removal of organic and inorganic pollutants, excess nutrients (like nitrogen and phosphorus), and heavy metals.

Again, μPBR would have the ability to offer controlled environmental conditions, ensuring optimal algae growth and effective removal of pollutants. The small-scale design would allow for high surface-area-to-volume ratios, promoting efficient light absorption and nutrient exchange. Additionally, μPBR could be equipped with sensors to monitor factors like pH, temperature, dissolved oxygen, and nutrient levels. Thus, these would provide real-time adjustments for maximum output. Through this level of control, μPBR could be fine-tuned to target specific pollutants in wastewater, achieving higher removal efficiency as compared to traditional, less-controlled treatment methods. Furthermore, potential of microalgal cultivation in a μPBR could be extended to methods like production of biofuels, animal feed, or other bio-products, creating a circular economy.

10.11 COMPARATIVE ANALYSIS BETWEEN PBR AND μPBR

A traditional PBR could be seen as a large-scale system, usually cylindrical, tubular, or flat-panel, designed for mass cultivation of microalgae, cyanobacteria, and other phototrophic organisms. It operates at liter to cubic meter volumes, using sunlight or artificial light, with aeration and CO_2 delivery systems. This could be optimized primarily for biomass production, biofuels, pigments, proteins, and carbon capture. In contrast, a μPBR is a miniaturized system, often chip-based or microfluidic, operating at microliter to milliliter volumes. It primarily helps in precise control of environmental parameters, supports high-throughput experiments, reduces reagent costs, and it primarily could be used for research, optimization, and metabolic studies. Table 10.1 lists tentative differences between a PBR and a μPBR.

TABLE 10.1
Comparative Analysis Between PRB and μPBR

Aspect	PBR	μPBR
Scale	Large-scale system (liters to cubic meters in volume)	Miniaturized system (milliliters in volume)
Design	Cylindrical, tubular, or flat-panel with complex infrastructure	Often chip-based or small transparent vessels with microfluidic channels
Light Source	Sunlight or artificial lamps	Programmable LED light sources with adjustable intensity and spectrum
Mixing & Aeration	Requires sparger, pumps, or bubbling for gas exchange and circulation	Gas exchange and mixing could be achieved through microfluidics and controlled channels
Integration	Standalone, industrial or pilot-scale setup	Compact, modular, easy to integrate with lab automation systems
Purpose	Mass cultivation of microalgae, cyanobacteria, or other phototrophic organisms	Research, optimization, and screening rather than large-scale production
Optimization	Biomass, biofuels, pigments, proteins, CO_2 capture	Precise control of light, pH, temperature, nutrients, and gas exchange

TABLE 10.1 (Continued)
Comparative Analysis Between PRB and μPBR

Aspect	PBR	μPBR
Control of Conditions	Less precise control compared to micro-scale	Highly precise and programmable environmental control
Throughput	Focus on large-scale efficiency and productivity	Enables high-throughput experiments with small sample volumes
Applications	Industrial biomass generation, product formation, CO_2 sequestration	Growth kinetics, metabolic engineering, stress response studies
Cost & Time	High resource demand due to large volumes	Reduces reagent cost and experimental time

10.12 CHALLENGES AND FUTURE PERSPECTIVES FOR μPBR

Thus, from the above discussion, it can be concluded that μPBR could offer promising avenues for microscale cultivation of photosynthetic organisms, but several challenges should be addressed for realizing its full potential. One major issue could be light distribution since spectral optimization and minimizing of shading in dense cultures might be crucial to ensure efficient photosynthesis. Similarly, factors like gas–liquid mass transfer could produce difficulties, particularly in delivering sufficient CO_2 and removing O_2 via membranes or bubbles. This could be attributed to the fact that bubble formation and management are particularly complex at small scales. Biofouling and operational stability again, could raise additional concerns, since maintaining biofilm thickness through antifouling coatings, flow reversal, or selective illumination could provide to be essentials for long-term usage. Scaling up of μPBR for industrial applications would require effective strategies, which involve replicating units while maintaining uniform flow and light conditions. Furthermore, the integration of analytical systems—such as on-chip sensors for optical density, fluorescence, oxygen, and pH would enable real-time monitoring and feedback control, enhancing process efficiency and automation. Therefore, a future with huge opportunities waits ahead for μPBR in ways which include hybrid illumination systems, optical fiber delivery, AI-driven control algorithms, and integrated extraction. Additionally, these systems might also contribute to sustainable resource recovery applications. Finally, addressing these challenges would be critical for transitioning μPBR from experimental tools to scalable, efficient bioprocessing platforms.

REFERENCES

AbdelJawad, N., Nagy, I. 2023. The impacts of urban sprawl on environmental pollution, agriculture, and energy consumption: Evidence from Amman city. *Revista de Gestao Social e Ambiental* 17:2 03347.

Abdel-Raouf, N., Al-Homaidan, A.A., Ibraheem, I.B.M. 2012. Microalgae and wastewater treatment. *Saudi Journal of Biological Sciences* 19:3 257–275. https://doi.org/10.1016/j.sjbs.2012.04.005

Afshar, A., Soleimanian, E., Akbari Variani, H., Vahabzadeh, M., Molajou, A. 2022. The conceptual framework to determine interrelations and interactions for holistic Water, Energy, and Food Nexus. *Environment, Development and Sustainability* 24:8 10119–10140. https://doi.org/10.1007/s10668-021-01858-3

Ajibade, F.O., Adelodun, B., Lasisi, K.H., et.al. 2021. Environmental pollution and their *socioeconomic impacts*. In: *Microbe Mediated Remediation of Environmental Contaminants*, 321–354. Woodhead Publishing, Elsevier.

Al Sharabati, M., Abokwiek, R., Al-Othman, A., et al. 2021. Biodegradable polymers and their nano-composites for the removal of endocrine-disrupting chemicals (EDCs) from wastewater: A review. *Environmental Research* 202 111694. https://doi.org/10.1016/j.envres.2021.111694

Banaduc, D., Simi, V., Cianfaglione, K., et al. 2022. Freshwater as a sustainable resource and generator of secondary resources in the 21st century: Stressors, threats, risks, management and protection strategies, and conservation approaches. *International Journal of Environment and Public Health* 19:24 16570. https://doi.org/10.3390/ijerph192416570

Dong, D., Sun, H., Qi Z., Liu, X. 2021. Improving microbial bioremediation efficiency of intensive aquacultural wastewater based on bacterial pollutant metabolism kinetics analysis. *Chemosphere* 265 129151. https://doi.org/10.1016/j.chemosphere.2020.129151

Dou, S., Xu, D., Zhu, Y., Keenan, R. 2023. Critical mineral sustainable supply: Challenges and governance. *Futures* 146 103101. https://doi.org/10.1016/j.futures.2023.103101

Fallahi, A., Rezvani, F., Asgharnejad, H., Nazloo, E. K., Hajinajaf N., Higgins, B. 2021. Interactions of microalgaebacteria consortia for nutrient removal from wastewater: A review. *Chemosphere* 272 129878. https://doi.org/10.1016/j.chemosphere.2021.129878

Fawell, J., Nieuwenhuijsen, M.J. 2003. Contaminants in drinking water. *British Medical Bulletin* 68 199–208. https://doi.org/10.1093/bmb/ldg027

Fuller, R., Landrigan, P.J., Balakrishnan, K., et al. 2022. Pollution and health: A progress update. *The Lancet Planetary Health* 6:6 535–547. https://doi.org/10.1016/S2542-5196(22)00090-0

Gentili, F.G. 2014. Microalgal biomass and lipid production in mixed municipal, dairy, pulp and paper wastewater together with added flue gases. *Bioresource Technology* 169 27–32. https://doi.org/10.1016/J.BIORTECH.2014.06.061

Javed, F., Aslam, M., Rashid, N., Shamair, Z., Khan, A.L., Yasin, M., et al. 2019. Microalgae based biofuels, resource recovery and wastewater treatment: A pathway towards sustainable biorefinery. *Fuel* 255 115826. https://doi.org/10.1016/J.FUEL.2019.115826

Jayaraman, J., Karthick, M., Yadav, A.S., Baskar, S., Kumaraswamy, J., Anish, M., et al. 2024. Wastewater treatment by algae-based membrane bioreactors: A review of the arrangement of a membrane reactor, physico-chemical properties, advantages and challenges. *RSC Advances* 14:47 34769–34790. https://doi.org/10.1039/D4RA04417G

Jin, Q., Wu, Q,. Shapiro, B.M., McKernan, S.E. 2022. Limited mechanistic link between the Monod equation and methanogen growth: A perspective from metabolic modelling. *Microbiology Spectrum* 10:2 1–10. https://doi.org/10.1128/spectrum.02259-21

Kalra, R., Gaur, S., Goel, M. 2021. Microalgae bioremediation: A perspective towards wastewater treatment along with industrial carotenoids production. *Journal of Water Process Engineering* 40 101794. https://doi.org/10.1016/j.jwpe.2020.101794

Landrigan, P.J., Fuller, R., Acosta, N.J.R., et al. 2018. The Lancet Commission on pollution and health. *The Lancet* 391:10119 462–512. https://doi.org/10.1016/S0140-6736(17)32345-0

Li, K., Liu, Q., Fang, F., et al. 2019. Microalgae-based wastewater treatment for nutrients recovery: A review. *Bioresource Technology* 291 121934. https://doi.org/10.1016/j.biortech.2019.121934

Lutzu, G.A., Ciurli, A., Chiellini, C., Di Caprio, F., Concas, A., Dunford, N.T. 2021. Latest developments in wastewater treatment and biopolymer production by microalgae. *Journal of Environmental Chemical Engineering* 9 104926. http://doi.org/10.1016/j.jece.2020.104926

Mairet. F., Baron. R. 2019. A physiologically structured equation to consider quota heterogeneity in the droop model. *IFAC Papers Online* 52:26 275–280. https://doi.org/10.1016/j.ifacol.2019.12.270

Mat Aron, N.S., Khoo, K.S., Chew, K.W., Veeramuthu, A., Chang, J.S., Show, P.L. 2021. Microalgae cultivation in wastewater and potential processing strategies using solvent and membrane separation technologies. *Journal of Water Process Engineering* 39 101701. https://doi.org/10.1016/J.JWPE.2020.101701

Molajou, A., Afshar, A., Khosravi, M., Soleimanian, E., Vahabzadeh, M., Variani, H.A. 2023. A new paradigm of water, food, and energy nexus. *Environmnetal Science and Pollution Research* 30:49 107487–107497. https://doi.org/10.1007/s11356-021-13034-1

Munzel, T., Hahad, O., Daiber, A., Landrigan, P.J. 2023. Soil and water pollution and human health: What should cardiologists worry about? *Cardiovascular Research* 119:2 440–449. https://doi.org/10.1093/cvr/cvac082

Munzel, T., Hahad, O., Lelieveld, J., et al. 2025b. Soil and water pollution and cardiovascular disease. *Nature Reviews Cardiology* 22:2 71–89. https://doi.org/10.1038/s41569-024-01068-0

Munzel, T., Kuntic, M., Lelieveld, J., et al. 2025a. The links between soil and water pollution and cardiovascular disease. *Atherosclerosis* 403 119160. https://doi.org/10.1016/j.atherosclerosis.2025.119160

Padder, F.A., Bashir, A., 2023. Scarcity of water in the twenty-first century: Problems and potential remedies. *Medalion Journal: Medical Research, Nursing, Health and Midwife Participation* 4:1 1–5. https://doi.org/10.59733/medalion.v4i1.66

Prasad, M.N.V., 2024. Bioremediation, bioeconomy, circular economy, and circular bioeconomy-Strategies for sustainability. In: *Bioremediation and Bioeconomy*, ed. M.N.V. Prasad, 3–32. Elsevier.

Saito, K., Cao, X., He, Y., Xu, Y., 2015. Progress in the molecular understanding of central regulation of body weight by estrogens. *Obesity* 23:5 919–926. https://doi.org/10.1002/oby.21099

Samuel, M. S., Selvarajan, E., Chidambaram, R, Patel H., Brindhadevi, K. 2021. Clean approach for chromium removal in aqueous environments and role of nanomaterials in bioremediation: Present research and future perspective. *Chemosphere* 284 131368. https://doi.org/10.1016/j.chemosphere.2021.131368

Shen, Y. 2014.Carbon dioxide bio-fixation and wastewater treatment via algae photochemical synthesis for biofuels production. *RSC Advances* 4 49672–49722. https://doi.org/10.1039/C4RA06441K.

Shukla, K., Agarwalla, S., Duraiswamy, S., Gupta, R.K. 2021. Recent advances in heterogeneous micro-photoreactors for wastewater treatment application. *Chemical Engineering Science* 235 116511. https://doi.org/10.1016/j.ces.2021.116511

Singh, V.P., Su, Q. 2022. Water-environment-energy-food nexus: Challenges and opportunities under climate change. *Indian Journal of Soil Conservation* 50:3 177–189.

Velugula-Yellela, S.R., Kohnhorst, C., Powers, D.N., et al. 2018. Use of high-throughput automated microbioreactor system for production of model IgG1 in CHO Cells. *Journal of Visualized Experiments* 28 58231. https://doi.org/10.3791/58231

Yadav, D., Rangabhashiyam, S., Verma, P., et al. 2021. Environmental and health impacts of contaminants of emerging concerns: Recent treatment challenges and approaches. *Chemosphere* 272 129492. https://doi.org/10.1016/j.chemosphere.2020.129492

Zhang, Y., Yuan, J., Ding, Y., Liu, B., Zhao, L., Zhang, S. 2021. Research progress on g–C3N4–based photocatalysts for organic pollutants degradation in wastewater: From exciton and carrier perspectives, *Ceramics International* 47:22 31005–31030. https://doi.org/10.1016/j.ceramint.2021.08.063

11 Advanced Oxidation and Photocatalytic Remediation Using Microreactors

Rashmi Acharya, Reema Nayak, Anubhav Naik, and Subhasish Mishra

11.1 INTRODUCTION

Water resources are experiencing a host of serious concerns and are indeed undergoing significant alterations as a result of the influx of a vast range of contaminants. These contaminants, originating from diverse sources such as agricultural, domestic, and industrial activities, are adversely affecting water quality (Gulfam-E-Jannat et al., 2023; Acharya et al., 2023; Mishra and Acharya, 2023). Chemical, distillery, paper, pulp, textile, food, and beverage sectors generate approximately 500 million liters of wastewater every day (Shamshad et al., 2025; Mansingh et al., 2018). The contaminated wastewater, with a wide range of pollutants mostly from the aforementioned sources, is severely impacting water resources. As a matter of fact, the discharge of untreated or partially treated wastewater injects harmful organic and inorganic substances into the aquatic environment, affecting both human life and the aquatic environment (Wear et al., 2021). So there exists a critical need and immediate requirement to develop effective as well as suitable treatment techniques for the removal of these pollutants. To remove these toxic contaminants from wastewater sources, various conventional techniques, including adsorption, reverse osmosis, solvent extraction, electro-coagulation, chemical precipitation, and ultrafiltration, have been carried out (Dutta et al., 2021; Mishra and Acharya, 2021; Acharya & Parida, 2022a). These traditional wastewater treatment technologies are becoming more complicated due to the introduction of complex pollutants owing to increased industrialization. Therefore, developing an effective and eco-friendly technique to treat wastewater is crucial for reducing pollutant levels in the environment (Razzak et al., 2022).

Advanced oxidation processes (AOPs) encompass a group of water purification methods that rely on the in-situ formation of highly reactive oxidants, including ozone (O_3), hydroxyl radicals ($^{\bullet}OH$), sulfate radicals ($SO_4^{\bullet-}$), and superoxide anions ($^{\bullet}O_2^-$) (Wang and Bai, 2017; Wang and Wang, 2018; Wang and Xu, 2012). These species possess high oxidative capability and react non-selectively with a wide array of

 DOI: 10.1201/9781003633761-11

organic pollutants, facilitating their transformation and mineralization (Mierzwa et al., 2018). Primarily, the AOPs based on hydroxyl radical (•OH), possessing an exceptionally high standard electrode potential (E_0= +2.80 V vs. SHE), are extensively studied due to their high electron affinity and rapid reaction kinetics (Cuerda-Correa et al., 2020).

Despite their efficacy, conventional AOP systems often face limitations, including low reaction efficiency, high energy consumption, and poor scalability. Microreactors are the cutting-edge systems that provide a revolutionary platform for improving AOP performance compared to traditional approaches (Miklos et al., 2018; Tanimu et al., 2017). They employ intricately designed microchannels, typically spanning lateral dimensions from a few micrometers to several millimeters, to precisely regulate chemical reactions and fluid dynamics. Their high surface-area-to-volume ratios and laminar flow conditions facilitate superior mass and heat transfer, resulting in reduced diffusion limitations and precise control over reaction kinetics (Dong et al., 2022; Dhamorikar et al., 2024). Their compact design, low reagent consumption, and minimal waste generation make them ideal for sustainable water treatment. Applications of microstructured reactors have been expanded in recent years due to their ease of operation and scalability (Gouran et al., 2022; Seo et al., 2025). To investigate the kinetics of diuron degradation, Khongthon et al. conducted several experiments using an electrochemical microreactor. Their results showed that the process's rate constant was 1–2 orders of magnitude higher than in a conventional reactor (Khongthon et al., 2016).

In this chapter, we first depict different types of AOPs employed for aquatic pollutant degradation using microreactors. Then, we focus on the photocatalytic removal of heavy metals from the aqueous environment with elucidation of the mechanism. The photocatalytic degradation of organic pollutants is addressed next by demonstrating the mechanism. At the end, the chapter is concluded with future research directions.

11.2 ADVANCED OXIDATION PROCESSES

11.2.1 UV-based AOPs

In this type of AOPs, ultraviolet light is utilized to directly photolyze water or activate an oxidant (namely, O_3, H_2O_2, Cl_2, SO_4^{2-}) to release •OH radicals, which decompose the pollutants into harmless products. This technique is again divided into the following categories.

11.2.1.1 UV/H_2O_2

The UV/H_2O_2 AOP enables efficient generation of hydroxyl radicals (•OH) via direct UV photolysis and •OH-mediated indirect photolysis, exhibiting a high quantum yield (Keen et al., 2016). The primary mechanism involves UV-induced cleavage of the O–O bond in hydrogen peroxide, consistent with the Haber–Weiss pathway, initiating free radical formation. Subsequent reactions propagate •OH generation and pollutant breakdown as per the following equations (Glaze et al., 1987).

$$H_2O_2 + h\nu \rightarrow 2\,HO^{\bullet} \tag{11.1}$$

$$H_2O_2 + 2HO^{\bullet} \rightarrow H_2O + HO_2^{\bullet} \quad (11.2)$$

$$H_2O_2 + HO_2^{\bullet} \rightarrow H_2O + O_2 + HO^{\bullet} \quad (11.3)$$

The $^{\bullet}OH$ and $^{\bullet}OOH$ species released from reactions undergo decomposition of pollutants. For instance, Verma et al. achieved 97% anatoxin-a removal under optimal conditions by applying a UV/H_2O_2 AOP-based technique. Necessary reaction parameters such as pH, oxidant dosage, water matrix, and UV transmittance critically influence treatment performance (Verma & Sillanpää, 2015). Single-capillary UV/H_2O_2 microreactors have been employed to effectively remove pollutants such as benzoylecgonine, isoxazole, and isothiazolinones by Russo et al. Figure 11.1(a) shows the photodegradation of mixtures of methylisothiazolinone (MIT), benzisothiazolinone (BIT), and isoxazole (ISOX) using UV254/H_2O_2 in MilliQ water at pH=6.0. The single capillary microreactor has demonstrated improved photo-kinetics as well as high-rate constant values for the reaction of $^{\bullet}OH$ with targeted pollutants. The rate constant values were estimated to be 2.09×10^9, 5.9×10^9, and 2.15×10^9 L/mol/s for the degradation of MIT, BIT, and ISOX, respectively (Russo et al., 2020; Russo et al., 2016).

11.2.1.2 UV/Ozone(O_3)

Ozone is a potent oxidant, but its reactivity increases significantly under UV exposure. This synergy is highly effective for treating waters rich in organic compounds. In this process, UV irradiation accelerates ozone (O_3) photolysis, producing hydrogen peroxide and subsequently hydroxyl radicals ($^{\bullet}OH$) (Sonntag et al., 2008; Lu et al., 2022).

$$O_3 + H_2O + h\nu(UV) \rightarrow O_2 + H_2O_2 \quad (11.4)$$

$$H_2O_2 + O_3 \rightarrow 2HO^{\bullet} + 3O_2 \quad (11.5)$$

These $^{\bullet}OH$ free radicals oxidize the pollutants to harm less products. A microporous tube-in-tube microchannel reactor for azo dye ozonation based on MR was invented by Gao et al. It has coaxial tubes that form an annular microchannel, with wastewater in the outer tube and ozone in the inner tube. Micropores allow ozone diffusion, significantly improving mass transfer and ozone utilization. Figure 11.1(b) shows the effect of annular channel width on decolorization efficiency and ozone utilization efficiency of AR14 dye. The decolorization efficiency shows a constant increment as the channel width of the microreactor is reduced from 1000 to 250 μm (Gao et al., 2012). While ozone can degrade certain hazardous compounds (such as pigments), it exhibits limited reactivity toward complex organic molecules. Thus $^{\bullet}OH$ based mechanisms are necessary for complete mineralization and enhanced biodegradability of these pollutants. However, only a small percentage of the produced H_2O_2 decomposes to $^{\bullet}OH$ due to cage recombinations, which restricts its full-scale application (Reisz et al., 2003).

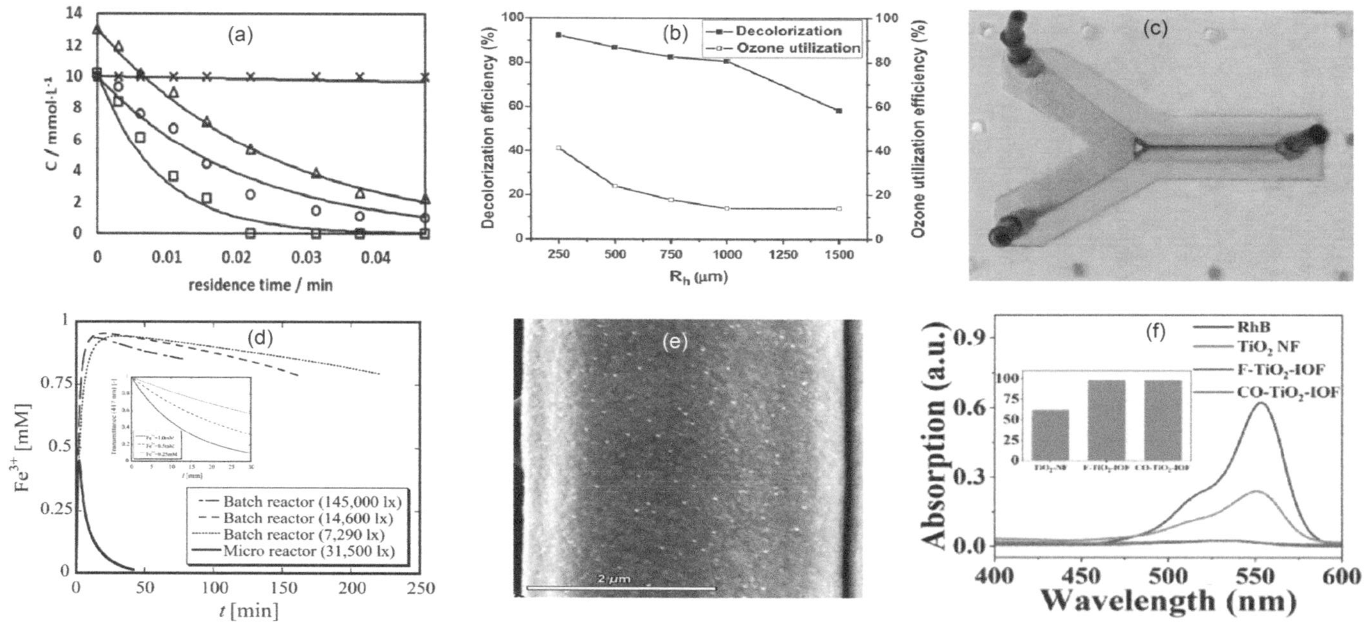

FIGURE 11.1 (a) The kinetic model for photodegradation of MIT (o), BIT (Δ), and ISOX (□) and H_2O_2 (×) using UV254/H_2O_2 (Reproduced from Russo et al., 2020), (b) the effect of annular channel width on decolorization and ozone utilization efficiency (Reproduced from Gao et al., 2012), (c) picture of the Y-shaped microreactor for continuous-flow Fenton-based azo dye decolorization (Reproduced from Rahimi et al., 2014), (d) comparison of the activity of conventional batch reactor with MR [reproduced from Shinozawa et al., 2020], (e) cross sectional SEM image of Au decorated TiO_2 layer deposited homogeneously onto the walls inside the SiO_2 microreactor (Reproduced from Vega et al., 2025), (f) RhB absorption and photocatalytic RhB degradation ability (inset) of the different catalysts (Reproduced from Huang et al., 2025b).

11.2.1.3 UV-Cl_2

UV, along with Cl_2, is another robust AOP carried out by the generation of $Cl^•$ and $^•OH$ free radicals owing to the photolysis of chlorine. $Cl^•$ [$E^0 = 2.47$ V] is more selective and highly oxidative than $^•OH$ radicals. The generation of these radicals resulted from the decomposition of hypochlorous acid (HOCl) and hypochlorite ion (ClO^-) (Kishimoto et al., 2019; Guo et al., 2022).

$$Cl_2 + h\nu \rightarrow 2^•Cl\left(\text{Photolysis of chlorine}\right) \tag{11.6}$$

$$HOCl + h\nu \rightarrow {}^•OH + {}^•Cl\left(\text{Photolysis of hypochlorous acid}\right) \tag{11.7}$$

$${}^•Cl + Cl^- \rightarrow Cl_2^{-•}\left(\text{Formation of dichloride radical}\right) \tag{11.8}$$

$${}^•Cl + H_2O \rightarrow ClOH^•\left(\text{Formation of reactive chlorine species}\right) \tag{11.9}$$

$${}^•OH + R\text{- }H \rightarrow R^• + H_2O\left(\text{Oxidation of organic contaminant}\right) \tag{11.10}$$

In low pH circumstances, this reaction typically outperforms UV/H_2O_2 systems. When trichloroethylene (TCE) is destroyed in a drinking water system utilizing a bench-scale experiment with an MP lamp, the UV-Cl_2 process performs better at pH 5 than UV-H_2O_2 (Wang et al., 2012). Nonetheless, the primary disadvantage of this procedure is the production of excess byproducts such as chlorate, perchlorate, and organo-chlorine species. Sometimes, in the UV/chlorine treatment process, the toxicity of the intermediates can occasionally even exceed that of raw textile wastewater (Sun et al., 2019).

11.2.1.4 UV/Persulfate ($S_2O_8^{2-}$)

Persulfate ions ($S_2O_8^{2-}$) can be activated by UV light to form sulfate radicals ($^•SO_4^-$), which are strong oxidants with a typical oxidation potential of E_0= +2.5 V to +3.1 V vs. SHE. These sulfate radicals can then combine with water or hydroxide to produce hydroxyl radicals (Ike et al., 2018).

$$S_2O_8^{2-} + h\nu \rightarrow 2\,SO_4^{-•} \tag{11.11}$$

$$SO_4^{-•} + H_2O \rightarrow HO^• + SO_4^{2-} + H^+ \tag{11.12}$$

$${}^•SO_4^- + OH^- \rightarrow {}^•OH + SO_4^{2-} \tag{11.13}$$

Sulfate radical-based AOPs have become more common in recent years because of their wider pH range of efficacy and greater stability than hydrogen peroxide. UV-C (100–380 nm) activation leads to the homolytic cleavage of peroxydisulfate ($S_2O_8^{2-}$). $S_2O_8^{2-}$ has more quantum yield than H_2O_2 (1.4 vs. 1.0), and its molar absorption is slightly greater, resulting in a higher rate of radical production when peroxodisulfate (PDS) is used as an oxidizing agent (Xiao et al., 2016). Recently, Son et al. have applied this method for the degradation of Bisphenol A from industrial wastewater. The combination of UV and persulfate was able to degrade completely in 30 min with a TOC removal efficiency of 41–72% (Kang et al., 2024).

11.2.2 Ozone-based AOPs

In water treatment, ozone has long been utilized as an oxidant and disinfectant. Ozonation itself is frequently regarded as an AOP or AOP-like process, as its reactions in actual aqueous solutions frequently involve the generation of •OH radicals (Merenyi et al., 2010). Numerous advantages of ozonation include low sludge generation, minimal installation cost, and no increase in effluent volume. The reaction of ozone with water or hydroxide ion can generate ample amount of hydroxyl radicals as shown below.

$$O_3 + H_2O \rightarrow 2HO^{\bullet} + O_2 \tag{11.14}$$

$$O_3 + HO^- \rightarrow O_2^{\bullet-} + HO_2^{\bullet} \tag{11.15}$$

$$O_3 + HO_2^{\bullet} \rightarrow 2O_2 + HO^{\bullet} \tag{11.16}$$

However, the rate of reaction between O_3 and many organic molecules is sluggish. AOPs increase the oxidative power of ozone by encouraging its breakdown into hydroxyl radicals (Buffle et al., 2006). There are various ozone-based AOPs, which are discussed below.

11.2.2.1 Ozone (O_3)/Hydrogen Peroxide (H_2O_2)

The direct interaction of ozone with hydrogen peroxide produces hydroxyl radicals in this pathway (Rekhate and Srivastava, 2020). This procedure is referred to as the peroxone process.

$$H_2O_2 + 2O_3 \rightarrow 2HO^{\bullet} + 3O_2 \tag{11.17}$$

Additionally, ozone reacts with the hydroperoxide anion ($^-HO_2$) to form $HO_2^{\bullet}$ precursors, which subsequently generate $HO^{\bullet}$

$$H_2O_2 \rightleftharpoons HO_2^- + H^+ \tag{11.18}$$

$$HO_2 - + O_3 \rightarrow HO_2^{\bullet} + O_2^{-\bullet} + O_2 \tag{11.19}$$

$$O_3 + HO_2^{\bullet} \rightarrow 2O_2 + HO^{\bullet} \tag{11.20}$$

Excess of H_2O_2 may lead to the inactivation of $HO^{\bullet}$ radicals. Therefore, before releasing the treated water into the receiving aquatic environment, the remaining H_2O_2 may need to be eliminated. Similarly, a low amount of H_2O_2 restricts the formation of $^{\bullet}OH$ due to incomplete oxidation resulting from competitive reactions (Hübner et al., 2015). For the peroxone process, the ideal molar ratio of H_2O_2 to O_3 (H_2O_2:O_3) is 0.5, with the ozone levels typically ranging from 1 to 20 mg/L (Katsoyiannis et al., 2011). However, recent research has demonstrated that the high competitive responses and ineffective radical production with ozone could restrict the benefits of its utilization in wastewater treatment (Pisarenko et al., 2012).

11.2.2.2 Ozone (O3)/Catalyst

Catalytic ozonation is characterized as homogeneous or heterogeneous according to catalyst solubility. Heterogeneous catalysts like metal oxides (Al_2O_3, MnO_2, TiO_2, FeOOH) are easily recoverable, but have lesser catalytic efficiency than homogeneous catalysts such as Mn^{2+}, Fe^{2+}, Co^{2+}, etc. (Kasprzyk-Hordern et al., 2003; Wang et al., 2019). Organic degradation occurs through both radical and non-radical mechanisms. Metal catalysts activate ozone by producing hydroxyl radicals (·OH) via surface hydroxyl groups, Lewis acid sites, and structural defects, while graphene and CNTs produce singlet oxygen through non-radical processes (Yu et al., 2020). Heterogeneous catalytic ozonation removes dissolved organics more effectively than homogeneous systems due to higher adsorption on the catalyst surface. This promotes ·OH utilization and reduces diffusion loss (Sun et al., 2015). To achieve optimal catalytic performance, organic and ozone adsorption must be balanced at the solid–liquid interface.

11.2.3 Fenton-based AOPs

These reactions include the catalytic breakdown of hydrogen peroxide by ferrous iron (Fe^{2+}). Fenton oxidation technologies are commonly utilized for wastewater treatment that involves the addition of Fenton reagent (Fe^{2+} and H_2O_2), which reacts to produce $^{\bullet}OH$ radicals, as shown below (Monteil et al., 2019).

$$Fe^{2+} + H_2O_2 \rightarrow Fe^{3+} + HO^{\bullet} + HO^{-} \tag{11.21}$$

$$HO^{\bullet} + H_2O_2 \rightarrow HO_2^{\bullet} + H_2O \tag{11.22}$$

$$2HO^{\bullet} \rightarrow H_2O_2 \tag{11.23}$$

A Y-shaped microreactor [shown in Figure 11.1(c)] was constructed for continuous-flow Fenton-based azo dye decolorization. The results demonstrated that 86% conversion was achieved in just 4.2 seconds, demonstrating microreactor efficiency (Rahimi et al., 2014). The classic Fenton process usually works best at an acidic pH (pH 2–4) because of the higher stability of Fe^{2+} and the increased reactivity of H_2O_2. Temperature, pH, Fe^{2+} concentration, and H_2O_2 dosage are all important operational parameters that influence the oxidation process significantly (Jayakumar et al., 2022). The maximum COD elimination efficiency can be observed when the molar ratio of H_2O_2/Fe^{2+} is between 150 and 250. Although homogeneous Fenton oxidation can effectively degrade organic pollutants, the precipitation of iron oxyhydroxides at higher pH values, as well as the production of iron sludge, limits its application in practical environments (Wang et al., 2016).

To overcome these limitations, the photo-Fenton method was adopted. The applications of UV or visible light in the presence of a catalyst considerably enhance the Fenton reaction. Additionally, the heterogeneous Fenton or Fenton-like processes can be performed across a wide range of pH, and the catalyst can be utilized circularly, avoiding the generation of iron sludge (He et al., 2016). Recently, Mishra et al. have designed a $NiFe_2O_4$/g-C_3N_4 hybrid catalyst for the effective degradation of Rhodamine B dye and Tetracycline antibiotic residue from wastewater samples. Within an hour of solar exposure, they reportedly achieved 84.32% and 98.6% of TC and RhB degradation, respectively (pH 9, 1 mL of H_2O_2, 0.4 g/L catalyst dose, 10 mg/L TCH, and 20 mg/L RhB). The Fe^{3+} ions can be reduced to Fe^{2+} ions by the electrons present in the CB of $NiFe_2O_4$. As a result, the Fe^{2+} ions break down the H_2O_2 to generate hydroxyl radicals. The holes in the VB of $NiFe_2O_4$ possess the requisite potential to create $^{\bullet}OH$ radicals via OH^- photo-oxidation reaction (Mishra et al., 2024b). The following equations demonstrate how these extremely potent free radicals can directly break down TCH and RhB molecules into smaller components,

$$Fe^{3+} + e_{CB}^{-} \rightarrow Fe^{2+} \tag{11.24}$$

$$Fe^{2+} + H_2O_2 \rightarrow Fe^{3+} + OH^{\cdot} \tag{11.25}$$

$$OH^{-} + h_{VB}^{+} \rightarrow OH^{\cdot} \tag{11.26}$$

$$O_2^{\cdot -}, OH^{\cdot}, h^{+} + RhB \rightarrow H_2O + CO_2 \tag{11.27}$$

For the degradation of carbofuran under visible light, Shinozawa et al. employed photo-Fenton AOP in a traditional batch reactor and compared the results with the microreactor system. The results shown in Figure 11.1d illustrate that the experiment carried out in the microreactor was 7.5 times faster than that in the conventional batch reactor (Shinozawa et al., 2020).

11.2.4 Photocatalysis

Photocatalysis is an AOP in which light of suitable energy is absorbed by the photocatalyst to generate electron-–hole pairs. These charged species react with water and dissolved oxygen after satisfying the required thermodynamic criteria to release reactive species such as $^{\bullet}O_2^-$and $^{\bullet}OH$ which undergo pollutant degradation reaction through a number of steps to produce harmless products. The prime merit of this process over other AOPs is that it uses visible light, which is a major component of the solar spectrum and hence can be harnessed cheaply. Over the past few decades, a lot of research has been done on the utilization of photo-active catalysts for oxidation reactions in water treatment. Photocatalytic reaction begins with the attachment of the pollutant molecules to the surface of the photocatalyst, which serves as the active site for subsequent chemical transformations. The reaction solution is exposed to a light source when the adsorption–desorption equilibrium is attained. In general, photocatalysis proceeds through the following five steps: (1) pollutant molecules are transferred to the photo catalyst's surface; (2) these molecules are then allowed to adsorb onto the catalyst's surface; (3) the catalyst is activated by irradiating it with light of suitable wave length to produce electron–hole pairs; (4) the photoinduced charged carriers are migrated to the surface of the photocatalyst; and (5) after satisfying thermodynamic criteria, the charge carriers undergo redox reactions with water and adsorbed O_2 to produce reactive species which undergo reaction with pollutant molecules (Khader et al., 2024; Rueda-Marquez et al., 2020; Mishra et al., 2025a).

Light irradiation can catalyze redox reactions over a wide range of materials popularly known as photocatalysts. These are semiconducting materials with an empty conduction band and a valence band rich in electrons. These electrons migrate from the valence band to the conduction band, resulting in holes, when photon energy exceeds the semiconductor's band-gap potential (Acharya and Parida, 2020a).

$$SC + h\nu \rightarrow e_{cb}^{-} + h_{vb}^{+} \tag{11.28}$$

The energetic charge carriers travel to the catalyst surface, where photogenerated holes help to oxidize the surface-bound water or hydroxide ion, resulting in the formation of hydroxyl radicals. The photo-induced electrons degrade the adsorbed oxygen molecules, producing superoxide radicals that can also contribute to the generation of $^{\bullet}OH$. The rate of recombination of photoinduced charge carriers determines how these reactive species will get produced. The lower recombination rate facilitates the generation of the reactive species. These highly reactive radicals further destroy many pollutants to completely mineralize them to CO_2 and H_2O (Mishra et al., 2025b; Khan et al., 2024).

$$h_{vb}^{+} + H_2O_{ads} \rightarrow {}^{\bullet}OH + H^{+} \tag{11.29}$$

$$h_{vb}^{+} + OH^{-}_{ads} \rightarrow {}^{\bullet}OH \tag{11.30}$$

$$e^{-}_{cb} + O_2 \rightarrow {}^{\bullet}O_2^{-} \tag{11.31}$$

$$O_2 + H^{+} \rightarrow HO_2^{\bullet} \tag{11.32}$$

$$HO_2^{\bullet} + HO_2^{\bullet} \rightarrow H_2O_2 + O_2 \tag{11.33}$$

$$H_2O_2 + {}^{\bullet}O_2^{-} \rightarrow {}^{\bullet}OH + O_2 + OH^{-} \tag{11.34}$$

A wide spectrum of aquatic pollutants, including dyes, drugs, and heavy metals, has been proven to be destroyed by photocatalytic reactions. Recently, a $Bi_4O_5I_2$/g-C_3N_4 hybrid photocatalyst has been applied for the simultaneous degradation of Cr (VI) and tetracycline from wastewater. This catalyst reported almost complete degradation of both the pollutants within 1 h of solar illumination owing to visible light activity, high charge transfer rate, and reduced recombination (Mishra et al., 2024a). Photocatalysis is a promising green technology due to its use of light energy and non-toxic catalysts, but challenges remain in achieving high quantum efficiencies and efficient catalyst separation.

The first application of a photocatalytic microfluidic reactor utilizing TiO_2 for the breakdown of MB dye was described by Lindstrom et al. The functionalized TiO_2 based microfluidic reactor fabricated using the deposition method enabled 99.9% of photocatalytic MB degradation (Lindstrom et al., 2007). Similar type of work has also been carried out by Vega et al., where they have prepared free standing SiO_2 microreactors functionalized with Au coated TiO_2 for photocatalytic applications. Figure 11.1(e) depicted the SEM image for the side view of a SiO_2 microreactor coated completely with a thin layer (~60 nm) of TiO_2 and decorated with Au nanoparticles identified as bright dots. The high density of free standing SiO_2 monolithic microreactor provides exceptionally high surface to volume ratio of ~7×10^5 m^2/m^3 (Vega et al., 2025). Most recently, Cheng and coworkers have demonstrated the novel preparation of TiO_2 inverse opal films in a microfluidic reactor using a limited planar rapid evaporation method. The as-prepared microreactor has demonstrated an outstanding activity for dye degradation since 95% of 5 mg/L rhodamine B (RhB) was degraded within 120 s. Figure 11.1(f) demonstrates the RhB absorption, and the inset shows the photocatalytic RhB degradation ability of the different catalysts. The results demonstrated that the TiO_2 inverse opal film-based microreactors have the minimum absorption and maximum photocatalytic degradation ability (Huang et al., 2025b).

11.3 PHOTOCATALYTIC ELIMINATION OF HEAVY METAL IONS

The widespread release of toxic heavy metal ions into aquatic environments, stemming from industrial, mining, and agricultural activities, represents a serious risk to human health and the environment. Among these, chromium (Cr (VI)), uranium (U (VI)), arsenic (As (III)), and mercury (Hg (II)) are of great concern owing to their pronounced toxicity, persistence, and bioaccumulation potential. These contaminants can lead to severe health effects, including carcinogenicity, nephrotoxicity, and neurotoxicity, even at low concentrations (Ray and Vashishth, 2024). As a result, the evolution of effective and sustainable methods for their removal has become a pressing environmental priority. Among the applied technologies, photocatalysis has emerged as an optimistic strategy for the remediation of heavy metal-tainted water owing to its capability to convert the toxic metal ions into harmless metal ions through a suitable redox reaction (Burakova et al., 2018). Unlike conventional treatment methods, photocatalysis offers advantages such as low energy input, minimal secondary pollution, and the potential for complete detoxification [Ahuja et al., 2023]. In this context, semiconductor-based photocatalysts can facilitate the reduction or oxidation of heavy metal ions through the generation of photogenerated electron–hole pairs (Serpone and Emeline, 2012). For instance, Cr (VI) can be converted into the less toxic Cr (III) species, while As (III) can be oxidized to the less mobile As (V), depending on the redox potential of the photocatalytic system. Recent research has focused on engineering advanced photocatalysts with enhanced light absorption, charge carrier separation efficiency, and surface reactivity to improve heavy metal ion removal performance (Zhang et al., 2023). Approaches such as the formation of heterojunctions, surface defect modification, and band structure modulation have been widely explored to enhance photocatalytic activity (Balapure et al., 2024). This study aims to explore a tailored photocatalyst system for the efficient removal of Cr (VI), U(VI), As (III), and Hg (II) under visible light irradiation. The photocatalytic removal of heavy metals is illustrated in the schematic diagram shown in Figure 11.2(a).

11.3.1 Photocatalytic Reduction of Cr (VI)

Cr (VI) present is a potent toxicant and known carcinogenic substance that possesses serious environmental and health hazards (Acharya et al., 2018; Tripathy et al., 2019). It typically enters the aquatic system through industrial discharge from

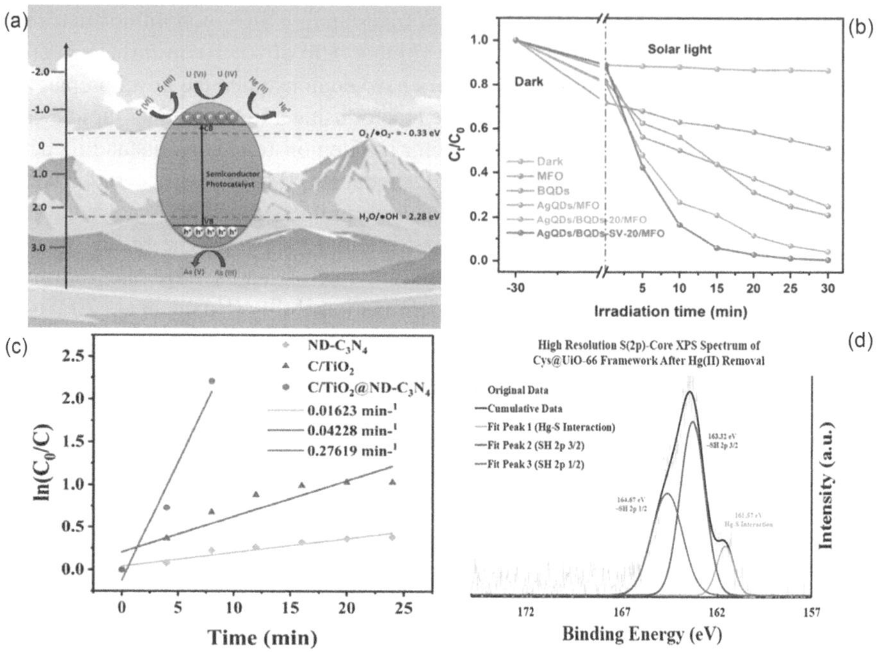

FIGURE 11.2 (a) Schematic diagram of photocatalytic elimination of heavy metal ions, (b) Photocatalytic degradation curves Cr(VI) over different reaction systems (Reproduced from He et al., 2025), (c) Photooxidation kinetics of As(III) from different photocatalysts [reproduced from Liu et al., 2023], (d) High-resolution XPS spectra of Cys@UiO-66 after Hg(II) removal (S(2p)-core after adsorption).

(Reproduced from Zahra et al., 2025.)

activities such as electroplating, leather tanning, textile dyeing, pigment production, and ore processing. Due to its high solubility in water and persistence in the environment, it contaminates both surface and groundwater sources (Prasad et al., 2021; Acharya & Naik, 2024; Acharya et al., 2024). In humans, chronic exposure to Cr (VI) contaminated water may pose serious health risks, including skin irritation, respiratory disorders, hepatic injury, renal impairment, and an elevated risk of lung and gastrointestinal cancers. Cr (VI) is readily taken up by the human body through ingestion or skin contact, and is reduced to Cr (III) through reactive oxygen species (ROS), causing intracellular damage. Its carcinogenic and mutagenic properties have led to its designation as a Group 1 carcinogen for humans by the International Agency for Research on Cancer (IARC) (Iyer et al., 2023; Acharya et al., 2022b).

At varying pH levels and concentrations, Cr (VI) exists in multiple oxyanion forms, which significantly influence its mobility and environmental behavior. Under highly acidic conditions (pH<1), chromic acid (H_2CrO_4) predominates. Within a moderately acidic range (pH 1–6), hydrogen chromate ($HCrO^{4-}$) and dichromate ($Cr_2O_7^{2-}$) ions are the dominant species. At pH levels above 6, chromate ions (CrO_4^{2-}) are thermodynamically stable (Stern et al., 2020; Acharya et al., 2024). These anionic forms of Cr (VI), particularly in neutral to alkaline environments, exhibit limited interaction with negatively charged soil surfaces due to electrostatic repulsion. Consequently, they remain highly mobile, readily leaching into groundwater and surface water systems, thereby posing a serious risk to aquatic ecosystems and human health (Liang et al., 2021; Acharya et al., 2021). In recognition of these threats, the United States Environmental Protection Agency (USEPA) has established stringent regulatory limits for Cr (VI), setting a maximum allowable concentration of 0.1 mg/L in surface water and 0.05 mg/L in drinking water (Deng et al., 2025; Acharya and Parida, 2020b). To reduce the hazardous effects of Cr (VI) on aquatic systems and downstream users, it is crucial to lower its concentration to a level below the regulatory limits prior to discharge into water bodies. Photocatalytic reduction of Cr (VI) to Cr (III) represents an efficient approach to minimize Cr (VI) toxicity due to its simplicity, efficiency, and eco-friendliness (Sharma et al., 2020; Acharya et al., 2025). Under visible light irradiation, photocatalysts generate electrons that reduce the highly toxic and mobile Cr (VI) to less toxic forms, less soluble Cr (III), which can be easily removed from water. This transformation significantly reduces Cr (VI) bioavailability and environmental risk, making photocatalysis a sustainable and practical approach for detoxifying Cr (VI) contaminated water (Wu et al., 2020; Acharya et al., 2017).

He et al. developed an AgQDs/BQDs-SV/MFO photocatalyst composed of ultrathin sheets, quantum dots, and engineered sulfur vacancies. The presence of sulfur vacancies enhanced light absorption, facilitated charge separation, and reduced electron transfer distance. The optimized AgQDs/BQDs-SV-20/MFO catalyst achieved 99.5% Cr (VI) degradation within 30 min under visible light using 0.6 g/L catalyst and a Cr (VI) concentration of 20 mg/L, demonstrating outstanding photocatalytic efficiency, as shown in Figure 11.2(b) (He et al., 2025). Zheng et al. fabricated Cu-MOF and Zn-MOF, among which Cu-MOF exhibited 2.5 times higher ability to remove Cr (VI) than Zn-MOF under sunlight irradiation. This increase in efficiency can be attributed to the wider light absorption range of Cu-MOF from 200 to 1100 nm, allowing it to effectively use the full spectrum of sunlight. In acidic conditions, Cr (VI) present in $Cr_2O_7^{2-}$ and $HCrO_4^-$ species are adsorbed onto Cu-MOF due to a positive surface charge at pH < 4.1. Upon sunlight irradiation, the Cu-MOF material undergoes photoexcitation, whereby electrons (e^-) are excited from the valence band (VB) to the conduction band (CB), generating corresponding holes (h^+) in the VB. These photogenerated holes subsequently oxidize water molecules, generating H_2O_2 and H^+ ions. Ultimately, Cr (VI) is efficiently converted to Cr (III) via the synergistic action of photogenerated electrons and H^+ ions. (Zheng et al., 2024).

11.3.2 Photocatalytic Reduction of U (VI)

Hexavalent uranium [U (VI)] is a toxic and radiologically hazardous pollutant, frequently detected in water bodies near uranium mining operations, nuclear sites, and industrial effluents (Zhu et al., 2025). In aqueous media, U (VI) predominantly exists as uranyl ions (UO_2^{2+}), which are highly soluble and mobile, enabling their rapid migration through soil and groundwater (Shan et al., 2025). Human exposure to U (VI) occurs via ingestion, inhalation, or dermal contact, with nephrotoxicity being the primary concern due to its strong affinity for renal tissues. Chronic exposure may also lead to oxidative stress, DNA damage, and accumulation in bones and liver, increasing the risk of long-term carcinogenic and mutagenic outcomes (Ma et al., 2020).

Photocatalytic reduction has gained prominence as a sustainable and efficient method for U (VI) detoxification. Under light irradiation, semiconductor photocatalysts generate electron–hole pairs, where the photogenerated electrons drive the stepwise reduction of U (VI) to insoluble tetravalent uranium (U (IV)) (Liu et al., 2021b). This process involves the initial adsorption of UO_2^{2+} onto the catalyst surface, followed by multi-electron transfer:

$$UO_2^{2+} + e^- \rightarrow UO_2^{+}\left(U(V)\right) + e^- \rightarrow U^{4+}\left(U(IV)\right) \tag{11.35}$$

Qiu et al. fabricated a $0.10BiVO_4/HCN$ heterojunction photocatalyst that achieved an impressive 98.5% reduction of U(VI) under visible light irradiation. The enhanced photocatalytic activity is primarily ascribed to the formation of an efficient heterojunction, which significantly enhances charge carrier separation and transfer, as confirmed by electrochemical impedance spectroscopy (EIS), transient photocurrent measurements, and photoluminescence (PL) analysis. Moreover, the conduction band positions of both $BiVO_4$ and HCN are more negative than the redox potentials of UO_2^{2+}/UO_2 (0.411 V) and UO_2^{2+}/U^{4+} (0.267 V vs. NHE), verifying the thermodynamic feasibility for photocatalytic U(VI) reduction (Qiu et al., 2025). Huang et al. synthesized two covalent organic frameworks (COFs), COF-AQ (anthraquinone-based) and COF-A (anthracene-based), for the photocatalytic removal of U(VI). COF-AQ exhibited a photoreaction rate constant 3.3 times higher than that of COF-A. This significant enhancement is attributed to the incorporation of the anthraquinone moiety, which not only increases the number of charge transfer pathways from two to four but also facilitates stronger adsorption of $[UO_2(H_2O)_2]^{2+}$ through hydrogen bonding interactions (Huang et al., 2025a). Chen et al. developed chitosan/polyacrylic (CS/PAA) hydrogel microspheres encapsulating $CdS/g\text{-}C_3N_4$ heterojunctions (CPCCN) integrated into a cation-exchange (CIEX) microreactor system for the efficient adsorption and photoreduction of U (VI). The microreactor is engineered to achieve dual enhancement in uranium extraction through synergistic mechanisms, including localized pH regulation, long-range electroosmotic flow, and light-driven reduction. Ion-exchange membranes within the reactor establish pH gradients that create spatially distinct acidic and basic microenvironments, which modulate U(VI) speciation and mobility. These conditions facilitate targeted transport of uranium ions toward the functionalized microsphere surfaces, thereby accelerating adsorption and enhancing binding efficiency. Additionally, the self-generated long-range flow disrupts adsorption–desorption equilibrium, ensuring continuous migration of U(VI) to active sites while conserving energy and minimizing environmental impact (Chen et al., 2025).

11.3.3 Photocatalytic Oxidation of As (III)

Arsenic contamination in water poses a serious environmental and public health threat, with arsenite [As (III)] being more toxic, mobile, and challenging to remove than arsenate [As (V)]. Major human-related sources of arsenic encompass mining activities, burning of fossil fuels, and the use of arsenic-based pesticides and wood preservatives (Sevak and Pushkar, 2024). In aqueous environments, As (III) predominantly exists in its neutral form as H_3AsO_3, which exhibits low affinity for conventional adsorbents (Liu and Qu, 2021). Within this framework, photocatalytic oxidation has gained recognition as a promising, sustainable, and alternative approach for arsenic removal (Paredes et al., 2023). Under exposure to UV or visible light, semiconductor photocatalysts produce electron–hole (e^-/h^+) pairs, which then generate ROS such as hydroxyl radicals ($^{\bullet}OH$) and superoxide anions ($^{\bullet}O_2^-$). These ROS, along with photogenerated holes, can oxidize As (III) into less harmful As (V) (Heiba et al., 2024). The oxidation mechanism involves both direct and indirect pathways: photogenerated holes (h^+) can directly oxidize As (III) adsorbed on the catalyst surface:

$$\mathrm{As(III)} + h^+ \rightarrow \mathrm{As(V)} \tag{11.36}$$

Simultaneously, water or hydroxide ions react with h^+ to form hydroxyl radicals ($^{\bullet}OH$):

$$h^+ + H_2O \rightarrow {}^{\bullet}OH + H^+ \tag{11.37}$$

$$h^+ + OH^- \rightarrow {}^{\bullet}OH \tag{11.38}$$

These $^{\bullet}OH$ species further promote the conversion of As (III) to the less toxic As (V), typically forming $H_2AsO_4^-$ or $HAsO_4^{2-}$, depending on the solution pH. This dual oxidative pathway underscores the potential of photocatalysis as an efficient and environmentally benign method for arsenic detoxification in water treatment applications (Yu et al., 2024).

Zhong et al. proposed a dual-functional photocatalytic approach utilizing bimetallic Fe/Ti-based metal–organic frameworks (MOFs) for the simultaneous redox transformation of As (III) under visible light irradiation. The photocatalyst demonstrated efficient charge carrier separation and intermetallic charge transfer, enabling concurrent oxidation of As (III) to As (V). In this system, Fe sites acted as electron mediators, promoting redox cycling, while Ti-oxo clusters served as active centers for light harvesting and ROS generation, particularly $^{\bullet}OH$ and $^{\bullet}O_2^-$ radicals. The As (III) oxidation process was predominantly governed by photogenerated holes and surface-bound hydroxyl radicals, effectively converting As (III) into the less toxic and more stable As (V) form. Additionally, the high surface area and porous framework of the MOF facilitated arsenic adsorption and activation, accelerating the redox conversion process (Zhong et al., 2025). In a separate study, Liu et al. developed a $C/TiO_2@ND\text{-}C_3N_4$

(CTN) heterojunction composite, achieving 95% As (III) removal with a kinetic rate constant of 0.27619 min^{-1}, approximately 6.5 and 17.0 times higher than that of individual C/TiO_2 and ND-C_3N_4 components, respectively, as shown in Figure 11.2(c). This superior photocatalytic activity was attributed to the formation of C–O–Ti interfacial bonds, which significantly enhanced charge separation and improved hole-driven oxidation. Furthermore, the heterojunction structure facilitated rapid adsorption of the oxidized As (V), synergistically boosting the overall arsenic removal efficiency (Liu et al., 2023).

11.3.4 Photocatalytic Reduction of Hg (II)

Mercury pollution, particularly in the form of divalent mercury ions [Hg (II)], represents a serious threat to the environment and public health due to its extreme toxicity, mobility, and tendency to bioaccumulate in aquatic food chains (Kayani and Mohammed, 2025). Major anthropogenic sources of Hg (II) include coal combustion, artisanal gold mining, chlor-alkali industries, and the dumping of mercury-containing waste (Xu et al., 2015). Once released into aquatic environments, Hg (II) can be converted into methylmercury (MeHg), a compound with pronounced neurotoxicity that accumulates in organisms and poses significant risks to both wildlife and humans (Regnell and Watras, 2019). Among various remediation techniques, photocatalytic reduction has emerged as an advanced and sustainable strategy for Hg (II) detoxification (Fu et al., 2020).

Mohanty et al. reported the synthesis of Au/SnS_2 quantum dots for the efficient photocatalytic reduction of Hg (II). The integration of gold nanoparticles into the SnS_2 matrix enhances localized surface plasmon resonance (LSPR), thereby improving light absorption and promoting effective charge separation. This plasmonic enhancement facilitates rapid electron transfer, significantly boosting photocatalytic performance. Under visible light irradiation, photogenerated electrons from SnS_2 reduce Hg^{2+} to elemental mercury (Hg^0), while photogenerated holes are efficiently scavenged by sacrificial agents, suppressing electron–hole recombination and improving redox efficiency (Mohanty and Kamali, 2024). In another study, Sharifzadeh et al. engineered defective Zr-based MOFs functionalized with $-NH_2$ and –SH groups to investigate their roles in Hg (II) capture and photoreduction. The incorporation of –SH functionalities provided strong thiol–Hg (II) coordination and facilitated rapid electron donation, thereby driving the reduction of Hg^{2+} to Hg^0, as evident from the XPS spectra provided in Figure 11.2(d). In contrast, the $-NH_2$ groups primarily enhanced Hg (II) adsorption via hydrogen bonding and weak electrostatic interactions but contributed less to the redox process. Additionally, structural defects in the MOF framework acted as electron-trapping centers, enhancing charge separation and promoting photocatalytic activity under light exposure (Zahra et al., 2025).

A comparative study of various photocatalysts evaluated under different experimental conditions for the degradation of heavy metal ions has been presented in Table 11.1.

TABLE 11.1
Removal Efficiency of Different Photocatalysts Under Optimum Conditions for the Elimination of Heavy Metal Ions

Sl. No.	Photocatalysts	Heavy Metal Ion	pH	Dose	Concentration	Time	Efficiency (%)	Ref
1	AgQDs/BQDs-SV-20/MFO	Cr (VI)		30 mg	20 mg/L	30 min	99.5	He et al., 2025
2	Cu-MOF	Cr (VI)	4	20 mg	10 mg/L	40 min	100	Zheng et al., 2024
3	0.10BiVO_4/HCN	U (VI)	6	25 mg	40 mg/L	60 min	98.5	Qiu et al., 2025
4	COF-AQ	U (VI)			200 mg/L	180 min	99.2	Huang et al., 2025a
5	CPCCN	U (VI)	5.5	2.5 mg	30 mg/L	120 min	100	Chen et al., 2025
6	NH2-MIL-88B(Fe0.6Ti0.4)	As (III)	2.0	20 mg	5 mg/L	360 min	83.54	Zhong et al., 2025
7	C/TiO_2@ND-C_3N_4	As (III)	7.0	20 mg	10 mg/L	24 min	95	Liu et al., 2023
8	Au/SnS_2 QDs	Hg (II)		0.5 mg/mL	10 mg/L	120 min	82.6	Mohanty and Kamali, 2024
9	Cys@UiO-66	Hg (II)	7.0	5 mg	500 mg/L	360 min	100	Zahra et al., 2025

11.4 PHOTOCATALYTIC DEGRADATION OF TOXIC ORGANIC COMPOUNDS

Toxic organic compounds such as dyes and antibiotics impart a significant adverse impact on the environment. Owing to their high stability and poor biodegradability, these tend to accumulate in ecosystems and enter the food chain. As a result, these cause ecological damage and chronic health issues [Vinayagam et al., 2023]. However, dyes are extensively applied in a wide range of industrial processes. For instance, dye-contaminated wastewater is accidentally released from various industries like textiles, leather, paper, plastic, pharmaceutical, food, etc., and contaminates the nearby natural aqueous water ecosystems (Gothwal and Shashidhar, 2015; Benkhaya et al., 2022; Abatan et al., 2014). It has been reported that about 700000 tons of dyes are annually produced worldwide, out of which 20% are released to the atmosphere through processing and manufacturing [Shetty et al., 2023]. These include xanthene dyes, anthraquinone dyes, indigoid dyes, triphenylmethane dyes, azo dyes, and phthalocyanine dyes. Similarly, antibiotics are released into water bodies due to their overuse and uncontrolled environmental discharge. Several kinds of antibiotics like quinolones, tetracycline, macrolides, sulfonamides, etc. are contributing to severe water pollution [Chen et al., 2020; Lu et al., 2020; de Ilurdoz et al., 2022]. Therefore, it is indispensable to remediate both these bioaccumulative and toxic pollutants from water ecosystems. They can be completely decomposed into harmless substances by semiconductor-mediated photocatalysis. The fundamental mechanism underlying the photocatalytic degradation of dyes and antibiotics and their subsequent removal through photocatalytic microreactor technologies was demonstrated in the following sections [Phoon et al., 2020; Sharma et al., 2023].

11.4.1 Basic Mechanism for Photoreduction of Antibiotics and Dyes

The degradation of antibiotics and dyes in semiconductor-mediated photocatalytic processes typically involves the generation of electron–hole pairs under light irradiation. These photoinduced charge carriers undergo desired redox reactions to produce ROS like hydroxyl ($^{\bullet}OH$), superoxide anions ($O_2^{\bullet -}$), etc., which effectively break down these organic pollutants into innocuous substances like CO_2 and water. Details of the photocatalytic pathway for the decomposition of organic pollutants were depicted in the following steps. These ROS attack the antibiotic or dye molecules, resulting in their oxidative degradation into a less toxic form or mineralized products such as CO_2, H_2O, and inorganic ions (Macedo et al., 2025; Liu et al., 2021a; Mishra et al., 2023).

11.4.1.1 Step I: Photon Absorption

When a semiconductor photocatalyst (SPC) is exposed to UV or visible light with energy equal to or more than its band gap energy (E_g), electrons are transferred to the CB of the semiconductor, whereas an equivalent number of holes are produced in the VB as shown in Equation 11.39 (Srivastava, 2024; Bai et al., 2022; Krivec et al., 2013).

$$SPC \xrightarrow{h\nu} eCB^{-} + hVB^{+} \tag{11.39}$$

11.4.1.2 Step II: Charge Carriers' Migration

These photo-induced electrons (e^-) and holes (h^+), acquiring the potentials of CB and VB, respectively, of the semiconductor, migrate to the surface to carry out surface redox reactions like oxygen reduction reaction (ORR) and water oxidation reaction (WOR) (Willis et al., 2023). However, they can recombine, releasing energy as heat or light, which reduces photocatalytic efficiency:

$$e_{CB}^- + h_{VB}^+ \rightarrow \text{heat / light}\left(\text{recombination}\right) \tag{11.40}$$

Strategies like band gap engineering, defect formation, and heterojunction construction are applied to minimize the recombination rate in order to accelerate the photocatalytic performance.

11.4.1.3 Step III: Generation of Reactive Oxygen Species

The photogenerated electrons (e^-) at the surface of the catalyst combine with dissolved oxygen molecules to undergo ORR for releasing $O_2^{\bullet-}$ as an ROS only when the CB potential becomes more negative than that of $O_2/O_2^{\bullet-}$ ($E^o_{O2/O2^{\cdot}}$ = –0.33 V vs. NHE at pH 7.0) redox couple. The ORR is shown in Equation 11.41.

$$O_2 + e^- \rightarrow O_2^{\bullet-}\left(E^o_{O2/O2} = -0.33\,\text{V vs.NHE at pH 7.0}\right) \tag{11.41}$$

Similarly, with more cathodic VB positions than the water oxidation potentials ($E^o_{H2O/.OH}$ = 2.29 V vs. NHE at pH 7.0 or $E^o_{OH^-/.OH}$ = 1.99 V vs. NHE at pH 7.0), holes undergo WOR to oxidize water or hydroxide ions to generate highly reactive hydroxyl radicals (•OH) as per the following equations.

$$H_2O + h_{VB}^+ \rightarrow \cdot OH + H^+\left(E^o_{H2O/.OH} = 2.29\,\text{V vs.NHE at pH 7.0}\right) \tag{11.42}$$

$$OH^- + h_{VB}^+ \rightarrow \cdot OH\left(E^o_{OH^-/.OH} = 1.99\,\text{V vs.NHE at pH 7.0}\right) \tag{11.43}$$

11.4.1.4 Step IV: Degradation of Organic Pollutants

The reactive species like hydroxyl radicals ((•OH), superoxide radicals ($\cdot O_2^-$) and holes (h^+) undergo oxidative degradation of dyes and antibiotics adsorbed on the catalyst surface through hydrogen abstraction, electron transfer, ring-opening reactions, and stepwise mineralization. Ultimately, these toxic organic compounds are decomposed into harmless molecules like CO_2 and H_2O. The degradation reaction can be presented as follows.

$$\text{Organic pollutant} + \cdot OH / h^+ / \cdot O_2^- \rightarrow CO_2 + H_2O \tag{11.44}$$

The plausible mechanism for photocatalytic pollutants degradation was presented in Figure 11.3.

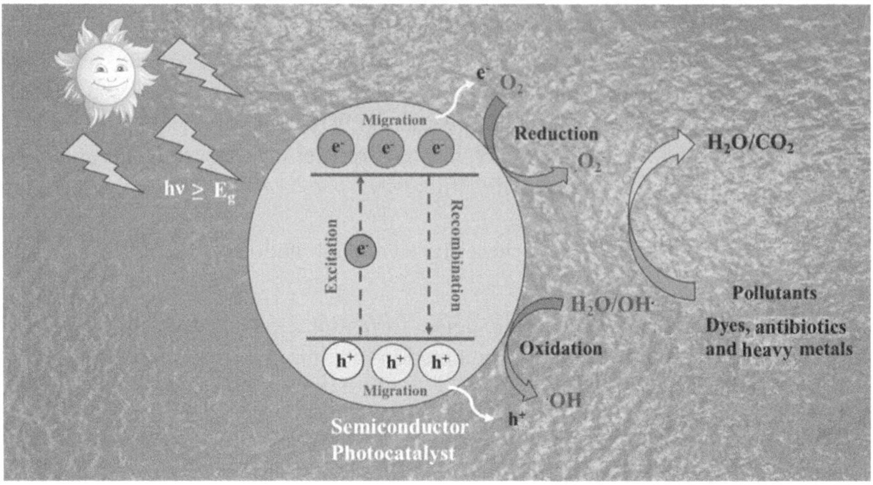

FIGURE 11.3 Schematic representations of semiconductor photocatalysis for the degradation of toxic organic compounds under solar light irradiation

11.4.2 Removal of Organic Dyes and Antibiotics through Photocatalytic Microwave Technologies

Numerous studies highlight TiO_2, ZnO, and other doped semiconductors used in microreactors as efficient photocatalysts in degrading antibiotics like tetracycline, sulfamethoxazole, and dyes such as methylene blue, rhodamine B, and methyl orange, showcasing the process's environmental significance (Singer et al., 2016; Russo, 2021). Among them, Li et al. developed an optofluidic microreactor incorporating TiO_2-coated fiberglass immersed in the microchannel. This design enhances mass transport by reducing diffusion length and inducing flow perturbations. When used for methylene blue degradation under UV irradiation, the reactor achieved a 2–3× higher reaction rate constant compared to conventional microreactors, and over 40% increase in degradation efficiency overall (Danner et al., 2019). Binjhade et al. designed a continuous-flow tubular microreactor using a TiO_2-coated quartz tube under UVC illumination to treat emerging pharmaceutical pollutants. The system degraded ciprofloxacin and metformin with ~92% efficiency individually and 85–90% efficiency in mixed solutions. After 9 h of operation, mineralization reached ~50% for ciprofloxacin and ~70% for metformin, with LC-MS/MS clarifying degradation pathways (Binjhade et al., 2025). Rabanimehr et al. developed a planar continuous-flow microreactor integrating chitosan nanofibers embedded with Bi_2WO_6/CNT/TiO_2 (15 wt. %) as a combined adsorptive–photocatalytic system. Under simulated sunlight, optimal removal of cephalexin was achieved as 99.2% and 92.4% TOC elimination at pH 4.8, retention time 256 s, light intensity 17.45 W/m^2, and initial cephalexin concentration of 29 mg/L (Ashrafi et al., 2024). Similarly, Jia et al. synthesized a type-II heterojunction CdS/ZnO (CSZ) nano dispersion using a one-pot precipitation method, aided by an anionic surfactant to enhance dispersion and photocatalytic activity. Under visible-light irradiation, the system degraded tetracycline (TC) with

94.4% efficiency at pH 7, showing reduced performance at pH values above or below neutral. The maximum degradation efficiency for Rhodamine B (RhB), methylene blue (MB), and methyl orange (MO) was found to be 99.6%, 95.9%, and 92.2%, respectively, within 120 min. The catalyst retained over 85% activity after 5 reuse cycles (Jia et al., 2022; Martin et al., 2023; Mahmoudi et al., 2024; Das et al., 2021; Latif et al., 2024).

Details on reaction conditions, operating parameters, and degradation efficiencies for photocatalytic degradation of various organic pollutants in microreactors are summarized in Table 11.2 (Akman and Aras, 2022; Agunbiade and Moodley, 2016; Roibu et al., 2025).

11.5 CONCLUSIONS AND FUTURE PROSPECTIVES

To summarize, AOPs combined with microreactor (MR) technology offer an outstanding approach to environmental remediation, particularly in the degradation of persistent organic compounds and heavy metal ions. Microreactors provide unrivalled benefits such as improved mass and photon transfer, precise control of reaction parameters, high surface-to-volume ratio, and compact design. When combined with photocatalysis, these systems make efficient use of light energy and catalyst surfaces, resulting in faster degradation kinetics and higher mineralization efficiency than traditional systems. Unlike conventional treatment methods, photocatalysis offers advantages such as low energy input, minimal secondary pollution, and the potential for complete detoxification. The synergism between the photocatalytic process and microfluidic reactors for the degradation of heavy metals (such as Cr, As, and U) and toxic organic compounds (dyes and antibiotics) was discussed using a variety of reactor designs that include single and multi-capillary and planar microreactors. More importantly, incorporating nanostructured photocatalysts such as TiO_2 into microreactors improves photocatalytic efficiency by enhancing charge separation and extending light absorption. The photocatalytic reaction mechanism has been thoroughly discussed with respect to organic pollutant and heavy metal ion degradation.

Despite tremendous advances, several difficulties and possibilities still exist for the broad deployment of photocatalytic AOPs in microreactors:

1. The creation of strong, visible light-active, novel photocatalytic materials with high quantum efficiency, stability, and recyclability should be the main emphasis of future research. Incorporation of recently discovered advanced functional materials (e.g., MOFs, COFs, Mxene) with modifications such as plasmonic enhancement, heterojunction creation, and defect engineering could be effective for achieving augmented activity.
2. Research and development are required in fluid dynamics, fabrication, and design strategies of the microreactor to fabricate reactors with optimal light distribution, enhanced residence time, and improved reactant dispersion for better degrading performance.
3. Theoretical modelling of the microfluidic processes and kinetic understanding of the chemical reactions taking place in a photocatalytic microreactor needs extensive research attention for further development.

TABLE 11.2
Photocatalytic Microreactor Technology for Removal of Toxic Organic Compounds.

Organic Pollutants	Photocatalytic System	pH	Catalyst Dose	Initial Pollutants Concentration (ppm)	Irradiation Time (min)	Efficiency	Ref.
Methylene blue (MB)	TiO_2 thin film coating/ TiO_2 NPs immobilized	~6–7	0.5–2 mg/cm^2	10–20 mg/L	10–20 min	100%	Danner et al., 2019
Ciprofloxacin and metformin	TiO_2 coated on quartz tube via modified dip-coating method	9	1 g/L	10 ppm	60 min	~92% for each drug individually; 85–90% in mixtures	Binjhade et al., 2025
Cephalexin (a β-lactam)	Chitosan/Bi_2WO_6/CNT/ TiO_2 nanofiber composite Plate-type planar microreactor	4.8	1.5 g/L	29 mg/L	120 min	~99.2%	Ashrafi et al., 2024
TC, RhB, MB and MO	CdS/ZnO nano system with surfactant aid	~7	0.5 g/L	10 ppm	120 min	TC ~ 94.4%; MB ~ 95.9% MO ~ 92.2% RhB ~ 100%	Jia et al., 2022
Ciprofloxacin	TiO_2 on quartz tube under UVC Continuous-flow tubular microreactor	9	0.7 g/L	5 ppm	60 min	~92%	Martin et al., 2023

Oxytetracycline	g-C_3N_4/Fe_3O_4	9	~ 0.7 g/L	30 mg/L	~60–80 min	~98%	Mahmoudi et al., 2024
RhB, Ciprofloxacin	$LaFeO_3$–MoS_2 nanocomposite	7	0.75 g/L	10 ppm	150 min	RhB ~96% Ciprofloxacin ~83%	Das et al., 2021
Tetracycline	$NiFe_2O_4$/CeO_2/GO nanocomposite	7	1 g/L	20 mg/L	90 min	~95%	Latif et al., 2024
Tetracycline (TC), Rhodamine B (RhB)	CuO/Fe_2O_3 nanocomposite	7	0.5 mg/L	10 ppm	40–80 min	TC ~88%, RhB ~96%	Akman and Aras, 2022
Ciprofloxacin	Zn-doped Cu_2O particles	7	0.75 mg/L	20 ppm	120 min	~94.6%	Agunbiade and Moodley, 2016
AP (acetaminophen)	TiO_2	9	0.1 g/L	50 mg/L	60 min	94%	Roibu et al., 2025
Sulfonamides	TiO_2 or ZnO based heterojunctions	9	0.5 g/L	10 mg/L	60 min	~90%	Srivastava, 2024
BPA	N-doped TiO_2	10	0.5 g/L	5 ppm	360 min	100%	Bai et al., 2022
Norfloxacin	TiO_2 or ZnO based heterojunctions	6	1.7 mg/L	30 ppm	90 min	~90.6%	Krivec et al., 2013
Methylene Blue dye	GO-SnO_2 nanocomposite	9	7.5 mg/L	16 ppm	60 min	~89%	Macedo et al., 2025
Ibuprofen	S-doped BiOBr	6	1.7 mg/L	30 ppm	45–90 min	~95.3%	Liu et al., 2021a

4. Combining photocatalysis with other AOPs (such as Fenton, ozonation, and electrochemical oxidation) in microreactors may result in synergistic effects that might increase the degradation rates and widen the range of degradable pollutants.
5. The current research focuses on model contaminants. Future research should verify microreactor performance in intricate, real-world matrices while taking competing species and catalyst deactivation into account.

The combination of microreactor engineering and photocatalytic research holds enormous promise for next-generation water treatment systems. Continued interdisciplinary collaboration will be indispensable to realize their full potential in tackling global environmental concerns.

ACKNOWLEDGMENT

The management of SOA deemed to be University, Bhubaneswar, is acknowledged with gratitude by the authors for their assistance in finishing the present investigation.

REFERENCES

Abatan, A., Obaigbena, A., and Ugwuanyi, E.D. et al. 2014. Metal oxide-based photocatalysts for the efficient degradation of organic pollutants. *J. Eng. Sci. Technol.* 5:543–554.

Acharya, R., Lenka, A., and Parida, K. 2021. Magnetite modified amino group based polymer nanocomposites towards efficient adsorptive detoxification of aqueous Cr (VI): A review. *J. Mol. Liq.* 337:116487.

Acharya, R., Martha, S., and Parida, K.M. 2017. Remediation of Cr (VI) using clay minerals, biomasses and industrial wastes as adsorbents. In *Advanced Materials for Waste Water Treatment*, ed. Shahid-ul-Islam, 129–170., John Wiley & Sons, Inc.

Acharya, R., Mishra, S., Acharya, L., and Parida, K. 2023. Graphitic carbon nitride (g-C_3N_4)-based photocatalysts for environmental applications. In *Two-Dimensional Materials for Environmental Applications*, ed. N. Kumar, R. Gusain, and S. Sinha Ray, 103–136. Springer International Publishing.

Acharya, R., Mishra, S., and Naik, A. 2024. Development of graphene-based photocatalysts for remediation of hexavalent chromium. In *Graphene-Based Photocatalysts. Advanced Structured Materials*, ed. M.R. Johan, M.N. Naseer, M. Ikram, A.A. Zaidi, Y. Abdul Wahab, vol. 217. Springer, Cham.

Acharya, R., and Naik, A. 2024. Adsorption of aqueous Cr (VI) onto UiO66NH_2 MOF: Isotherm, thermodynamics and mechanism studies. *Mater. Today Proc.* https://doi.org/10.1016/J.MATPR.2024.01.003

Acharya, R., Naik, B., and Parida, K.M. 2018. Adsorption of Cr (VI) and textile dyes on to mesoporous silica, titanate nanotubes, and layered double hydroxides. In *Nanomaterials in the Wet Processing of Textiles*, ed. Shahid ul-Islam, B.S. Butola, 219–260. John Wiley & Sons, Inc.

Acharya, R., and Parida, K. 2020a. A review on TiO_2/g-C_3N_4 visible-light-responsive photocatalysts for sustainable energy generation and environmental remediation. *J Environ Chem Eng.* 8:103896.

Acharya, R., and Parida, K. 2020b. A review on adsorptive remediation of Cr (VI) by magnetic iron oxides and their modified forms. *Biointerface Res. Appl. Chem.* 10:5266–5272.

Acharya, R., and Parida, K. 2022a. Valorization of agricultural wastes as low-cost adsorbents towards efficient removal of aqueous Cr(VI). In *Handbook of Biomass Valorization for Industrial Applications*, ed. Shahid-ul-Islam, Aabid Hussain Shalla, Salman Ahmad Khan, 507–530. John Wiley & Sons, Inc..

Acharya, R., Pati, S., and Parida, K. 2022b. A review on visible light driven spinel ferrite-g-C_3N_4 photocatalytic systems with enhanced solar light utilization. *J. Mol. Liq.* 357:119105.

Acharya, R., Sillanpää, M., and Al-Farraj, S. 2025. A comprehensive review on sequestration of aqueous Cr (VI) over graphene based adsorbents. *J. Hazard. Mater. Adv.* 18:100670.

Agunbiade, F.O., and Moodley, B. 2016. Occurrence and distribution pattern of acidic pharmaceuticals in the Msunduzi River, South Africa. *ET&C and Environ. Toxicol. Chem.* 35:36–46.

Ahuja, T., Brighu, U., and Saxena, K. 2023. Recent advances in photocatalytic materials and their applications for treatment of wastewater: A review. *J. Water Process Eng.* 53:103759.

Akman, B., and Aras, O. 2022. Usability, durability and regeneration of Ag/ZnO coated microreactor for photocatalytic degradation of methylene blue. *J. Mol. Struct.* 1251:132003.

Ashrafi, M., Farhadian, M., Nazar, A.R.S., Hajiali, M., and Noorbaksh, A. 2024. The tetracycline degradation in a photocatalytic fuel cell microreactor using a ZnO nanorod/ Bi_2MoO_6/ZIF-67 photocatalyst responsive to visible light. *Energy Convers. Manag.* 312:118565.

Bai, X., Chen, W., Wang, B., Sun, T., Wu, B., and Wang Y. 2022. Photocatalytic Degradation of Some Typical Antibiotics: Recent Advances and Future Outlooks. *Int. J. Mol. Sci.* 23:8130.

Balapure, A., Dutta, J.R., and Ganesan, R. 2024. Recent advances in semiconductor heterojunctions: A detailed review of the fundamentals of photocatalysis, charge transfer mechanism and materials. *RSC Appl. Interfaces.* 1:43–69.

Benkhaya, S., M'rabet, S., Lgaz, H., El Bachiri, A., and El Harfi, A. 2022. Dyes: Classification, pollution, and environmental effects. In *Dye Biodegradation, Mechanisms and Techniques*, ed S.S. Muthu, A. Khadir, Sustainable Textiles: Production, Processing, Manufacturing & Chemistry, 1–50.

Binjhade, R., Mondal, R., and Mondal, S. 2025. Photocatalytic degradation of ciprofloxacin and metformin in a continuous-flow tubular reactor. *React. Chem. Eng.* 10:1577–1586.

Buffle, M.-O., Schumacher, J., Meylan, S., Jekel, M., and von Gunten, U. 2006. Ozonation and advanced oxidation of wastewater: Effect of O_3 dose, pH, DOM and HO-scavengers on ozone decomposition and HO generation. *Ozone: Sci. Eng.* 28:247–259.

Burakova, A.E., Galunina, E.V., and Burakovaa, I.V. et al. 2018. Adsorption of heavy metals on conventional and nanostructured materials for wastewater treatment purposes: A review. *Ecotoxicol. Environ. Saf.* 148:702–712.

Chen, D., Cheng, Y., Zhou, N. et al. 2020. Photocatalytic degradation of organic pollutants using TiO_2-based photocatalysts: A review. *J. Cleaner Prod.* 268:121725.

Chen, L., Fu, L., Feng, K. et al. 2025. Ion exchange-based microreactors harness self-generated pH gradients and long-range flow for dual-enhanced uranium extraction. *Adv. Funct. Mater.* e09597.

Cuerda-Correa, E.M., Alexandre-Franco, M.F., and Fernández-González, C. 2020. Advanced oxidation processes for the removal of antibiotics from water. An overview. *Water* 12:102.

Danner, M.C., Robertson, A., Behrends, V., and Reiss, J. 2019. Antibiotic pollution in marine environments: A review. *Sci. Total Environ.* 664:793–804.

Das, S., Dutta, S., Mistry Tama, A., and Basith, M.A. 2021. Nanostructured $LaFeO_3$-MoS_2 for efficient photodegradation and photocatalytic hydrogen evolution. *Mater. Sci. Eng. B* 271:115295.

de Ilurdoz, M.S., Sadhwani, J.J., and Reboso, J.V. 2022. Antibiotic removal processes from water & wastewater for the protection of the aquatic environment—A review. *Water Process Eng.* 45:102474.

Deng, X., Wu, Z., Yin, B., Lei, J., Bian, Z., and Li, H. 2025. Bifunctional 2D K-SnS for one-pot photocatalytic reduction of Cr(VI) and in situ ion-exchange recycle of product Cr(III). *ACS EST Engg.* 5:1430–1440.

Dhamorikar, R.S., Lade, V.G., Kewalramani, P.V., and Bindwal, A.B. 2024. Review on integrated advanced oxidation processes for water and wastewater treatment. *J. Ind. Eng. Chem.* 138:104–122.

Dong, G., Yang, M., and Chen, B. et al. 2022. Advanced oxidation processes in microreactors for water and wastewater treatment: Development, challenges, and opportunities. *Water Res.* 211:118047.

Dutta, D., Arya, S., and Kumar, S. 2021. Industrial wastewater treatment: Current trends, bottlenecks, and best practices. *Chemosphere* 285:131245.

Fu, K., Liu, X., Yu, D., Luo, J., Wang, Z., and Crittenden, J.C. 2020. Highly efficient and selective Hg(II) removal from water using multilayered $Ti_3C_2O_x$ MXene via adsorption coupled with catalytic reduction mechanism. *Environ. Sci. Technol.* 54:16212–16220.

Gao, M., Zeng, Z., Sun, B., Zou, H., Chen, J., and Shao, L. 2012. Ozonation of azo dye Acid Red 14 in a microporous tube-in-tube microchannel reactor: Decolorization and mechanism. *Chemosphere* 89:190–197.

Glaze, W.H., Kang, J.-W., and Chapin, D.H. 1987. The chemistry of water treatment processes involving ozone, hydrogen peroxide and ultraviolet radiation. *Ozone Sci. Eng.* 9:335–352.

Gothwal, R., and Shashidhar, T. 2015. Antibiotic pollution in the environment: A review. *Clean: Soil Air Water* 43:479–489.

Gouran, A., Mohadesi, M., and Falsafi, M. 2022. Removal of dyes from aqueous solution using ultraviolet/hydrogen peroxide/titanium dioxide process in a rotating microreactor. *Int. J. Environ. Sci. Technol.* 19:5421–5436.

Gulfam-E-Jannat, S., Golui, D., and Islam, S. et al. 2023. Industrial water demand and wastewater generation: Challenges for Bangladesh's water industry. *ACS ES&T Water* 3:1515–1526.

Guo, K., Wu, Z., Chen, C., and Fang, J. 2022. UV/Chlorine process: An efficient advanced oxidation process with multiple radicals and functions in water treatment. *Acc. Chem. Res.* 55:286–297.

He, J., Liu, L., and Fu, N. et al. 2025. Construction of sulfur-vacancy-modified 0D/ 2D S-scheme heterojunction for enhancing photocatalytic Cr(VI) reduction. *ACS Materials Lett.* 7:566–575.

He, J., Yang, X.F., Men, B., and Wang, D.S. 2016. Interfacial mechanisms of heterogeneous Fenton reactions catalyzed by iron-based materials: A review. *J. Environ. Sci.* 39:97–109.

Heiba, H.F., Bullen, J.C., Kafizas, A., Petit, C., Jiang, D., and Weiss, D.J. 2024. Role of the Sn-TiO_2/Ti-SnO_2 heterojunction in enhancing the photocatalytic oxidation of arsenite (AsIII) through the promotion of charge carrier lifetime. *ACS Appl. Mater. Interfaces.* 16:69239–69252.

Huang, Q., Zhang, H., Yang, H. et al. 2025a. Site engineering of covalent organic frameworks to increase charge transfer channels and provide hydrogen bond toward enhanced photocatalytic reduction of U(VI). *Appl. Catal. B: Environ. and Eng.* 362:124721.

Huang, Z., Wang, Y., Ye, M., and Cheng, H. 2025b. Fabrication of a TiO_2 inverse opal film-modified photocatalytic microreactor for highly efficient degradation of dyes. *Langmuir* 41:8138–8143.

Hübner, U., Zucker, I., and Jekel, M. 2015. Options and limitations of hydrogen peroxide addition to enhance radical formation during ozonation of secondary effluents. *J. water reuse desalin.* 5:8–16.

Ike, I.A., Linden, K.G., Orbell, J.D., and Duke, M. 2018. Critical review of the science and sustainability of persulphate advanced oxidation processes. *Chem. Eng. J.* 338: 651–669.

Iyer, M., Anand, U., and Thiruvenkataswamy, S., et al. 2023. A review of chromium (Cr) epigenetic toxicity and health hazards. *Sci. Total Environ.* 882:163483.

Jayakumar, V., Govindaradjane, S., Rajamohan, N., and Rajasimman, M. 2022. Biosorption potential of brown algae, Sargassum polycystum, for the removal of toxic metals, cadmium and zinc. Environ. *Sci. Pollut. Control Ser.* 29:41909–41922.

Jia, K., Liu, G., Lang, D.N. et al. 2022. Fast photodegradation of antibiotics and dyes by an anionic surfactant-aided CdS/ZnO nanodispersion. *New J. Chem.* 46:11303–11314.

Kang, J., Choi, J., Lee, D.,and Son, Y. 2024. UV/persulfate processes for the removal of total organic carbon from coagulation-treated industrial wastewaters. *Chemosphere* 346:140609.

Kasprzyk-Hordern, B., Ziółek, M., and Nawrocki, J. 2003. Catalytic ozonation and methods of enhancing molecular ozone reactions in water treatment. *Appl. Catal. B Environ.* 46:639–669.

Katsoyiannis, I.A., Canonica, S., and von Gunten, U. 2011. Efficiency and energy requirements for the transformation of organic micropollutants by ozone, O_3/ H_2O_2 and UV/ H_2O_2. *Water Res.* 45:3811–3822.

Kayani, K.F., and Mohammed, S.J. 2025. Mercury in aquatic environments: Toxicity and advances in remediation using covalent organic frameworks. *Mater. Adv.* 6:3371.

Keen, O.S., Love, N.G., Aga, D.S., and Linden, K.G. 2016. Biodegradability of iopromide products after UV/H_2O_2 advanced oxidation. *Chemosphere* 144:989–994.

Khader, E.H., Muslim, S.A., and Mohammed, T.J., et al. 2024. Recent advances in photocatalytic advanced oxidation processes for organic compound degradation: A review. *Desalination Water Treat.* 318:100384.

Khan, S., Noor, T., Iqbal, N., and Yaqoob, L. 2024. Photocatalytic dye degradation from textile wastewater. *ACS Omega* 9:21751.

Khongthon, W., Jovanovic, G., Yokochi, A., Sangvanich, P., and Pavarajarn, V. 2016. Degradation of diuron via an electrochemical advanced oxidation process in a microscale-based reactor. *Chem. Eng. J.* 292:298–307.

Kishimoto, N. 2019. State of the art of UV/chlorine advanced oxidation processes: Their mechanism, byproducts formation, process variation, and applications. *J. Water Environ. Technol.* 17:302–335.

Krivec, M., Žagar, K., Suhadolnik, L., Čeh M., and Drazic, G. 2013. Highly efficient TiO_2-based microreactor for photocatalytic applications. *ACS Appl. Mater. Interfaces* 5:9088–9094.

Latif, M., Nawaz, R., and Hammad Aziz, M. et al. 2024. Evaluation of tetracycline photocatalytic degradation using $NiFe_2O_4$/CeO_2/GO nanocomposite for environmental remediation: In silico molecular docking, Antibacterial performance, degradation pathways, and DFT calculations. *Sep. Purif. Technol.* 351:128074.

Liang, J., Huang, X., and Yan, J. et al. 2021. A review of the formation of Cr(VI) via Cr(III) oxidation in soils and groundwater. *Sci. Total Environ.* 774:145762.

Lindstrom, H., Wootton, R., and Iles, A. 2007. High surface area titania photocatalytic microfluidic reactors. *AIChE J.* 53:695–702.

Liu, H., Yu, Z., and Xiao, M. et al. 2023. Synergistic photocatalytic oxidation and adsorption boost arsenic removal by in-situ carbon-doped TiO_2 and nitrogen deficiency C_3N_4 heterojunction. *J. Environ. Manag.* 348:119218.

Liu, R., and Qu, J. 2021. Review on heterogeneous oxidation and adsorption for arsenic removal from drinking water. *J. Environ. Sci.* 110:178–188.

Liu, Y., Hu, Z., and Yu, J.C. 2021a. Photocatalytic degradation of ibuprofen on S-doped BiOBr. *Chemosphere* 268:130376.

Liu, Y., Wu, S., Liu, J., Xie, S., and Liu, Y. 2021b. Synthesis of g-C_3N_4/TiO_2 nanostructures for enhanced photocatalytic reduction of U(VI) in water. *RSC Adv.* 11:4810–4817.

Lu, H., Li, Q., and Feng, W. 2022. Application progress of O_3/UV advanced oxidation technology in the treatment of organic pollutants in water. *Sustainability* 14:1556.

Lu, Z.-Y., Ma, Y.-L., Zhang, J.-T., Fan, N.-S., Huang, B.-C., and Jin, R.-C., J. 2020. A critical review of antibiotic removal strategies: Performance and mechanisms. *Water Process Eng.* 38:101681.

Ma, M., Wang, R., Xu, L., Xu, M., and Liu, S. 2020. Emerging health risks and underlying toxicological mechanisms of uranium contamination: Lessons from the past two decades. *Environ. Int.* 145:106107.

Macedo, H.E., Lehner, B., Nicell, J.A., Khan, U., and Klein, E.Y. 2025. Antibiotics in the global river system arising from human consumption. *PNAS Nexus* 4(4):pgaf096.

Mahmoudi, K., Farzadkia, M., Rezaei Kalantary, R., Reza Sobhi, H., Yeganeh, M., and Esrafili, A. 2024. Efficient removal of oxytetracycline antibiotics from aqueous media using UV/g-C_3N_4/Fe_3O_4 photocatalytic process. *Heliyon* 10:e30604.

Mansingh, S., Acharya, R., Martha, S., and Parida, K.M. 2018. Pyrochlore $Ce_2Zr_2O_7$ decorated over rGO: A photocatalyst that proves to be efficient towards the reduction of 4-nitrophenol and degradation of ciprofloxacin under visible light. *Phys. Chem. Chem. Phys.* 20:9872–9885.

Martin, M.V., Eusebi, A.L., Manassero, A., Berli, C.L.A., and Satuf, M.L. 2023. Degradation of binary mixtures of water pollutants in a photocatalytic microreactor: Competitive kinetics with explicit radiation absorption effects. *J. Environ. Chem. Eng.* 11:110929.

Merenyi, G., Lind, J., Naumov, S., and von Sonntag, C. 2010. Reaction of ozone with hydrogen peroxide (peroxone process): A revision of current mechanistic concepts based on thermokinetic and quantum-chemical considerations. *Environ. Sci. Technol.* 44:3505–3507.

Mierzwa, J.C., Rodrigues, R., and Teixeira, A.C. 2018. UV-hydrogen peroxide processes. In *Advanced Oxidation Processes for Waste Water Treatment*, ed. S.C. Ameta & R. Ameta, 13–48. Academic Press, Elsevier.

Miklos, D.B., Remy, C., Jekel, M., Linden, K.G., Drewes, J.E., and Hübner, U. 2018. Evaluation of advanced oxidation processes for water and wastewater treatment–A critical review. *Water Res.* 139:118–131.

Mishra, S., Acharya, L., Marandi, B., Sanjay, K., and Acharya, R. 2024a. Boosted photocatalytic accomplishment of 3D/2D hierarchical structured $Bi_4O_5I_2$/g-C_3N_4 p-n type direct Z-scheme heterojunction towards synchronous elimination of Cr(VI) and tetracycline. *Diam. Relat. Mater.* 142:110834.

Mishra, S., Acharya, L., Sharmila, S., Sanjay, K., and Acharya, R. 2024b. Designing g-C_3N_4/$NiFe_2O_4$ S-scheme heterojunctions for efficient photocatalytic degradation of Rhodamine B and tetracycline hydrochloride. *Appl. Surf. Sci. Adv.* 24:100647.

Mishra, S., and Acharya, R. 2021. Photocatalytic applications of graphene based semiconductor composites: A review. *Materials Today: Proceedings* 35:164–169.

Mishra, S., and Acharya, R. 2023. Recent updates in modification strategies for escalated performance of Graphene/MFe_2O_4 heterostructured photocatalysts towards energy and environmental applications. *J. Alloys Compd.* 960:170576.

Mishra, S., Acharya, R., and Parida, K. 2023. Spinel-ferrite-decorated graphene-based nanocomposites for enhanced photocatalytic detoxification of organic dyes in aqueous medium: A review. *Water* 15:81.

Mishra, S., Marandi, B., Sanjay, K., and Acharya, R. 2025b. Graphitic carbon nitride loaded $Bi_4O_5I_2$ for elevated photocatalytic tetracycline degradation. *J. Met. Mater. Miner.* 35:2261.

Mishra, S., Tudu, B.L., Mishra, N., Sanjay K., and Acharya, R. 2025a. Fabrication of oxygen vacancy modified 2D–2D g-C_3N_4/$ZnFe_2O_4$ heterostructures for amplifying photocatalytic methyl orange degradation and hydrogen production. *Mater. Adv.*, Advance Article. https://doi.org/10.1039/D5MA00533G.

Mohanty, A., and Kamali, K. 2024. Au/SnS_2 hybrid quantum dots for surface-enhanced raman scattering-based monitoring and photoreduction of Hg (II) ions. *ACS Appl. Nano Mater.* 7:3326–3338.

Monteil, H., Echaud, Y.P., Oturan, N., and Oturan, M.A. 2019. A review on efficiency and cost effectiveness of electro- and bio-electro-Fenton processes: Application to the treatment of pharmaceutical pollutants in water. *Chem. Eng. J.* 376:119577.

Paredes, M.Y., Martinez, L.P., and Barja, B.C. et al. 2023. Efficient method of arsenic removal from water based on photocatalytic oxidation by a plasmonic–magnetic nanosystem. *Environ. Sci.: Nano.* 10:166–177.

Phoon, B.L., Ong, C.C., Mohamed Saheed, M.S., et al. 2020. Conventional and emerging technologies for removal of antibiotics from wastewater. *J. Hazard. Mater.* 400:122961.

Pisarenko, A.N., Stanford, B.D., Yan, D., Gerrity, D., and Snyder, S.A. 2012. Effects of ozone and ozone/peroxide on trace organic contaminants and NDMA in drinking water and water reuse applications. *Water Res.* 46:316–326.

Prasad, S., Yadav, K.K., and Kumar, S. et al. 2021. Chromium contamination and effect on environmental health and its remediation: A sustainable approaches. *J. Environ. Manag.* 285:112174.

Qiu, J., Zhu, Y., Wang, B. et al. 2025. Enhanced removal of aqueous uranium: Photocatalytic U(VI) reduction over bismuth vanadate/hydrothermal carbon nitride under visible light. *J. Environ. Chem. Eng.* 13:115009.

Rahimi, M., Aghel, B., Sadeghi, M., and Ahmadi, M. 2014. Using Y-shaped microreactor for continuous decolorization of an Azo dye. *Desalination water treat.* 52:5513–5519.

Ray, S., and Vashishth, R. 2024. From water to plate: Reviewing the bioaccumulation of heavy metals in fish and unraveling human health risks in the food chain. *Emerg. Contam.* 10:100358.

Razzak, S.A., Faruque, M.O., and Alsheikh, Z. et al. 2022. A comprehensive review on conventional and biological-driven heavy metals removal from industrial wastewater. *Environ. Adv.* 7:100168.

Regnell, O., and Watras, C.J. 2019. Microbial mercury methylation in aquatic environments: A critical review of published field and laboratory studies. *Environ. Sci. Technol.* 53:4–19.

Reisz, E., Schmidt, W., Schuchmann, H.P., and Von Sonntag, C. 2003. Photolysis of ozone in aqueous solutions in the presence of tertiary butanol. *Environ. Sci. Technol.* 37:1941–1948.

Rekhate, C.V., and Srivastava, J.K. 2020. Recent advances in ozone-based advanced oxidation processes for treatment of wastewater- A review. *Chem. Eng. J. Adv.* 3:100031.

Roibu, A., Udroiu, R., Dinu, A., and Andronic, L. 2025. Microfluidic platform for screening the activity of immobilized photocatalysts for degradation of water pollutants in flow. *RSC React. Chem. Eng.* 10:2345–2355.

Rueda-Marquez, J.J., Levchuk, I., Ibañez, P.F., and Sillanpää, M. 2020. A critical review on application of photocatalysis for toxicity reduction of real wastewaters. *J Clean Prod.* 258:120694.

Russo, D. 2021. Kinetic modeling of advanced oxidation processes using microreactors: Challenges and opportunities for scale-up. *Appl. Sci.* 11:1042.

Russo, D., Cochran, K.H., and Westerman, D. et al. 2020. Ultrafast photodegradation of isoxazole and isothiazolinones by UV254 and UV254/H_2O_2 photolysis in a microcapillary reactor. *Water Res.* 169:115203.

Russo, D., Spasiano, D., and Vaccaro, M. et al. 2016. Investigation on the removal of the major cocaine metabolite (benzoylecgonine) in water matrices by UV254/H_2O_2 process by using a flow microcapillary film array photoreactor as an efficient experimental tool. *Water Res.* 89:375–383.

Seo, E., Park, J., and Cho, M.S. et al. 2025. Water purification of organic pollutants using photocatalytic microreactors integrating Au/TiO_2/carbon cloth under solar irradiation. *Chem. Eng. J. Adv.* 24:100854.

Serpone, N., and Emeline, A.V. 2012. Semiconductor photocatalysis past, present, and future outlook. *J. Phys. Chem. Lett.* 3:673–677.

Sevak, P., and Pushkar, B. 2024. Arsenic pollution cycle, toxicity and sustainable remediation technologies: A comprehensive review and bibliometric analysis. *J. Environ. Manag.* 349:119504.

Shamshad, J., and Ur Rehman, R. 2025. Innovative approaches to sustainable wastewater treatment: A comprehensive exploration of conventional and emerging technologies. *Env. sci., Adv.* 4:189–222.

Shan, X., Yang, X., Liu, F. et al. 2025. Photoreduction and extraction of U(VI) from mining wastewater using magnetically recoverable $ZnIn_2S_4$-based Z-scheme photocatalysts. *Chem. Eng. J.* 517:164483.

Sharma, M., Rajput, D., and Kumar, V. et al. 2023. Photocatalytic degradation of four emerging antibiotic contaminants and toxicity assessment in wastewater: A comprehensive study. *Environ. Res.* 231:116132.

Sharma, N., Dey, A.K., and Sathe, A.K. et al. 2020. Highly efficient visible-light-driven reduction of Cr(VI) from water by porphyrin-based metal–organic frameworks: Effect of band gap engineering on the photocatalytic activity. *Catal. Sci. Technol.* 10:7724–7733.

Shetty, S.S., Sonkusare, S., Naik, P.B., and Madhyastha, H. 2023. Environmental pollutants and their effects on human health. *Heliyon* 9:E19496.

Shinozawa, Y., Heggo, D., Ookawara, S., and Yoshikawa, S. 2020. Photo-fenton degradation of carbofuran in helical tube microreactor and kinetic modeling. *Ind. Eng. Chem. Res.* 59:3811–3819.

Singer, A.C., Shaw, H., Rhodes, V., and Hart, A. 2016. Review of antimicrobial resistance in the environment and its relevance to environmental regulators. *Front. Microbiol.* 7:1728.

Sonntag, C.v. 2008. Advanced oxidation processes: Mechanistic aspects. *Water Sci. Technol.* 58:1015–1021.

Srivastava, S.K. 2024. Recent advances in removal of pharmaceutical pollutants in wastewater using metal oxides and carbonaceous materials as photocatalysts: A review. *RSC Appl. Interfaces* 1:340–429.

Stern, C.M., Hayes, D.W., Kgoadi, L.O., and Elgrishi, N. 2020. Emerging investigator series: Carbon electrodes are effective for the detection and reduction of hexavalent chromium in water. *Environ. Sci.: Water Res. Technol.* 6:1256–1261.

Sun, J., Kong, D., and Aghdam, E. et al. 2019. The influence of the UV/chlorine advanced oxidation of natural organic matter for micropollutant degradation on the formation of DBPs and toxicity during post-chlorination. *Chem. Eng. J.* 373:870–879.

Sun, Q., Wang, Y., Li, L., Bing, J., Wang, Y., and Yan, H. 2015. Mechanism for enhanced degradation of clofibric acid in aqueous by catalytic ozonation over MnOx/SBA-15. *J. Hazard. Mater.* 286:276–284.

Tanimu, A., Jaenicke, S., and Alhooshani, K. 2017. Heterogeneous catalysis in continuous flow microreactors: A review of methods and applications. *Chem. Eng. J.* 327:792–821.

Tripathy, S.P., Subudhi, S., Acharya, R., Acharya, R., Das, M., and Parida, K. 2019. Adsorptive removal of Cr(VI) onto UiO-66-NH_2 and its determination by radioanalytical techniques. *J. Radioanal. Nucl. Chem.* 322:983–992.

Vega, D., Gupta, Y., Soler, L., Llorca, J., and Rodríguez, A. 2025. Novel silicon oxide microtube microstructures for photocatalytic reactions: Application to hydrogen photoproduction. *Chem. Eng. J. Adv.* 23:100827.

Verma, S., and Sillanpää, M. 2015. Degradation of anatoxin-a by UV-C LED and UV-C LED/H_2O_2 advanced oxidation processes. *Chem. Eng. J.* 274:274–281.

Vinayagam, V., Palani, K.N., and Ganesh, S. et al. 2023. Recent developments on advanced oxidation processes for degradation of pollutants from wastewater with focus on antibiotics and organic dyes. *Environ. Res.* 240:117500.

Wang, B., Zhang, H., Wang, F., et al. 2019. Application of heterogeneous catalytic ozonation for refractory organics in wastewater. *Catalysts* 9:241.

Wang, D., Bolton, J.R., and Hofmann, R. 2012. Medium pressure UV com bined with chlorine advanced oxidation for trichloroethylene destruction in a model water. *Water Res.* 46:4677–4688.

Wang, J.L., and Bai, Z.Y. 2017. Fe-based catalysts for heterogeneous catalytic ozonation of emerging contaminants in water and wastewater. *Chem. Eng. J.* 312:79–98.

Wang, J.L., and Wang, S.Z. 2018. Activation of persulfate (PS) and peroxymonosulfate (PMS) and application for the degradation of emerging contaminants. *Chem. Eng. J.* 334:1502–1517.

Wang, J.L., and Xu, L.J. 2012. Advanced oxidation processes for wastewater treatment: Formation of hydroxyl radical and application. *Crit. Rev. Environ. Sci. Technol.* 42:251–325.

Wang, N.N., Zheng, T., Zhang, G.S., and Wang, P. 2016. A review on Fenton-like processes for organic wastewater treatment. *J. Environ. Chem. Eng.* 4:762–787.

Wear, S.L., Acuña, V., McDonald, R., and Font, C. 2021. Sewage pollution, declining ecosystem health, and cross-sector collaboration. *Biological Conservation* 255:109010.

Willis, D.E., Sheets, C.E., and Worbington, M.R. et al. 2023. Efficient chemical-free degradation of waterborne micropollutants with an immobilized dual-porous TiO_2 photocatalyst. *ACS EST.* 3:1694–1705.

Wu, S.S., Hou, W.C., and Wang, D.K. 2020. Photocatalytic reduction of Cr(VI) by graphene oxide materials under sunlight or visible light: The effects of low-molecular-weight chemicals. *Environ. Sci.: Nano.* 7:2399–2409.

Xiao, Y., Zhang, L., Zhang, W., Lim, K.-Y., Webster, R.D., and Lim, T.-T. 2016. Comparative evaluation of iodoacids removal by UV/persulfate and UV/H_2O_2 processes. *Water Res.* 102:629–639.

Xu, J., Bravo, A.G., Lagerkvist, A., Bertilsson, S., Sjöblom, R., and Kumpiene, J. 2015. Sources and remediation techniques for mercury contaminated soil. *Environ. Int.* 74:42–53.

Yu, G., Wang, Y., Cao, H., Zhao, H., and Xie, Y. 2020. Reactive oxygen species and catalytic active sites in heterogeneous catalytic ozonation for water purification. *Environ Sci Technol.* 54:5931–5946.

Yu, Z., Xiao, M., and Liu, H. et al. 2024. FeOOH modified carbon doped TiO_2 p-n heterojunction wrapped Fe_3O_4 for highly efficient removal of As(III) through photocatalytic oxidation and adsorption. *J. Water Process Eng.* 61:105314.

Zahra, S., Sayed, A.A.R., and Ali, M. 2025. Functionalization of defective Zr-MOFs for water decontamination: Mechanistic insight into the competitive Roles of $-NH_2$ and –SH sites in the removal of Hg(II) ions. *ACS Appl. Mater. Interfaces.* 17:17726–17740.

Zhang, P., Jin, Y., Li, M., Wang, X., and Zhang, Y. 2023. Dual heterojunction-based Au@TiO_2 photoelectrode exhibiting efficient charge separation for enhanced removal of organic dye under visible light. *RSC Sustain.* 1:622–630.

Zheng, J., Sun, L., Xue, Y., Ye, L., and Fan, Q. 2024. Construction of pillared-layer metal–organic frameworks as an all-visible-light switchable photocatalyst for aqueous Cr(VI) reduction. *Inorg. Chem.* 63:15841–15850.

Zhong, X., Xu, J., Song, S., Li, H., and Hu, B. 2025. Synchronous photocatalytic redox conversion of chromium(VI) and arsenic(III) by bimetallic Fe/Ti metal–organic frameworks. *Inorg. Chem.* 64:1443–1454.

Zhu, Y., Chen, S., Li, Z., Li, H., Shabanab, M., and Chen, C. 2025. Nanoscale zero-valent iron composites for uranium-contaminated water treatment and environmental remediation: A review. *Environ. Sci.: Nano.* 12:20–40.

12 Hybrid and Integrated Approaches to Microreactor-Based Remediation

Akshay Kumar Dey, Sinchon Roy, Anik Das, Monisankar Roy, Sankha Chakrabortty, Arindam Mandal, and Biswajit Kamila

12.1 INTRODUCTION

The increasing concern on water pollution around the world, due to industrial activities, agricultural runoff, and the presence of contaminants of emerging concern (CECs) necessitates the development of innovative and efficient remediation technologies. Traditional methods of wastewater treatment often face different problems in effective removal of pollutants with complex mixtures, particularly in endocrine-disrupting chemicals (EDCs) such as 17 β-estradiol (E2) and 17 α-ethinylestradiol (EE2) (Diório et al. 2024). Microreactors have become a powerful tool in environmental engineering for making it possible to speed up processes and improve water treatment. Microreactors having small channels with diameters between 10 and 1000 μm are showing many built-in benefits for chemical and environmental processes. These include much better heat and mass transfer because of high surface-area-to-volume ratios and short diffusion distances. This fact leads to a faster rate of reaction, better selectivity, and less byproduct formation (Yoshihisa et al. 2007). The proper control over reaction conditions in micro-channels also reduces the use of reagents and energy waste. This also includes the ideas of green chemistry and sustainable development. Microreactor-based technologies are also efficient for making them easy to scale up or down to handle higher level of treatment and types of pollutants (Adama and Shing 2020, Prashant et al. 2018, Gorges et al. 2004). Microreactors have manifested a huge number of uses such as making nanoparticles, organic compounds, and heterogeneous catalysis. However, this type of reactor technology may not work well on its own for treating complex wastewater. These types of limitations may encompass challenges associated with the effective elimination of a wide range of pollutants, management of a huge load of substantial pollutants, and the attainment of complete mineralization of resistant organic compounds. To address various challenges and maximize the advantages of microreactor technology in environmental

 DOI: 10.1201/9781003633761-12

remediation, the integration of microreactors with other advanced treatment processes has garnered considerable attention (Arturo et al. 2019, Aiden et al. 2018).

The chapter describes the idea of hybrid and integrated approaches for the remediation using microreactors. This chapter demonstrates the synergistic effects and improved performance due to such integrations, and it also manifests the different combinations of microreactors with other physical, chemical, and biological treatment technologies. The chapter includes the fundamentals of these hybrid systems and gives instances of how they have been used to remove various types of pollutants. This chapter also discusses the difficulties and prospects for the future in this quickly developing field. The focus will be on how these integrated systems can offer more sustainable, efficient, and robust solutions for tackling the complex water pollution challenges of the 21st century.

12.1.1 Microreactors: A Short Overview

A microstructured reactor (or microchannelled) reactor is an apparatus in which chemical reactions are carried out in a confinement with usual lateral dimensions less than 1 mm; the most usual type of such confinement are microchannels (Diório et al. 2024, Elizabeta et al. 2022). Microreactors are researched in micro process engineering, along with other equipment (e.g., micro heat exchangers) in which physical processes take place. The microreactor is typically a continuous flow reactor (compare with/to a batch reactor). Microreactors have the potential to provide numerous benefits compared with traditional scale reactors, some of which include enhanced energy efficiency, reaction rate and yield, safety, reliability, scalability, on-site/on-demand manufacture, and significantly finer process control, as depicted in Figure 12.1 (Helena et al. 2022).

12.1.2 Microreactors: Fundamentals and Advantages

Microreactors are devices with microscale channels, typically with dimensions ranging from a few micrometers to a few millimeters. This small scale confers several advantages:

- **High Surface-Area-to-Volume Ratio**: This leads to enhanced mass and heat transfer, accelerating reaction rates and improving efficiency.

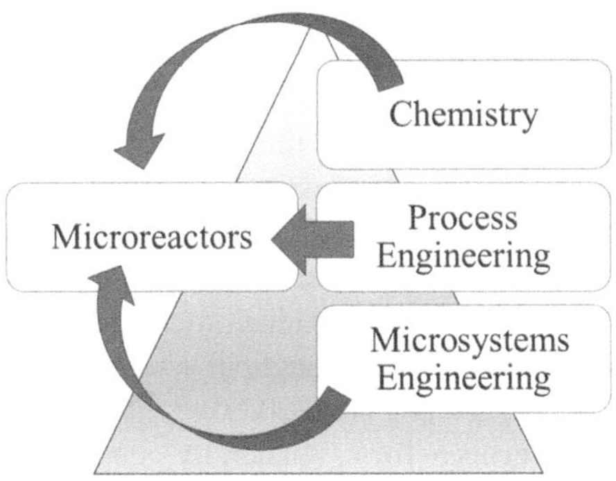

FIGURE 12.1 Interdisciplinary approach of microreactor technology.

- **Precise Control**: Microreactors have the potential to accurately control temperature, pressure, and flow rates. They also ensure consistent and optimized reaction conditions.
- **Laminar Flow**: The microreactors consisting of a typically laminar flow regime show well-defined and quasi-predictable flow patterns. In laminar zone microreactors, the laminar flow also makes it advantageous for some particular reactions.
- **Rapid Mixing**: Uniform reaction conditions within microreactors can be achieved by small volume-specified efficient mixing.
- **Compact Size and Portability**: Microreactors are typically small and portable; therefore, they can be used for mobile applications and on-site treatment.
- **Safety**: Accidents and exposure to hazardous materials are less in microreactors due to their small volumes.
- **Reduced Waste**: Microreactors can create more sustainable processes by reducing waste production and reagent consumption.

Because of these features, microreactors are appealing for a number of uses, such as the synthesis of chemicals, the creation of nanoparticles, and—more and more—environmental remediation (Helena et al. 2022, Qasim et al. 2024, Ahmed et al. 2023).

12.2 TYPES OF MICROREACTORS

Based on materials, design, and intended application, microreactors are classified in some common types; those include:

- **Channel Microreactors**: These are the most common type, consisting of microchannels etched or machined into a substrate material. The channels can have various geometries, such as straight, serpentine, or interconnected networks, depending on the specific application (Bugra and Aras 2022, Pil and Kim 2010, Norbert et al. 2011).
- **Packed-Bed Microreactors**: These microreactors contain a packed bed of catalyst particles within the microchannels. They are used for heterogeneous catalytic reactions, where the catalyst is in a different phase from the reactants (Luka et al. 2016, Sanaa et al. 2020, Lu and Yue 2025, Andrew et al. 2007).
- **Membrane Microreactors**: These microreactors incorporate a membrane, which can be used for separation, extraction, or controlled addition of reactants (Lun et al. 2005, Xiaoyao and Li 2013, Seelam et al. 2013).
- **Droplet Microreactors**: These microreactors generate and manipulate tiny droplets of fluid, which act as individual microscale reaction vessels. They are particularly useful for high-throughput screening and single-cell analysis (Linbo et al. 2020, Cheung et al. 2009).
- **Homogeneous Microreactors**: Generally curved micro-channels of inner radius 200 μm, channel length of 1600 mm, and reaction volume of 260 μL are now used as appropriate dimensions for homogeneous reactors.

The reactor geometry is optimized to employing tensile stress analysis so that sufficient flow stability is provided throughout the reactions are reported (Barbora 2025, Matic et al. 2013). It was carried out through esterification of acetic acid with n-butanol in the presence of arenesulphonic acid groups at 60 °C in silicon monolith-based microreactors. The micro-channels of length 1000 mm were prepared as macro-pores and micro-pores of diameter 30–50 μm through chemical etching technique. A residence time of 155 seconds with an intensification factor of ~180 over conventional reactors for high flow rates up to 90ml/min was achieved (Zhongjian et al. 2023).

- **Heterogeneous Microreactors**: Parallel, brief micro-channels of 300 mm length, 400 μm depth, and distinct inlets of 200 μm width, for reactant supply and catalyst charging. Recently, a try to conduct acid-catalyzed liquid-liquid reaction at elevated temperatures and reduced channel diameter has been employed for two-phase reactions (heterogeneous catalysis). Maximum loading volume of catalyst is 29 μL and individual weirs (40 × 20 μm) are utilized for catalyst retention in the channel flow. Another inlet for gas phase is given to separate the two-phase flow. The heated section may be extended up to 47 mm length based on the length of the halo region etched between silicon channels. Some T-junctions are constructed with manifold section per inlet so that there is lower pressure drop through the channels and resulting higher stability of flow. In heterogeneous catalysis, solid catalysts, as a packed bed that occupies 60% channel length, cause additional pressure drop which would be less than that through the channels. Following the occurrence of reactions, the product trickles out through a single outlet without further distribution. Flow regimes, *viz.*, (i) Rivulet flow, (ii) Dry patch falling film, and (iii) Full falling film flow have been reported (Thomas 2013, Haocui et al. 2009, Xingjun et al. 2015). Common FFMR employs stainless steel or polymethylmethacrylate (PMMA) as wall material for applications at high pressure. Scale-up of FFMR involves straightforward numbering-up technique or change in channel geometry (Mashuri et al. 2020, Chun et al. 2011). Low flow rates and curved profile for liquid film created provide high mass transfer within the channels. Complete falling film on a flat surface is the prevailing flow pattern, and a liquid-side mass transfer coefficient (kL) substantially relies on molecular diffusion.
- **Falling Film Microreactors**: Falling film microreactor (FFMR) is one of the ordinary and most usual category of microreactors for concerted liquid-gas interphase reactions (Pohar and Plazl 2009, Zhang et al. 2009). This basically works through the accumulation of gravity-assisted wetted thin liquid phase over the solid surface. For long microchannels, a slit may be included at the inlet side for better flow-liquid contacting process, which is based on the film-penetration theory of mass transfer that results in low residence time. Accordingly, the mass transfer coefficient is calculated as follows (Equation 12.1): (Haocui et al. 2009).

$$k_L = \frac{uD}{nL} \tag{12.1}$$

where

u—film surface velocity (based on liquid film thickness and kinetic viscosity)
D—gas diffusivity
L—length of the microchannel

- **Membrane microreactors**: They utilize the synergistic performance of both membranes and micro-channels for specific applications in biochemical catalysis and aim for complete reactant conversion (Thomas 2013). Ceramic (aluminum oxide) membranes of pore size 0.2 μm, thickness ranging from 5 to 100 μm and tortuosity of 2–2.5 for very short lengths are fabricated using anodization technique. These are integrated with silicon through chemical etching to form porous surfaces with a surface-area-to-volume ratio measuring ~100 m^2/cm^3 (Hu et al. 2014, Luyang et al. 2014). Heterogeneous catalysts are impregnated over them with a thickness in nanoscale. A graphite gasket (thickness 2 mm) supports the membrane inside the body of the reactor module (Diório et al. 2024, Ahmed et al. 2023).

The broad classification of microreactors is depicted in Figure 12.2 (Krishna and Biswas 2016).

12.2.1 Materials for Construction of Microreactors

Microreactors may be made of a range of materials, depending on the application and conditions of operation. Typical materials for the construction of microreactors are:

✓ **Silicon**: Silicon is a common material for microfabrication because its processing methods are well established.
✓ **Glass**: Glass has good chemical stability and optical clarity and so is well suited to photochemical reactions.
✓ **Polymers**: Polymers are cheap to process and produce and so are good choices for low-cost or disposable microreactors.
✓ **Metals**: Metals have good thermal and mechanical strength, and thus they can be used for high-temperature or high-pressure conditions.
✓ **Ceramics**: Ceramics are chemically and thermally stable, and hence they can be used for extreme chemical conditions (Qasim et al. 2024, Ahmed et al. 2023).

12.3 FUNDAMENTALS OF MICROREACTORS FOR ENVIRONMENTAL REMEDIATION

Microreactors provide a novel platform for performing chemical and physical operations at a microscale, which results in a number of benefits that are important to environmental remediation. Appreciation of these basic features is important in order to understand why they should be coupled with other treatment technologies (Soomin et al. 2018, Fang et al. 2015).

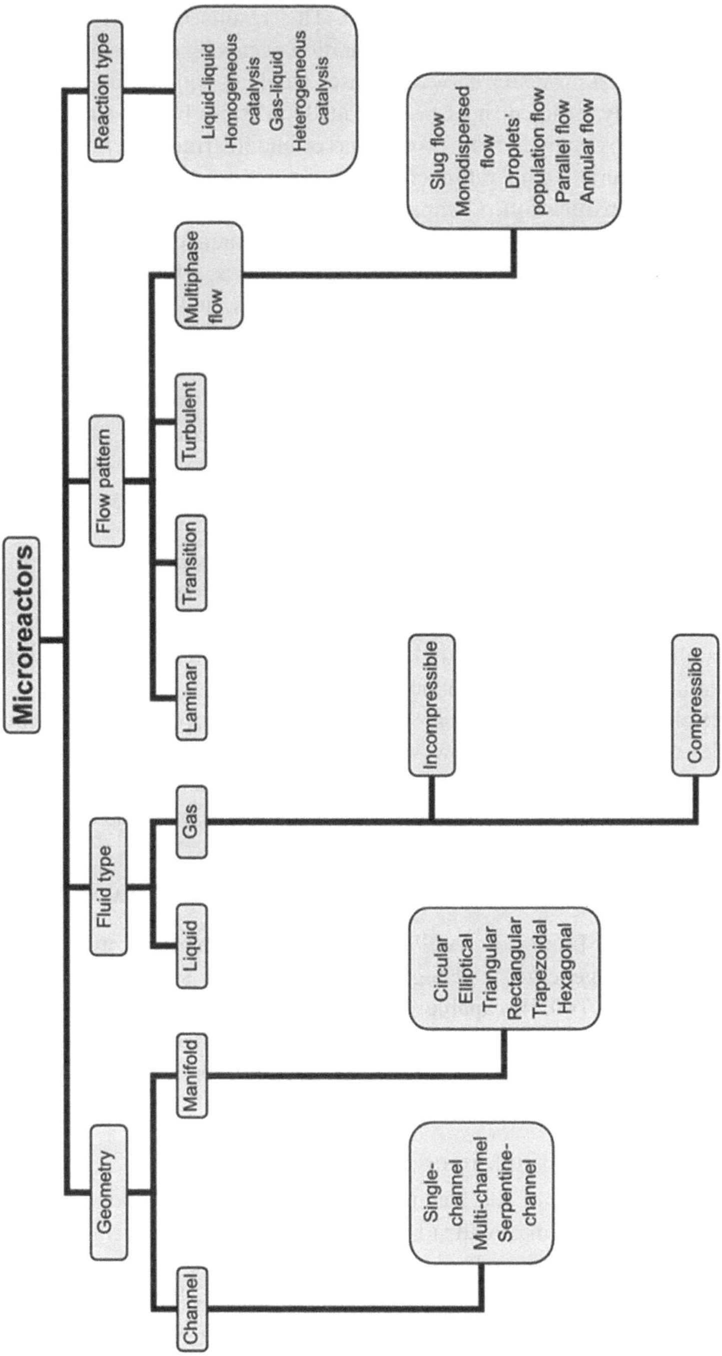

FIGURE 12.2 Classification of microreactors.

- **Increased Mass and Heat Transfer**: Fast diffusion and convective transport are made possible by the small microchannel dimensions, which results in a high surface-area-to-volume ratio. This results in improved contact between reactants and catalysts or separation agents by drastically lowering the rates of mass transfer between phases (such as liquid-solid and liquid-gas). Similarly, efficient heat transfer made possible by the short thermal diffusion paths enables both exothermic or endothermic reaction management and accurate temperature control.
- **Process Intensification**: Compared to conventional batch or macroscale reactors, microreactors are a tool for process intensification because of their superior mass and heat transfer characteristics, which enable them to achieve higher throughput, shorter reaction times, and smaller equipment sizes. This is very useful for wastewater treatment, in which a large amount of water must be treated efficiently.
- **Lower Reagent and Energy Consumption**: The control of reaction conditions and diffusion limitations in microreactors is frequently coupled with increased selectivity and lower byproduct generation. This means lesser consumption of reagents and possibly lower energy input for mixing and separation.
- **Scalable and Modular Systems**: Microreactors are simply integrated into scalable systems, which can be flexibly scaled up or down for varying treatment capacities by merely adding or subtracting reactor units. Such inherent scalability is a major benefit for decentralized water treatment processes or for adjusting to variable pollutant loads.
- **Multiple Operating Modes**: Microreactors are operated in multiple modes such as continuous flow, segmented flow, and droplet microfluidics, providing the versatility of working with various reactions and separation processes for the removal of pollutants (Xi et al. 2020, Marzieh et al. 2024).

12.3.1 Advanced Oxidation Processes (AOPs) in Microreactors

Advanced Oxidation Processes (AOPs) are a class of chemical treatment methods that involve the generation of highly reactive oxygen species (ROS), primarily hydroxyl radicals (•OH), capable of oxidizing and mineralizing a wide range of organic pollutants. The integration of AOPs with microreactor technology offers significant potential for enhancing their efficiency and applicability (Diório et al. 2024, Elizabeta et al. 2022).

12.3.1.1 Photocatalysis in Microreactors

- Photocatalysis, one of the leading AOPs, employs semiconductor photocatalysts like titanium dioxide (TiO_2) to produce ROS when irradiated with visible light or UV. Microreactors are the best platform to be used for photocatalytic degradation because they have a high surface-area-to-volume ratio, which makes photocatalyst immobilization efficient and improves light penetration (Xi et al. 2020, Marzieh et al. 2024, Jiayi and Zeng 2017, Wang et al. 2021).

- **Increased Light Distribution**: Microchannels' small size can enhance the light distribution in the reactor, which will ensure a greater portion of the photocatalyst surface is exposed to light.
- **Efficient Mass Transfer of Pollutants to Catalyst Surface**: The improved mass transfer in microreactors ensures the delivery of pollutants to the photocatalyst surface where the oxidation reactions take place.
- **Continuous Flow Operation**: Continuous flow photocatalytic treatment is made possible by microreactors, which leads to higher throughput and easier integration with downstream processes.

Acetamiprid degradation and E2 and EE2 degradation are two instances of photocatalytic AOPs in conjunction with microreactors (Diório et al. 2024, Elizabeta et al. 2022). The design and functionality of photocatalytic microreactors for pollutant destruction have also been enhanced through the use of CFD modeling (Krishna and Biswas 2016).

12.3.1.2 Fenton and Photo-Fenton Processes in Microreactors

The Fenton process utilizes ferrous ions (Fe^{2+}) as the catalyst for decomposing hydrogen peroxide (H_2O_2) into hydroxyl radicals. The photo-Fenton process is an improvement of that, using UV or visible light irradiation to facilitate regeneration of Fe^{2+} and potentially the production of further reactive oxygenated species (ROS). Microreactors can provide greater control of reaction conditions and better mixing in such processes (Krishna and Biswas 2016, Jiayi and Zeng 2017).

12.3.1.3 Ozonation and Peroxone Processes in Microreactors

Ozonation makes use of ozone (O_3) as a potent oxidant, whereas the peroxone process uses a combination of ozone and hydrogen peroxide to produce hydroxyl radicals more effectively. Microreactors may help ensure the effective dissolution and mass transfer of ozone into the liquid phase for increasing the oxidation process. Integrated ozone microreactor technology has been established for decontaminating reclaimed potable water and greywater.

12.3.2 Hybrid Systems Integrating Microreactors with Physical Separation Processes

Combining microreactors with physical separation techniques can lead to synergistic effects, where the microreactor enhances the efficiency of the separation process or vice versa. Few hybrid systems are discussed in the following sections.

12.3.2.1 Membrane Separation Microreactors

Combining microreactors with membrane separation techniques like microfiltration, ultrafiltration, nanofiltration, and reverse osmosis has the potential to provide a robust route for concurrent reaction and separation.

- **Reactive Membranes**: Catalytic materials or reactive agents can be functionalized in membranes to incorporate them into microchannels to allow both pollutant degradation and separation in one unit.

- **Improved Mass Transfer to the Membrane Surface**: The environment in the microreactor may improve the mass transfer of pollutants to the membrane surface, enhancing separation efficiency as well as minimizing membrane fouling.
- **Compact Design and Continuous Operation**: Membrane microreactors are perfect for on-site water treatment because of their compact design and continuous operation. An illustration of this approach is the tube-in-tube membrane microreactor used to break down E2 and EE2 via the UV/H_2O_2/TiO_2 process (Diório et al. 2024).

12.3.2.2 Adsorption Microreactors

Adsorption is a highly effective method for eliminating a wide variety of contaminants from water. Because of the microenvironment's superior mass transfer characteristics and large surface area, adding adsorbent materials to microreactor channels can boost adsorption efficiency.

- **Packed Bed Microreactors**: Microchannels containing adsorbent substances, such as metal-organic frameworks (MOFs), zeolites, or activated carbon, can provide a large surface area for the pollutants' adsorption and efficient removal.
- **Flow-Through Adsorption**: In microreactors, the continuous flow operation creates the effective contact between the adsorbent material and the contaminated water.
- **Integration with Regeneration Processes**: Theoretically in-situ or ex-situ regeneration techniques for adsorbent materials can be combined with microreactors for making the treatment process more environmentally friendly.

12.3.3 Integrated Systems Combining Microreactors with Biological Treatment Processes

For the removal of some pollutants, biological treatment techniques like bioremediation employing microorganisms are both economical and ecologically benign. Control and efficiency can be improved by integrating microreactors with biological processes.

12.3.3.1 Biofilm Microreactors

Microreactors can be designed to support the development of biofilms on the packed materials within the channels or on the channel walls. Such biofilms can later break down organic contaminants as the water passes through the microreactor.

- **High Surface Area for Biofilm Development**: The high surface-area-to-volume ratio in microreactors offers sufficient space for development of biofilms.
- **Controlled Microenvironment for Microbial Activity**: Microreactors provide complete control over flow rate, nutrient delivery, and temperature.

These types of microreactors can be also used to optimize microbial activity and pollutant degradation.

- **Minimized Biofilm Detachment**: The closed space in microchannels of microreactors can reduce biofilm detachment, resulting in more stable and effective bioremediation.

12.3.3.2 Microfluidic Bioreactors for Toxicity Assessment

Although not used directly for remediation, microfluidic systems can be utilized to quantify the toxicity of contaminants and the efficacy of various treatment processes on microbes. Such data may be useful to maximize integrated bioremediation strategies (Diório et al. 2024).

12.4 THE CHALLENGE OF WATER POLLUTION AND THE ROLE OF MICROREACTORS

Large amounts of wastewater containing organic dyes have been produced as a result of the increase in pollution from various industries, especially the textile and dye sectors. These effluents pose a risk to the environment. The environment and human health are significantly impacted by untreated wastewaters, which are a combination of household and industrial wastes. Numerous pollutants and toxins found in untreated wastewater can contaminate soil, water, and air, endangering both human health and the environment. Phenols, pesticides, fertilizers, detergents, and synthetic dyes are examples of organic pollutants in wastewater that can lead to eutrophication, which lowers the oxygen content of water bodies and endangers aquatic life. Since contaminated water is the cause of 80% of illnesses and 50% of child fatalities worldwide, wastewater pollution also poses a major threat to human health. Every year, 300–400 million tons of trash are dumped into the world's oceans by factory plants, and wastewater pollution causes environmental issues. Release of untreated industrial wastewaters into water bodies places the health of communities around these polluted water sources at serious risk (Donia et al. 2019).

Harmful substances in industrial waste consist of heavy metals, solvents, and poisonous sludge, which result in acute and chronic health issues among humans, including neurological disease, respiratory disorders, and other types of cancer. Water shortage is a significant issue among humans nowadays, and it is imperative to meet this demand sustainably with eco-friendly methods without destroying the environment. Wastewater treatment technologies and water purification processes are significant solutions applied in industry. Traditional wastewater treatment plants are equipped to eliminate these TOCs; yet traditional methods only remove part of these chemicals and leave TOCs in wastewater treatment plant effluents released into surface water. The majority of organic contaminants present in waterways are not biodegradable in the ecosystem and are harmful to human and animal health. Microreactors provide an attractive alternative or addition to conventional wastewater treatment processes. Their primary advantages, such as high surface-area-to-volume ratios, accurate control over reaction conditions, and improved mass transfer, best position them to cope with the water pollution challenge. Through the creation of a controlled environment for chemical reactions, microreactors are able to enhance the effectiveness and efficiency of pollutant removal (Marzieh et al. 2024).

The results and discussions of few case studies that elucidate the application of hybrid and advanced microreactors in various types of environmental remediation are discussed in the following sections.

12.4.1 Case Study 1: Application of Rotating Photoreactor to Determine Reactive Species Like Acetamiprid

A reticulated batch mode was switched on to the immobilized layer of TiO_2 and activated under UV-A light to investigate the photocatalytic degradation of acetamiprid. [8] The authors basically investigated the influential potential of various interfering compounds by determining the concentration of ROS generated during the entire photocatalytic process. The authors reported that the TiO_2 P25 layer was activated under UV-A radiation to give superoxide anion radicals and singlet oxygen species that played a crucial role to degrade acetamiprid. When the pure TiO_2 was modified using UVC, the novel TiO_2-UVC showed a remarkable marginal decrease in the UV-DRS plotted bandgap than the pure TiO_2.

The authors also reported the impact of NaN_3 and p-benzoquinone (p-BQ) towards the photodegradation, the role of H_2O_2, isopropanol, and EDTA as potent scavengers against electrons, hydroxyl radicals, and holes, respectively. The presence of H_2O_2 along with NaN_3 and p-BQ suppresses the degradation efficiency slightly as the latter two are considered as an effective superoxide radical as well as singlet oxygen scavengers. The authors reported that 1 mM of p-BQ (as $\bullet O_2^-$ scavenger) is more effective in suppressing the degradation efficiency than I mM of NaN_3 (as $^1O_2$ scavenger). Also the interference of N_3^- anion with observed species is questionable as it tends to form azide radicals ($N_3\bullet$) by reacting with •OH radicals (Tamara et al. 2020). Taking into account the other scavengers, 3 mM of H_2O_2 decreases the degradation efficiency highly compared to 1 mM H_2O_2, 5 mM H_2O_2 with TiO_2-UVC photocatalyst. Subsequently, 5 mM isopropanol (compared to 1 mM isopropanol, 3 mM isopropanol, and TiO_2-UVC) and 3 mM EDTA (compared to 1 mM EDTA, 5 mM EDTA, and TiO_2-UVC) were highly able to suppress the degradation efficiency of the photocatalyst. Finally, the authors aligned the experimental results into a pseudo-second order kinetics and the RMSD model with k value of 2.78×10^3 min^{-1} for TiO_2-UVC at a pH of 6.5 and 8 W UV-A lamp.

Hence, the required qualitative and quantitative determinations of the crucial intermediate, ROS for determining the photodegradation mechanism were successfully done by the authors to improve the process efficiency and a real-world application toward the photocatalytic technologies.

12.4.2 Case Study 2: Functioning of a Tube-in-Tube Membrane Microreactor for UV/H_2O_2/TiO_2 Assisted E2/EE2 Degradation

The reactor functioned without any contaminants, circulating pure water at both laminar and turbulent flow rates. H_2O_2 was injected using either radial permeation (RP) or upstream injection (UI) techniques to assess concentration profiles and steady-state conditions. Radial permeation facilitated a consistent distribution of the oxidant throughout the reactor's length, leading to a uniform concentration profile. The concentration of the oxidant gradually rose to the steady-state value ($t/\tau \geq 80$) (Diório et al. 2024). Membrane titration of minimal oxidant quantities improves phase contact with the numerous pores that serve as dosage points. The membrane effect on oxidant dosage is particularly noticeable under turbulent flow conditions, as it requires a longer duration to reach steady-state, and the profile for H_2O_2 achieved through radial permeation mode is smoother compared to those from upstream injection mode. The enhancement of phase contact in RP, due to the extremely high dosage points, promotes oxidant uniformity along the reactor's length and facilitates degradation reactions.

The research sought to identify the ideal flow rate for achieving the highest removal efficiency of contaminants E2 and EE2 in a photocatalytic process. Turbulent flow produced the greatest removal rates, with E2 and EE2 being eliminated at rates of 2.87 $\mu M\ min^{-1}$ and 2.91 μM/min, respectively (Diório et al. 2024). The optimal concentration of H_2O_2 strikes a balance between the generation of HO• and the reduction of the scavenging effect, thereby enhancing both efficiency and cost-effectiveness. Additionally, the study investigated the influence of oxidant dosage on removal rates across laminar, turbulent, UVC irradiation, and dark conditions. The study also examined the impact of stock solution concentration on estrogen removal rate in a reactor using UVC irradiation. Results showed that UVC light, combined with oxidant, enhanced estrogen removal through the formation of hydroxyl radicals, indicating no significant variation in steady-state conditions. Also the estrogens such as β-estradiol is highly influenced by pH in terms of the interfacial interaction between the photocatalyst and the particular contaminant. According to the authors' report, pH 6.7 showed the highest concentration profile, while pH 5 had the highest removal rate. Lower pH values should lower E2/EE2 degradation, but the interfacial electrostatic repulsion, opposite to the aforesaid, between contaminants and the catalyst surface can also affect removal efficiency (Diório et al. 2024).

12.4.3 Case Study 3: Modeling and Kinetic Assessment of a Continuous Flow Photocatalytic Microreactor (CFPM) in Degrading Selected Pollutants

The research by Jamil et al. centers on determining the ideal flow rate and quantity of lights necessary for the photodegradation of pollutants within a CFPM (Diório et al. 2024). The pollutant stem residence time is influenced by the flow rates as well as the contact duration between the photocatalyst and the pollutant, thereby maximizing interaction. The degradation efficiency observed at various flow rates was recorded at 88, 70, 60, and 54 %, demonstrating the relation between degradation efficiency and the flow rate is inversed. Additionally, the study explored the impact of flow rate on the abasement of both intermediates and parent compounds, with a focus on measuring the Total Organic Carbon (TOC). The findings indicated that the TOC removal of mineralized Plasmocorinth B (PB) dye diminished from 26 to 1% as the flow rate increased. Furthermore, the research assessed the influence of light intensity on the degradation process, revealing that the degradation efficiency of the PB dye under various illuminations was reported to be the highest (95 %) with 6 lamps in contrast

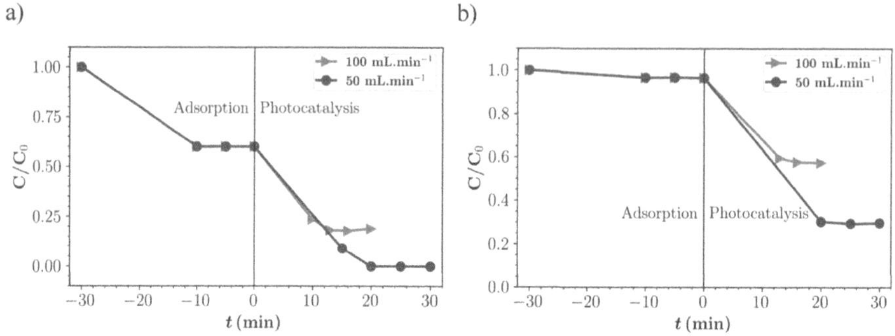

FIGURE 12.3 IBU degradation efficiency in (a) deionized water with different illuminations and (b) ISO medium with different flow rates.

(Image reproduced with permission from Jamil et al. 2024.)

to the lowest (55%) with only 1 lamp. The TOC removal efficiency was also evaluated under varying LED lamp configurations, indicating an increase in TOC removal efficiency from 21 to 26% as the number of illuminating lamps was raised from 1 to 3, followed by a decrease from 20% to 3% as the number of lamps was further escalated from 4 to 6. Authors could not find the specific reasons behind the inconsistencies in the results and urged to necessitate additional investigation.

The study also evaluated the suitability of the designed reactor for the photocatalytic degradation of three most commonly found pharmaceuticals, *viz.*, Ibuprofen (IBU), Sulfamethoxazole (SMX), and Diclofenac (DCF) in deionized water as well as in ISO medium. Authors conducted experiments to determine the optimum light intensity and flow rate for the degradation process. The results showed that the photocatalyst was maximum efficient for IBU and for that, the degradation efficiency was 100% and 82% in deionized water under different lamps, while the efficiency in ISO medium was 71% and 62% at different flow rates. The efficiency decreased in ISO medium could be attributed to the interference of various inorganic ions in the solution, which impeded pollutant adsorption and could react with hydroxyl radicals. The study highlights the importance of flow rate and light intensity in pharmaceutical degradation processes as depicted in Figure 12.3.

12.4.4 Case Study 4: CFD Modeling of Fe-N-TiO_2 Thin Film Microreactor Assisted Photooxidative Removal of Diclofenac Amide (DCFA) Removal

An improved Multiphysics CFD modeling coupled with radiation field modeling and species transport was brought in front by Yusuf et al. to investigate the dark adsorption-desorption as well as the apparent reaction rate constants for the photooxidative degradation of DCFA under UV radiation (Marzieh et al. 2024). The authors ranged the initial DCFA concentration within 0.73–1.7 ppm and that of the recirculated flow rate within 0.15–3.73 L/h. The chosen model restrict both the kinetic constants to be <0.03 that typically helped in the revelation of mass transfer limitations within the kinetic regime. The simulation of DCFA in both the dark and light field helped in the derivation of incident radiation (G(y)) profile while both the radiation field assisted Fe-N-TiO_2 immobilization helped to project the DCFA degradation rate law. The velocity field of the immobilization produced the maximum G value of 300 W/m^2 that indicates the compact morphology of the microreactor system comparable to the pure sputtered slid TiO_2. The authors also indicated the possibility of limited mass transfer upon reactant adsorption impact may be due to the non-porous nature of the thin film.

12.4.5 Case Study 5: Formation of Carbonized TiO_2-coated Melamine Foam (CTiMF) Loaded 3D Framework Microreactor for Improved Solar-irradiated Remediation of Cotton Dying Wastewater

Lu et al. put effort into the fabrication of CTiMF that involves coating pristine melamine foam (PMF) with polydopamine (PDA), which effectively bridges the MF and TiO_2 NPs (Marzieh et al. 2024). The original brown TiMF reduces in size and turns black, indicating the formation of CTiMF. They followed a strategical biomimetic polydopamine coating process to immobilize TiO_2 on to the carbonized melamine foam fibrils (Shu et al. 2018). The foam binder improved the thermal stability of the TiO_2 nanoparticles on the melamine framework. The system exhibited remarkable photocatalytic degradation efficiency of 98% towards RhB dye only in 60 minutes of direct solar irradiation and the authors reported that it was at least 16–20 times higher than the pristine nanoparticle. The carbonized N-doped foam showed a superior light absorbing capability along with the inhibited electron-hole pair recombination that marked up for such an efficiency.

12.4.6 Case Study 6: A Responsive Visible Light Driven Tetracycline Degradation through Photocatalytic Fuel Cell Microreactor (PFCM) Loaded with ZnO Nanorod/Bi_2MoO_6/ZIF-67 Photocatalyst

A promising PFCM microreactor was fabricated by Ashrafi et al. in which the researchers loaded it with ZnO nanorod/Bi_2MoO_6/ZIF-67 photocatalyst to investigate its photodegradation efficiency towards a pharmaceutical pollutant, namely, tetracycline (TTC) (Marzieh et al. 2024). The authors reported that the three component photocatalyst was synthesized by depositing those as photoanode through a viable drop-casting method. The FTO substrate reduced the effective bandgap to 1.61 eV which previously was reported to be 1.74 V. At an optimal pH of 7, TTC concentration of 20 ppm, initial flow rate of 120 μL/min and an irradiation intensity of 7.9 mW/cm^2, the photocatalyst showed an excellent TTC removal efficiency of 97.16% and the TOC was calculated to be 92.4%. The Langmuir-Hinshelwood kinetic model allows the degradation system to have a maximum power density up to 35.1 mW/cm^2 and the fluid phase mass transfer resistance allows the photocurrent density to be 1.4 mA/cm^2. The removal efficient showed a breaking point after 1450 minutes of operation at which the time-on-stream removal efficiency was dropped below 80%.

12.4.7 Case Study 7: Photo-assisted Dual-function Capturing and Removal of Organic Contaminants Present in Wastewater through a Self-propelled MOF Nanomotors

The very advanced motion capabilities at a nanoscale perimeter enables self-propelled MOF nanomotors to exhibit high potential towards various environmental and biological remediations. A multifunctional MOF-625 based nanomotor has been prepared by Liu and Zhao (2021) to execute a dual-function capturing of aqueous pollutants and their subsequent degradation. In this study, the authors have focused on the self-propulsion of the nanomotor that is primarily attributed to the cobalt-assisted catalytic decomposition of H_2O_2. Even in very less quantity and concentration, H_2O_2 acts as a potent chemical fuel. The further addition of Fe_3O_4 into the system enhanced the magnetic properties of the system, which in turn, helped an effortless recovery of the nanomotor. The authors employed malachite green (MG) dye to investigate the dual-functionality of adsorption as well as photocatalytic degradation of the resulting nanomotor. The authors chose a very low concentration (0.1%) of H_2O_2 as in slightly higher concentration it tends to drive the nanomotor in a Brownian-like motion. As mentioned earlier, the catalytic co-doping of cobalt enhances the Fenton-like photo-disintegration of H_2O_2, in which Co^{2+} acts as active catalytic sites. The authors reported that the nanomotor possessed maximum velocity upon the injection of 0.4% H_2O_2 as compared to 0.05%, 0.1%, 0.2% and 0.3% of the same species. In contract, the moving distance was observed to be the lowest with a 0.5% H_2O_2 dosage. While investigating the MG dye degradation efficiency, the authors started with an initial dye concentration of 40 ppm. The blank set up could show only up to 38.21 ppm of degraded dye, followed by MOF-525, MOF-525-Co, MOF-Fe_3O_4, while the maximum removal efficiency was showed by MOF-Co-Fe_3O_4 from an initial 40 ppm to 5.71 ppm in dark to 2.58 ppm in light irradiation along a contact time of 2h. The non-Co-doped composites were unable to exhibit an up-to-the-mark removal efficiency in comparison to the Co-doped composites and authors attributed this difference to the catalytic activity between H_2O_2 and Co^{2+}. This phenomena clearly demonstrates the self-propulsion of the nanomotor so as to enhance the removal of the targeted pollutant. Also it is evident that the darker medium only supports the adsorption process while the lighter medium supports both the adsorption and photocatalytic degradation and hence resulted in an improvised dye removal efficiency (Zhiwei et al. 2022). The authors also performed a magnetization of all the as-prepared composites and reported that the saturation magnetization of MOF-525-Co-Fe showed a highest value of 14.49 emu/g as obtained by the VSM hysteresis loop. This will help in an excellent magnetic compliance, suitable for controllability and separation. The overall facility of the as-prepared nanomotor indicates the possibility of a viable industrial application of the same.

We can conclude that microreactors have proven effective in the degradation of various pollutants, including organic dyes, pharmaceuticals, and pesticides. The high surface-area-to-volume ratio in microreactors enhances the contact between pollutants and reactive species, leading to improved degradation rates. The design and operation of microreactors also play a critical role in their performance. The immobilization of photocatalysts within microreactors is a key advantage, as it prevents catalyst loss and allows for continuous operation. The integration of microreactors with other treatment processes, such as advanced oxidation processes (AOPs), has also shown promise. Factory plants release 300–400 million tons of garbage annually into the world's seas, and wastewater pollution causes environmental problems. AOPs can efficiently break down a variety of contaminants by producing extremely reactive species, such as hydroxyl radicals. AOPs and microreactors can work in concert to produce synergistic effects by utilizing the improved mass transfer and reaction control that microreactors provide (Diório et al. 2024, Elizabeta et al. 2022). Notwithstanding the benefits, using microreactors for environmental remediation is not without its difficulties. The possibility of clogging is one difficulty, especially when handling wastewater with high suspended solid concentrations. Further research is needed to develop microreactor designs and operating conditions that minimize clogging and fouling.

In summary, a variety of pollutants have been removed by integrating microreactors with other treatment technologies. Some of the important ones are: (Amit et al. 2011, Diório et al. 2024, Elizabeta et al. 2022, Zhiwei et al. 2022, Marzieh et al. 2024, Huan et al. 2023, Hunge et al. 2017).

- **Chemical Synthesis**: Microreactors are utilized for the synthesis of organic and inorganic compounds, such as fine chemicals, pharmaceuticals, and nanomaterials.
- **Organic Dyes Removal**: Microreactors have been applied in the remediation of contaminated water with organic dyes. A Y-shaped microreactor obtained 86% decolorization of an azo dye in 4.2 seconds based on the Fenton process. Magnetic torus microreactors have also been demonstrated to be capable of removing dyes.
- **Synthesis of Nanoparticles**: The control provided by microreactors is essential for the synthesis of nanoparticles with controlled size, shape, and composition.
- **Catalysis**: Microreactors are applied to investigate and improve catalytic reactions, homogeneous and heterogeneous.
- **Analysis**: Microreactors may be coupled with analysis techniques for fast and sensitive chemical analysis.
- **Pharmaceuticals and Other Contaminant Removal**: Microreactors have shown promise in the removal of pharmaceuticals and EDCs.
- **Energy**: Microreactors are being considered for use in fuel cells, solar cells, and biofuel production.
- **Environmental Remediation**: Microreactors are employed for degradation of contaminants in wastewater and polluted soil.
- **Biotechnology**: Microreactors are employed for cell culture, enzyme catalysis, and other biotechnological operations.
- **Elimination of Emerging Contaminants**: Microreactors are being considered to eliminate emerging contaminants, such as mineral contaminants.

12.5 LIMITATIONS OF INDIVIDUAL REMEDIATION TECHNOLOGIES

Before discussing hybrid and integrated approaches, it is crucial to acknowledge the limitations of individual remediation technologies when used in isolation.

- **Adsorption**: Adsorption uses a solid adsorbent to remove impurities from water by attaching them to its surface. Though useful for some impurities, adsorption is not effective and has some drawbacks, such as: (Marzieh et al. 2024)
  - **Finite capacity**: Adsorbents possess a limited capacity, and when saturated, they must be regenerated or replaced.
  - **Regeneration issues**: Regeneration of adsorbents can be expensive and can create secondary waste.
  - **Ineffectiveness for certain pollutants**: Adsorption can be ineffective for all forms of pollutants, especially those which are highly soluble or non-polar.
- **Advanced Oxidation Processes (AOPs)**: AOPs generate very reactive radicals, e.g., hydroxyl radicals (OH•), to degrade and oxidize pollutants. Although AOPs can degrade a wide variety of organic pollutants, they too have some limitations: (Diório et al. 2024, Elizabeta et al. 2022)
  - **High energy consumption**: Some AOPs, such as those based on UV radiation, can be energy-intensive.
  - **Chemical expenses**: The cost of chemicals, such as ozone or hydrogen peroxide, may be high.
  - **Byproduct formation**: AOPs can occasionally produce harmful byproducts.
  - **Selectivity**: AOPs can react with contaminants and other components of the water matrix because they are generally non-selective.
- **Membrane Filtration**: Semi-permeable membranes are used in membrane filtration processes, such as reverse osmosis (RO), ultrafiltration (UF), microfiltration (MF), and nanofiltration (NF) to remove impurities from water. These have the following limitations: (Diório et al. 2024, Elizabeta et al. 2022, Luyang et al. 2014).
  - **Fouling**: Pollution can contaminate the membranes, reducing their longevity and performance.
  - **High pressure demands**: High pressure is required for some membrane processes, particularly RO, which raises energy consumption.
  - **Incomplete removal**: Membrane filtration might not remove all pollutants, particularly small or dissolved ones.
  - **Concentrate generation**: Membrane processes produce a concentrated waste stream that must be treated or disposed of.
- **Biological Treatment**: Biological treatment employs microbes to break down organic pollutants. Restrictions are: (Juan et al. 2018, Yoshihisa et al. 2008).
  - **Slow kinetics**: Biological processes may be slow, particularly for recalcitrant contaminants.
  - **Sensitivity to conditions**: Microorganisms are sensitive to conditions around them, including pH, temperature, and the presence of toxic compounds.

- **Sludge generation**: Sludge is produced through biological treatment, which must be treated or dumped.
- **Space needs**: Biological treatment systems frequently demand much space.

12.6 CHALLENGES AND FUTURE PERSPECTIVES

- The impressive development in hybrid and integrated microreactor-based remediation has faced various challenges that must be overcome for their extensive usage: (Juan et al. 2018, Yoshihisa et al. 2008, Milica et al. 2023, Raffaella et al. 2021).
- **Scaling-Up**: Scaling down from laboratory-scale microreactor systems to industrial-scale implementation is still a major challenge. The establishment of cost-efficient and effective numbering-up or scaling-out microreactor units is essential.
- **Channel Clogging and Membrane Fouling**: Fouling and clogging of microchannels may happen in the case of membrane or particulate-containing systems, causing performance decline and higher maintenance needs. It is important to address these problems.
- **Complexity of Integration**: The design and operation of the entire system may become more complicated when different treatment technologies are integrated. It's critical to maximize the compatibility and integration of different components.
- **Economic Effectiveness**: For long-term industry adoption, microreactor-based systems' initial capital and operating costs must be on par with or less than those of conventional treatment methods.
- **Long-Term Durability and Stability**: Under continuous operation, the materials and packaged components inside the microreactors must maintain their long-term stability and durability.
- **Real Wastewater Matrix Effects**: Analyzing hybrid and integrated microreactor systems' performance requires the use of complex real wastewater matrices, which may contain a variety of interfering substances.

Looking towards the future, several promising research directions and perspectives can be identified: (Yufei et al. 2016, Juan et al. 2018, Yoshihisa et al. 2008, Zahhad et al. 2020).

- **Development of Novel Microreactor Designs**: Investigating novel reactor designs and microchannel geometries to improve contact between phases, light distribution, and mass transfer.
- **Integration of Smart Materials and Sensors**: Utilizing in-situ sensors and smart materials with responsive qualities to monitor and manage the treatment process in real time.
- **Hybrid Systems with Biological Processes**: Additional research into combining microreactors with cutting-edge biological treatment techniques like enzyme-based bioremediation or microbial fuel cells.

- **Nanomaterial-Enhanced Microreactors**: Enhancing performance in microreactor systems by incorporating nanomaterials with specific characteristics (such as increased photocatalytic activity or high adsorption capacity).
- **Modeling and Simulation**: Optimizing the design and functioning of integrated microreactor systems through the use of sophisticated CFD and reactive transport modeling.
- **Life Cycle Assessment**: Carrying out thorough life cycle analyses to appraise the economic and environmental viability of remediation technologies based on hybrid microreactors.

12.7 CONCLUSION

The environmental remediation technology based on hybrid and integrated microreactors has shown remarkable capability to minimize the increasing rate of water pollution. Their synergistic effects are manifesting sustainability, more efficiency and flexibility to eliminate the broad-spectrum pollutants including CECs by incorporating the distinct benefits of microreactors with other emerging technologies. World researchers and scientists are continuously focusing on the exact path for the broad application of these innovative technologies for the treatment of wastewater and the remediation of the environment accepting the existing challenges with scaling-up, integration complexity and cost-effectiveness. Probably, microscale reactors and macro-scale processes will be integrated in the future of water treatment technologies to incorporate hybrid systems that combine the best features of both fields for achieving a more sustainable and greener future.

ACKNOWLEDGMENT

The authors are thankful to the Department of Chemical Engineering, University of Calcutta; Department of Chemistry, Brainware University; and School of Chemical Engineering, KIIT Deemed to be University for the successful completion of this manuscript.

REFERENCES

Adama, A. B., and Shing, W. H. (2020) "Design, fundamental principles of fabrication and applications of microreactors." *Processes* 8, no. 8: 891.

Ahmed, Y., Garlisi, C., Peralta, R., Moreira, M., Puma, G. L., and Palmisano, G. (2023) "Multiphysics computational fluid dynamics (CFD) modelling of diclofenac amide removal by photocatalytic oxidation on Fe-TiO2/N-TiO2 thin films microreactor." *Chemical Engineering Science* 274: 118662.

Aiden, P., Hodgson, Z., and Merk, B. (2018). "Advanced micro-reactor concepts." *Progress in Nuclear Energy* 107: 61–70.

Amit, B., Devi, S. S., and Chakrabarti, T. (2011) "Azo dyes: Past, present and the future." *Environmental Reviews* 19: 350–371.

Andrew, R. B., Mason, B. P., Sylvester, K. T., and McQuade, D. T. (2007) "Improving solid-supported catalyst productivity by using simplified packed-bed microreactors." *Angewandte Chemie-International Edition in English* 46, no. 10: 1698.

Arturo, M., Plouffe, P., Patience, G. S., and Roberge, D. M. (2019) "Experimental methods in chemical engineering: Micro-reactors." *The Canadian Journal of Chemical Engineering* 97, no. 10: 2578–2587.

Barbora, K. (2025) "A critical review on applications of microreactors for the treatment of polluted water with organic dyes." *Science of The Total Environment* 969: 178897.

Bugra, A., and Aras, O. (2022) "Usability, durability and regeneration of Ag/ZnO coated microreactor for photocatalytic degradation of methylene blue." *Journal of Molecular Structure* 1251: 132003.

Cheung, S. H., Bandyopadhyay, A., Bose, S., and Weitz, D. A. (2009) "Double emulsion droplets as microreactors for synthesis of mesoporous hydroxyapatite." *Chemistry of Materials* 21, no. 22: 5548–5555.

Chun-Xia, Z., He, L., Qiao, S. Z., and Middelberg, A. P. (2011) "Nanoparticle synthesis in microreactors." *Chemical Engineering Science* 66, no. 7: 1463–1479.

Diório, A., Bergamasco, R., and Vieira, M. F. (2024). "Performance of tube-in-tube membrane microreactor in E2/EE2 degradation via UV/H2O2/TiO2 process." *Desalination and Water Treatment* 320: 100852.

Donia, F., Lee, A. F., Wilson, K., Jalili, R., and Caruso, R. A. (2019) "Printing approaches to inorganic semiconductor photocatalyst fabrication." *Journal of Materials Chemistry A* 7, no. 18: 10858–10878.

Elizabeta, Z. I., Povijač, K., Gilja, V., Tomašić, V., and Gomzi, Z. (2022) "Photocatalytic degradation of acetamiprid in a rotating photoreactor-Determination of reactive species." *Catalysis Communications* 169: 106474.

Fang, H., Chen, G., Yu, Y., Zhou, Y., Zheng, Y., and Hao, S. (2015) "The synthesis of condensed C-PDA–gC 3 N 4 composites with superior photocatalytic performance." *Chemical Communications* 51, no. 31: 6824–6827.

Gorges, R., Meyer, S., and Kreisel, G. (2004) "Photocatalysis in microreactors." *Journal of Photochemistry and Photobiology A: Chemistry*, 167, no. 2–3: 95–99. https://doi.org/10.1016/j.jphotochem.2004.04.004

Haocui, Z., Chen, G., Yue, J., and Yuan, Q. (2009) "Hydrodynamics and mass transfer of gas–liquid flow in a falling film microreactor." *AIChE Journal* 55, no. 5: 1110–1120.

Helena, B., Kamenická, B., Weidlich, T., Beneš, L., Vlček, M., Lacina, P., and Švec, P. (2022) "Application of Raney Al-Ni alloy for simple hydrodehalogenation of Diclofenac and other halogenated biocidal contaminants in alkaline aqueous solution under ambient conditions." *Materials* 15, no. 11: 3939.

Hu, L., Wei, H., Zhang, Y., Zhang, S., and Li, B. (2014). "TiO2/carbon paper composite materials with hierarchically porous structure for photocatalysis." *Materials Letters*, 119: 88–91.

Huan, Z., Liu, Y., Liu, N., and Kang, S. (2023) "Understanding the interface properties of photocatalytic reactors for rational engineering applications." *Chemical Engineering Journal* 472: 145057.

Hunge, Y. M., Mahadik, M. A., Patil, V. L., Pawar, A. R., Gadakh, S. R., Moholkar, A. V., Patil, P. S., and Bhosale, C. H. (2017) "Visible light assisted photoelectrocatalytic degradation of sugarcane factory wastewater by sprayed CZTS thin films." *Journal of Physics and Chemistry of Solids* 111: 176–181.

Jamil, Q., Žener, B., Putar, U., and Matoh, L. (2024). "Continuous flow photocatalytic reactor for degradation of selected pollutants: Modeling, kinetics, mineralization rate, and toxicity assessment." *Heliyon*, 10, no. 21.

Jiayi, Q., and Zeng, H. (2017) "Photocatalysts fabricated by depositing plasmonic Ag nanoparticles on carbon quantum dots/graphitic carbon nitride for broad spectrum photocatalytic hydrogen generation." *Applied Catalysis B: Environmental* 209: 161–173.

Juan Carlos, C., Nair, V., Kuna, E., and Łomot, D. (2018) "Development of photocatalyst coated fluoropolymer based microreactor using ultrasound for water remediation." *Ultrasonics Sonochemistry* 41: 297–302.

Krishna, M. R., and Biswas, S. (2016) " Micro-reactors – An overview." *The Scitech Journal* 3, no. 4: 31–41.

Linbo, L., Xiang, N., and Ni, Z. (2020) "Droplet-based microreactor for the production of micro/nano-materials." *Electrophoresis* 41, no. 10–11: 833–851.

Liu, M., and Zhao, K. (2021). "Engineering active micro and nanomotors." *Micromachines* 12, no. 6: 687.

Lu, Z., and Yue, J. (2025) "Packed bed microreactors for sustainable chemistry and process development." *Chemistry* 7, no. 2: 29.

Luka, S., Pohar, A., Likozar, B., and Čeh, M. (2016) "Mechanism and kinetics of phenol photocatalytic, electrocatalytic and photoelectrocatalytic degradation in a TiO2-nanotube fixed-bed microreactor." *Chemical Engineering Journal* 303: 292–301.

Lun, Y. K., Zhang, X., Lau, W. N., and Aranda, R. M. (2005) "Experiments and modeling of membrane microreactors." *Catalysis Today* 110, no. 1–2: 26–37.

Luyang, H., Wei, H., Zhang, Y., Zhang, S., and Li, B. (2014) "TiO2/carbon paper composite materials with hierarchically porous structure for photocatalysis." *Materials Letters* 119: 88–91.

Marzieh, A., Farhadian, M., Solaimany Nazar, A. R., Hajiali, M., and Noorbaksh, A. (2024) "The tetracycline degradation in a photocatalytic fuel cell microreactor using a ZnO nanorod/Bi2MoO6/ZIF-67 photocatalyst responsive to visible light." *Energy Conversion and Management* 312: 118565.

Mashuri, S., Izati, S., Ibrahim, M. L., Kasim, M. F., Mastuli, M. S., Rashid, U., Abdullah, A. H., and Islam, A. (2020) "Photocatalysis for organic wastewater treatment: From the basis to current challenges for society." *Catalysts* 10, no. 11: 1260.

Matic, K., Zagar, K., Suhadolnik, L., Ceh, M., and Drazic, G. (2013) "Highly efficient TiO2-based microreactor for photocatalytic applications." *ACS Applied Materials & Interfaces* 5, no. 18: 9088–9094.

Milica, S. M., Šekuljica, N., Dajić, A., Mihajlović, M., Marković, M., Knežević-Jugović, Z., and Mijin, D. (2023) "Microfluidic technology for macro systems: Removal of textile dyes from wastewater in a microreactor." In *E3S Web of Conferences*, vol. 436, p. 10003. EDP Sciences.

Norbert, K., Gottsponer, M., and Roberge, D. M. (2011) "Scale-up concept of single-channel microreactors from process development to industrial production." *Chemical Engineering Journal* 167, no. 2–3: 718–726.

Pil, P. C., and Kim, D. P. (2010) "Dual-channel microreactor for gas– liquid syntheses." *Journal of the American Chemical Society* 132, no. 29: 10102–10106.

Pohar, A., Plazl, I., and Žnidaršič-Plazl, P. (2009) "Lipase-catalyzed synthesis of isoamyl acetate in an ionic liquid/n–heptane two-phase system at the microreactor scale." *Lab on a Chip* 9, no. 23: 3385–3390.

Prashant, L. S., Gumfekar, S. P., Bhanvase, B. A., Sonawane, S. H., and Pimplapure, M. S. (2018) "A review on microreactors: Reactor fabrication, design, and cutting-edge applications." *Chemical Engineering Science* 189: 431–448.

Qasim, J., Žener, B., Putar, U., and Matoh, L. (2024) "Continuous flow photocatalytic reactor for degradation of selected pollutants: Modeling, kinetics, mineralization rate, and toxicity assessment." *Heliyon* 10, no. 21: 833–846.

Raffaella, T., Bersano, A., and Segantin, S. (2021) "Review of nuclear microreactors: Status, potentialities and challenges." *Progress in Nuclear Energy* 138: 103822.

Sanaa, H., Aristodemou, E., Manos, G., Al-Salem, S. M., and Constantinou, A. (2020) "Modelling of packed bed and coated wall microreactors for methanol steam reforming for hydrogen production." *RSC Advances* 10, no. 68: 41680–41692.

Seelam, P. K., Huuhtanen, M., and Keiski, R. L. (2013) "Microreactors and membrane microreactors: Fabrication and applications." *Handbook of Membrane Reactors* 2: 188–235.

Shu, H., Lu, X., Li, Z., Ravishankar, H., Wang, J., and Wang, X. (2018) "A biomimetic approach towards the synthesis of TiO2/carbon-clay as a highly recoverable photocatalyst." *Journal of Photochemistry and Photobiology A: Chemistry* 351: 131–138.

Soomin, S., Jung, P. H., Park, J., Chae, D., Huh, D., Byun, M., Ju, S., and Lee, H. (2018) "Customizable 3D-printed architecture with ZnO-based hierarchical structures for enhanced photocatalytic performance." *Nanoscale* 10, no. 46: 21696–21702.

Tamara, S. J., Firak, D. S., Ribeiro, R. R., and Peralta-Zamora, P. (2020) "Use of scavenger agents in heterogeneous photocatalysis: Truths, half-truths, and misinterpretations." *Physical Chemistry Chemical Physics* 22, no. 27: 15723–15733.

Thomas, W., ed. (2013) *Microreactors in organic chemistry and catalysis*. John Wiley & Sons.

Wang, Y., Zhang, S. Y. W., and Guan, Y. (2021) "A 3D printed synergistic aerogel microreactor toward stable and high-efficiency photocatalytic degradation." *Materials Today Chemistry* 22: 100566.

Xi, L., Li, Z., Liu, Y., Tang, B., Zhu, Y., Razal, J. M., Pakdel, E., Wang, J., and Wang, X. (2020) "Titanium dioxide coated carbon foam as microreactor for improved sunlight driven treatment of cotton dyeing wastewater." *Journal of Cleaner Production* 246: 118949.

Xiaoyao, T., and Li, K. (2013) "Membrane microreactors for catalytic reactions." *Journal of Chemical Technology & Biotechnology* 88, no. 10: 1771–1779.

Xingjun, Y., Zhang, Y., Du, L., Liu, J., and Yao, J. (2015) "Review of the applications of microreactors." *Renewable and Sustainable Energy Reviews* 47: 519–539.

Yoshihisa, M., Ichimura, T., Ohba, N., Kumada, S., Sakeda, K., Suzuki, T., Tanibata, H., and Murata, T. (2007) "Recent progress on photoreactions in microreactors." *Pure and Applied Chemistry* 79, no. 11.

Yoshihisa, M., Ohba, N., Kumada, S., Sakeda, K., Suzuki, T., and Ichimura, T. (2008) "Photocatalytic reactions in microreactors." *Chemical Engineering Journal* 135: S303–S308.

Yufei, G., Li, J., Yuan, Y., Li, L., Zhang, M., Zhou, C., and Lin, Z. (2016) "A rapid microwave-assisted thermolysis route to highly crystalline carbon nitrides for efficient hydrogen generation." *Angewandte Chemie* 128, no. 47: 14913–14917.

Zahhad, A., Mohamed, E., Hammad, A. S., Radwan, A., Memon, S., El-Shazly, A. H., and El-Kady, M. F. (2020) "Dye removal with magnetic graphene nanocomposite through micro reactors." *International Journal of Solar Thermal Vacuum Engineering* 2, no. 1: 79–94.

Zhang, H., Chen, G., Yue, J., and Yuan, Q. (2009). "Hydrodynamics and mass transfer of gas–liquid flow in a falling film microreactor." *AIChE Journal*, 55, no. 5: 1110–1120.

Zhiwei, W., Liu, Z., Huang, J., Chen, Y., Su, R., He, J., and Lv, G. (2022) "Zr6O8-porphyrinic MOFs as promising catalysts for the boosting photocatalytic degradation of contaminants in high salinity wastewater." *Chemical Engineering Journal* 440: 135883.

Zhongjian, T., Shi, H., Zheng, Y., and Cao, Y. (2023) "A 3D homogeneous microreactor with high mixing intensity at wide Re range for MOF preparation and POCT application." *Chemical Engineering Journal* 476: 146481.

13 Microreactors for Renewable Energy Production

Prasenjit Chakraborty and Ramesh Kumar

13.1 INTRODUCTION

Global energy policies must prioritize increasing the proportion of low-carbon energy for achieving their climate variation objectives (Aziz et al. 2024; Li et al. 2024). A number of reasons, including population expansion, economic and technological improvements, and the growing energy demand brought on by fast industrialization and infrastructure development, have contributed to the sharp increase in energy consumption during the past 50 years (Vishwakarma et al. 2023). Considerable attempts have been made to encourage the use of renewable energy in order to satisfy this increasing apprehension.

Technologies that can provide baseload power with less carbon are crucial for reaching zero carbon footprint, which enhances the production of renewable energy. Renewable energy sources can meet present demands without jeopardizing the needs of future energy. Renewable energy sources (RES) such as solar, hydrogen, geothermal, ocean energy, wind, and hydroelectricity are regarded as viable ways of energy generation. Nonetheless, their sporadic nature complicates the provision of a reliable energy supply, necessitating alternate sources to serve as key load or baseload suppliers. Due to its zero-carbon emissions, there is a growing transition toward renewable energy (Panwar et al. 2011).

Renewable energy sources generated 29% of the world's electricity in 2020, compared to 27% the previous year (Dey et al. 2022). While industrial bioenergy consumption increased by 3%, biofuel consumption dropped by the same amount due to lower demand for oil and mixed biofuels. The most rapid annual rise in renewable power output since the 1970s will occur in 2021, when it is projected to exceed 8300 TWh, an increase of more than 8%. Despite falling demand for all other fuels in 2020, the usage of renewable energy climbed by 3%. An increase of nearly 7% in the production of energy from sources of renewable energy was the main factor (Dey et al. 2022). The shifting of fossil fuel-based energy to renewable energy sources signifies a massive societal and economic upheaval that goes well beyond the realm of revolution of the energy sector. Integrating renewable energy sources with microreactors can yield a carbon-neutral, dependable, and advanced energy system (Kutscher et al. 2018; Zhou et al. 2015).

Microreactors are developing as novel nuclear energy technology capable of delivering low-carbon electricity. Their development probability is enhanced by their

DOI: 10.1201/9781003633761-13

capacity to penetrate new markets where large-scale energy technologies are impractical (Black et al. 2023; Testoni et al. 2021). These compact reactors, often yielding between 1 MWe and 20 MWe, may produce electricity utilizing advanced non-light water reactor and light water reactor configurations (Thomas and Ramana 2022; Ingersoll, 2009). The highest power outputs of 50 MWe may be attained. They can supply power, generate process heat for immediate consumption, or both (Dittmeyer et al. 2017). The obstacles pertaining to production and operational expenses, together with legal and institutional impediments, are surmounted. Microreactors possess the capacity to address unfulfilled demands in energy portfolios when fossil fuel constraints are apparent (Ekinci et al. 2024). Microreactors may also gain competitiveness in regions without extensive electrical networks, where fuel transportation is difficult or costly, where economies of scale are inapplicable, or where spatial limitations or particular climatic circumstances render renewable energy impractical (Yadav et al. 2023; Yan et al. 2020). Current technical progress in microreactors design presents both opportunities and challenges for their incorporation into energy portfolios. Advancements in microreactors can support national low-carbon initiatives and enhance resilience goals in regions where extreme weather and unforeseen catastrophes endanger safety, protection, and financial security. The importance of microreactors increases amid worldwide apprehensions regarding power safety and durability.

Utilizing microreactors to bolster adaptability can augment preparedness and restoration from difficulties, thereby ensuring operational continuity in vital sectors such as important government services, energy-dependent industries, critical medical amenities, and rapid connectivity for remote workers (Mehla et al. 2022). Despite ongoing economic challenges hindering the competitiveness of microreactors against other carbon-free sources of energy, certain cost forecasts indicate that microreactors could be competitive in markets in which big nuclear technologies are unfeasible (Yadav et al. 2023). The microreactors are used in the production of various renewable energies like biodiesel, bio-oil, and hydrogen. The use of microreactors in the transesterification process of biodiesel manufacture allows for a faster rate of chemical reactions, shorter reaction times, and faster synthesis of products (Madhawan et al. 2018). The earliest known microreactors utilized for the production of biodiesel were capillary type. Subsequently, numerous progressive microreactors were constructed utilizing various methodologies and ingredients.

The current chapter addresses the generation of renewable energy through microreactors. The employment of various microreactors and their design approaches toward renewable energy production, particularly in bio-fuel production and hydrogen generation are discussed. Technical benefits, financial challenges, and future scope of microreactors have been explored.

13.2 DESIGN AND CONSTRUCTION PROCESS OF MICROREACTORS

Microreactors differ from traditional chemical reactors in that they are based on fundamental concepts that govern their design and operation. These tiny devices have larger surface-area-to-volume ratios despite lower operational volumes, making scaling crucial. Reaction conditions may be precisely controlled, and heat and mass transport techniques are generally made more effective, attributed to this scaling down (Bugay et al. 2024). The type of flow and stirring mechanism have a significant impact on microreactor operation, which affects the quality of the product and response consistency. Optimal response results necessitate an in-depth comprehension of patterns of flow due to the dynamic nature of the flow of fluid in microchannels. A low-cost production approach that enables three-dimensional designs and the use of suitable substances is necessary to develop a productive and portable microreactor. The reactor design precludes the use of conventional thin film processes due to the device's need for a continuous flow of reactants and products. The integration and modification of conventional thin film processes make it usable for proper use (Holladay et al. 2004). To facilitate a comprehensive understanding of microreactor fabrication, detailed analysis of the choice of materials, the selection of microreactors of wall-coated and packed bed, different manufacturing schemes, stability assessments, making of catalysts and their assessment, and performance evaluations are needed (Yadav et al. 2023).

13.2.1 Choice of Materials

Chemical suitability, suitable thermal characteristics, and structural stability are the three most important criteria to consider while selecting fabrication materials. Silicon, metals, polymer, and low-temperature co-fired ceramics are the four key groups based on which materials are selected. Polymeric substances have been widely employed in microreactors as relevant materials. The choosing methods of materials is challenging due to the manufacturing process that may alter the characteristics of the materials. Microreactors possess the capability to be produced using cost-effective and large volume systems, nevertheless, considerations about selection of materials, design, and production must prioritize the price to develop a profitable equipment. Ceramic oxides, metals, and organic polymers can also act as supports of catalyst for micro conduits. Ceramics, on the other hand, are frequently employed as supporting substance of catalysts in chemical reactions. Polymers, quartz, metal alloys and glass are among the materials that can be used to construct microreactors (Zhu et al. 2022; Wang et al. 2021). Microreactors made of silicon are ideal because of the material's great strength, ease of fabrication through micromachining, and inertness (Pattekar et al. 2001). A silicon chip-constructed microreactor device with one channel layout and a 400 μm × 1000 μm cross-section composed of a bed microchannel packed with a Cu/ZnO catalyst was effectively designed and experimented by Pattekar and Kothare (2004). A maximum 88.19% of conversion of methanol to hydrogen gas which produces 9.48W of power for use in fuel cell was obtained by the fabricated microreactor.

13.2.2 Construction Methodologies of Microreactors

Numerous microchannel manufacturing methods are used, including lithography, etching process (wet and dry), laser cutting, micro-drilling and micro-milling, micromachining, and wire sawing. The construction of microreactors is affected by multiple parameters such as the requisite total dimensions, aspect ratio, cross-section,

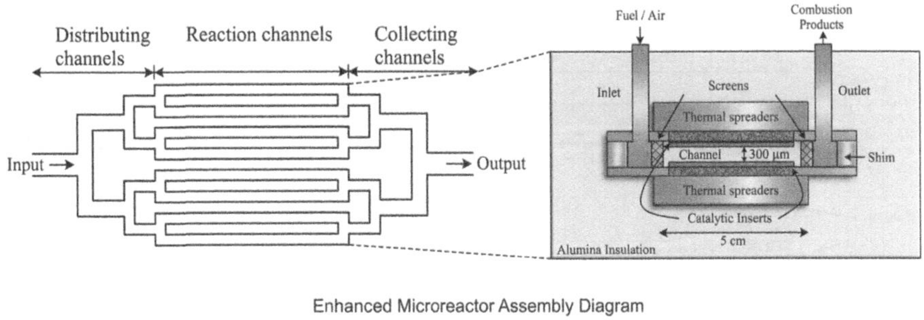

FIGURE 13.1 (a) Single phase microreactor with heat and fluid flow, (b) Structure of single phase microreactor.

(Taken from Bugay et al. 2024.)

geometric limitations which facilitate the materials flow under certain circumstances, internal wall construction that can optimize mixing and improve the activity of the catalyst, pressure and temperature specifications of various materials utilized, and suitability of the catalyst components and inner surface of the microchannel substances.

An autothermal microreactor of three-dimensional printed uniform porous framework was created by Zheng et al. (2022) to facilitate combustion processes for effective and continuous hydrogen production that ensured a steady supply of hydrogen in fuel cells. Sanz et al. (2016) fabricated a microreactor of stainless steel by micromilling technique for steam reforming of methanol. Higher transformation of methanol was reported through the prepared microreactor. The SLM approach for microreactor design was designed by Liu et al. (2020) to create metal catalysts with various porosities and shapes for use in steam reforming. They showed that the total surface area accessible for loading of catalyst is increased when various metal particles adhere to the surface of the porous catalytic that is produced using SLM. A potential method for the economical fabrication of complex microreactors suitable with numerous substances is required for additive manufacturing. In order to create high-tech microreactors that are suitable with different materials, researchers are concentrating on finding inexpensive methods (Figure 13.1).

13.2.3 Design and Synthesis of Catalyst

Catalyst performance is highly dependent on its chemical and physical characteristics, which are in turn related to the method employed for preparation, making the preparation step crucial. Many catalyst parameters, including surface area, crystal arrangement, porosity, and size of the particle are sensitive to the manufacturing technique. The selectivity, activity of catalyst, and stability of the compound can be greatly affected by these parameters (Yadav et al. 2023). Thus, it is crucial to choosing a proper catalyst preparation process with attention to guarantee the required characteristics for a certain proposal. The efficiency of microreactors used for production of hydrogen is greatly affected by the catalyst layer; hence, it is constructed

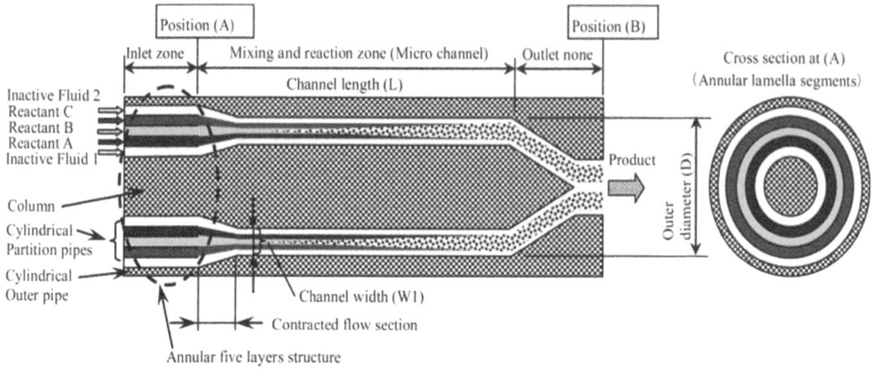

FIGURE 13.2 Schematic of microreactor flow

(Taken from Bugay et al. 2024.)

with precision. Different methods exist for depositing this catalyst layer onto the surface of the microchannel inside the microreactor. The tolerance to heat and chemical flexibility of the material are key factors in determining the most effective preparation technique (Okonkwo and Otor 2021). Pre- and post-coating are two primary ways to coat the microreactor. The catalyst is deposited directly into the microchannel before the microreactor is fully constructed, known as pre-coating. Contrarily, post-coating entails adding of coating onto a microreactor that has already been created. Complete cleaning of microchannel is essential for accurate employing of the catalyst.

According to Tajrishi et al. (2018), the microchannel must be pretreated before coating is applied in order to greatly enhance the catalyst's durability. A novel catalyst $CuFeMg/Al_2O_3$ was created by Zhang et al. (2021). This structure demonstrated exceptional catalytic activity and stability throughout extended reactions. They found that, compared to the conventional catalyst, the produced catalyst resulted in a 2.5-fold increase in hydrogen generation rate. Over the course of the 100-h test, the resulting gas concentration and methanol transformation rate were both maintained at 85% (Zhang et al. 2021) (Figure 13.2).

13.3 MICROREACTOR TYPES FOR RENEWABLE ENERGY

Microreactor technology has been suggested to facilitate the renewable energy generation process within a brief timeframe. This is available in diverse forms and configurations, engineered to facilitate effective reaction and mixing. The generation of renewable energy using microreactors can increase the product synthesis rate by reducing reaction time and producing a high rate of reaction. Microreactors utilized for the production of hydrogen, bio-oil or biodiesel, and methanol were initially known to be capillary microreactors. Subsequently, numerous sophisticated microreactors were constructed utilizing diverse methodologies and materials.

13.3.1 Microtube-Based Microreactor

Microtube technology has garnered significant interest recently due to its capacity for greater mass transfer, effective heat dissipation, and lower diffusion length. Silva et al. (2010) conducted the process of transesterifying soybean oil into fatty acid ethyl esters. The reaction was conducted by them in a microtube reactor with 0.76 mm of inner diameter and volume of 36.5 mL. Significant yields using conditions of 398 K temperature, 1:20 molar ratio (oil and ethanol), and 20 MPa pressure were achieved. An increase in the microreactor's capacity for biodiesel production was investigated by Billo et al. (2015). They developed a higher scale scheme of microreactor system that reduced the reaction time needed from hours to minutes, also increased the capability of biodiesel production per reactor volume. The biodiesel production process using a 0.508 mm inner diameter and height of 1.2 m microtube reactor was conducted through methanol and palm oil transesterification employing catalyst KOH (Kaewchada et al. 2016). A substantial rise in %FAME was observed at low residence duration and the catalyst amount increased from 5 to 10 mg/g, according to their research. Using catalyst loading of 10 mg/g, 6:1 methanol-to-oil ratio and a 5 s residence duration produced 97.1% FAME (Kaewchada et al. 2016).

13.3.2 Membrane-Based Microreactor

The membrane process is renowned for its large surface area per unit volume, great selectivity, and capacity to regulate constituent mixing among two phases. Membrane reactors are typically categorized based on four criteria: the membrane reactor nature like catalytic or inert, reactor configuration, membrane varieties such as inorganic, organic, and porous involved in the reaction, and the specific reactions (esterification and dehydrogenation) taking place within the membrane reactor. Membrane process of separation is a prevalent method for achieving high pure hydrogen (typically exceeding 99.99%) in industrial applications. It is attractive to integrate the hydrogen purification process with the generating reactors to create compact equipment for microreactors. In the process of producing biodiesel, the membrane either keeps the unreacted triglycerides inside the membrane or extracts glycerol from the resulting stream (Saleh et al. 2010; Guerreiro et al. 2006). Achmadin et al. (2010) created a membrane microreactor for methanol and triolein transesterification which has a 63.5 mm diameter. The membrane's pores serve as a reaction site, achieving an 80% conversion at 35 °C temperature and reaction duration of 19 minutes. Greater recovery of hydrogen in traditional systems requires very high temperatures (300 °C or higher) and very high pressures (typically more than 250 psi).

Microreactors' ability to accommodate complicated shapes and small feature sizes has led to the fabrication of very thin membranes made of metal, which in turn enable higher recovery of hydrogen at low pressures.

13.3.3 Oscillatory Flow Reactors

One innovative type of reactor that is gaining popularity is the oscillatory flow reactor (OFR), which uses tubes with orifice plate baffles that are evenly spaced. An oscillating motion appears over the total system fluid flow, providing a flow arrangement that aids in competent mass transfer and blending. Larger residence times can be attained because the degree of mixing does not depend on total flow (Harvey et al. 2003). Radial mixing is improved by the regulated motion of oscillation. A novel oscillatory baffled reactor (mesoscale) of 5 mm in diameter and a sequence of tubes measuring 340 mm in length was developed by Phan et al. (2011) to constantly evaluate the biodiesel production. The results showed that production of methyl ester by the binary-phase reaction in the reactor. Production of biodiesel through jatropha oil transesterification was investigated by (Highina and Bugaje, 2014) using a vertical tube arrangement stainless-steel reactor consisting 1.4 m height and 0.1 m inner diameter. They reported that at a ratio of 6:1 (methanol:oil), and NaOH (1 wt%) as catalyst, the oscillatory based baffled reactor scheme showed jatropha oil transformation of 94% in 5 min.

13.3.4 Packed Bed Microreactor

The microchannel of a packed bed reformer is densely packed with tiny catalyst materials (Zhuang et al. 2020; Chein et al. 2011). In spite of its reputation for large efficiency with compact size, packed bed reactors display notable variations in temperature and a high-pressure drop (Karim et al. 2005). Zhuang et al. (2020) created a packed bed microreactor that distributed fluids in channels using a split inlet. In the packed bed reactor, the catalyst particles consisted of CuO, ZnO, and Al_2O_3. Higher conversion of methanol (94%) was achieved using developed packed bed microreactor and catalyst consisting ZnO, CuO, and Al_2O_3. A typical fixed bed reactor with packed bed plate type microreactor was compared in a study by Wang et al. (2013). They observed slow activity in the reactor inlet with a gradient distribution of catalyst but greater activity throughout the microreactor channel. Wang et al. (2023) developed a unique approach to low-scale dispersed hydrogen generation using steam methane reforming process utilizing a combined particle packed bed design. According to their design, every channel of the structure's multiple channel frame has a large thermal conductivity and is packed with Ni/Al_2O_3 catalyst elements. A hybrid packed bed was suggested as a solution to the challenges of generation of hydrogen in tiny scale, including resistance of high mass and heat transfer of the current fixed bed reactors.

13.4 TECHNOLOGICAL ADVANCES, FUTURE POSSIBILITIES, AND CHALLENGES

Innovative design of microreactor provides numerous benefits to consumers. Due to detailed technical infrastructure obligation in their manufacturing process, microreactors are gaining significant interest. Reactivity monitoring is a crucial technical characteristic that is necessary for the secure and effective operation of microreactors. The developments in technology and benefits of microreactors make them unique. Due to their automated and remote-controlled schemes, these reactors can run safely in distant locations without constant human oversight. Additionally, microreactors provide an excellent solution to generate power in the event of a power loss or natural catastrophe. Conventional ISO containers are designed to accommodate a number of microreactors, making their transportation easier (Peakman et al. 2018). On the other hand, there may be certain regulatory hurdles associated with transporting fuel-based microreactors. Due to their adaptability, low cost, and ability to increase energy security and resilience, microreactors are becoming increasingly popular.

Microreactors have the potential to solve energy needs in inaccessible or remote regions due to their mobility and ease of installation. Their rapid startup and ability to begin power generation, in contrast to massive nuclear power plants, allows for substantial reduction in both time and expenses (Morones-García and François, 2024; Peakman and Lindley, 2023; Testoni et al. 2021). Despite the products are easily transportable for the conventional container design of microreactors, it has various regulatory hurdles. Economic evaluations assess the benefits and expenses associated with microreactors. Microreactors are economically superior to generators powered by diesel. Microreactors are a more cost-effective long-term solution than diesel generators because of their large initial price and fuel consumption (Xue et al. 2021).

Microreactors are typically more affordable when it comes to the cost of producing power (Peakman and Lindley, 2023). The compact dimensions and industrial manufacturing capabilities of microreactors contribute to cost reduction. Furthermore, maintenance and operation expenses of microreactors can be decreased and its designs and distribution conditions have an impact on prices. The manufacturing process of the microreactor is the main difficulty. The performance of the microreactor can be affected by fabrication issues such as catalyst dislocation and improper loading of catalyst (Qing et al. 2020). Furthermore, because ceramic-based materials are fragile, conventional machining methods are inappropriate for their use (Li et al. 2022). Higher porosity of catalyst is recognized to be advantageous for enhancing the efficiency of the microreactor though it also poses some restrictions on the microreactor's construction. Another major obstacle to the widespread use of fuel cell devices in steam reforming is their high cost (Herdem et al. 2019). The energy and power density of microreactors may require additional effort for upgrading. Nonetheless, it has considerable problems like irregularity and inadequate transportation, which complicate its incorporation into the power system (Tazvinga et al. 2014). Creative approaches are necessary to address these challenges. To optimize solar energy utilization, it can be integrated with additional techniques for energy transformation to enhance its total competence and applicability.

Consequently, in order to increase hydrogen generation, extensive research into systems based on many microreactors is required. Methanol is the ideal fuel for microreactor based steam reforming. Microreactors can use a variety of fuels and applied in the renewable energy sector. In the coming years, there will be plenty of chances to investigate the tremendous potential of microreactors for use in the renewable energy sector. In order to overcome the conventional power generation restrictions, a lot of research has been done toward scalability in renewable energy like hydrogen, bio-ethanol, and bio-oil production utilizing microreactors.

13.5 CONCLUSION

Technological developments in microreactors for use in the renewable energy sector are abridged in this chapter. The article delves into different microreactors and the methods used in their design to generate renewable energy, with a focus on bioethanol, bio-oil and hydrogen production. Future possibilities, financial hurdles, and technical advantages and challenges of microreactors have all been discussed. This study offers useful information into microreactor function by meticulously analyzing multiple factors, thereby enhancing researchers' understanding of the method of reforming or conversion reaction in microreactors. Present chapter offers useful information into microreactor function by meticulously analyzing multiple factors, thereby enhancing researchers' understanding of the various renewable energy generation method in microreactors. As a result, researchers are able to develop and improve microreactors for particular purposes utilizing the insight learned from this chapter. The chapter underscores the significance of designing microreactor and optimizing heat usage to attain outstanding results. In general, microreactor process is highly promising and has the potential for application in numerous renewable energy sectors, thus leading us to a greener and sustainable society in future.

REFERENCES

Achmadin, L. M., Misri, G., Mohammad, N., Siswa, S., Young, J. Y. Membrane microreactor in biocatalytic transesterification of triolein for biodiesel production. *Biotechnol. Bioprocess Eng.* 15 (2010), 911–916.

Aziz, S., Ahmed, I., Khan, K., Khalid, M. Emerging trends and approaches for designing net-zero low-carbon integrated energy networks: A review of current practices. *Arab. J. Sci. Eng.* 49 (2024), 6163–6185.

Billo, R. E., Oliver, C. R., Charoenwat, R., Dennis, B. H., Wilson, P. A., Priest, J. W., Beardsley, H. A cellular manufacturing process fora full scale biodiesel microreactor. *J. Manuf. Syst.* 37 (2015), 409–416. https://doi.org/10.1016/j.jmsy.2014.07.004

Black, G., Shropshire, D., Araújo, K., van Heek, A. Prospects for nuclear microreactors: A review of the technology, economics, and regulatory considerations. *Nucl. Technol.* 209 (Suppl. S1) (2023), S1–S20.

Bugay, C. A., Caballas, M. C., Mercado, S. B., Rubio, J. F., Serote, P. K., Villarte, P. N., Rubi, R. V. C. A review of microreactors for process intensification. *Eng. Proc.* 67(1) (2024), 21.

Chein, R. Y., Chen, Y. C., Lin, Y. S., Chung, J. N. Experimental study on the hydrogen production of integrated methanol-steam reforming reactors for PEM fuel cells. *Int. J. Therm. Sci.* 50 (2011), 1253–1262, https://doi.org/10.1016/j.ijthermalsci.2011.03.001

Dey, S., Sreenivasulu, A., Veerendra, G. T. N., Rao, K. V., & Babu, P. A. (2022). Renewable energy present status and future potentials in India: An overview. *Innov. Green Dev.* 1(1), 100006.

Dittmeyer, R., Boeltken, T., Piermartini, P., Selinsek, M., Loewert, M., Dallmann, F., Kreuder, H., Cholewa, M., Wunsch, A., Belimov, M. et al. Micro and micro membrane reactors for advanced applications in chemical energy conversion. *Curr. Opin. Chem. Eng.* 17 (2017), 108–125.

Ekinci, F., Guzel, M. S., Acici, K., Asuroglu, T. The future of microreactors: Technological advantages, economic challenges, and innovative licensing solutions with blockchain. *Appl. Sci.* 14(15) (2024), 6673.

Guerreiro, L., Castanheiro, J. E., Fonseca, I. M., Martin-Aranda, R. M., Ramos, A. M., Vital, J. Transesterification of soybean oil over sulfonic acid functionalised polymeric membranes. *Catal. Today* 118(1-2) (2006), 166–171. https://doi.org/10.1016/j.cattod.2005.12.012

Harvey, P. A., Mackley, R. M., Seliger, T. Process intensification of biodiesel production using a continuous oscillatory flow reactor. *J. Chem. Technol. Biotechnol.* 78(2–3) (2003), 338–341. https://doi.org/10.1002/jctb.782

Herdem, M. S., Sinaki, M. Y., Farhad, S., Hamdullahpur, F., An overview of the methanol reforming process: Comparison of fuels, catalysts, reformers, and systems. *Int. J. Energy Res.* 43 (2019), 5076–5105, https://doi.org/10.1002/er.4440

Highina, K. B., Bugaje, U. B. Production of biodiesel from Jatropha oil using continuous oscillatory baffed reactor arrange ment and homogenous catalyst. *World J Chem Sci Eng* 1 (2014), 1–6.

Holladay, J. D., Wang, Y., Jones, E. Review of developments in portable hydrogen production using microreactor technology. *Chem. Rev.* 104(10) (2004), 4767–4790.

Ingersoll, D. T. Deliberately small reactors and the second nuclear era. *Prog. Nucl. Energy* 51 (2009), 589–603.

Kaewchada, A., Pungchaicharn, S., Jaree, A. Transesterification of palm oil in a microtube reactor. *Can. J. Chem. Eng.* 94(5) (2016), 859–864. https://doi.org/10.1002/cjce.22464

Karim, A., Bravo, J., Gorm, D., Conant, T., Datye, A. Comparison of wall-coated and packed-bed reactors for steam reforming of methanol. *Catal. Today* 110 (2005), 86–91, https://doi.org/10.1016/j.cattod.2005.09.010

Kutscher, C. F., Milford, J. B., Kreith, F. *Principles of Sustainable Energy Systems*, 3rd ed. CRC Press, Boca Raton, FL, 2018, p. 655.

Li, K., Zhou, Y., Huang, X., Xiao, H., Shan, Y. Low-carbon development pathways for resource-based cities in China under the carbon peaking and carbon neutrality goals. *Environ. Sci. Pollut. Res.* 31 (2024), 10213–10233.

Li, X., Fu, D., Xu, W., Yuan, D., Jiang, X., Fu, T., Chu, X., Gao, Y., Zhou, W., Zhou, W. Fabrication of high aspect ratio ceramic micro-channel in diamond wire sawing as catalyst support used in micro-reactor for hydrogen production. *Int. J. Hydrog. Energy* 47 (2022), 35123–35135, https://doi.org/10.1016/j.ijhydene.2022.08.082

Liu, J., Gao, Y., Fan, Y., Zhou, W. Fabrication of porous metal by selective laser melting as catalyst support for hydrogen production microreactor. *Int. J. Hydrog. Energy* 45 (2020), 10–22, https://doi.org/10.1016/j.ijhydene.2019.10.173

Madhawan, A., Arora, A., Das, J., Kuila, A., Sharma, V. Microreactor technology for biodiesel production: A review. *Biomass Convers. Biorefinery*, 8(2) (2018), 485–496.

Mehla, S., Gudi, R. D., Mandaliya, D. D., Hisatomi, T., Domen, K., Bhargava, S. K. Additive manufacturing as the future of green chemical engineering. In *Additive Manufacturing for Chemical SciencesA and Engineering*. Springer Nature, Singapore, 2022, pp. 239–307. https://doi.org/10.1007/978-981-19-2293-0_8

Morones-García, E., François, J. L. Nuclear microreactors: Status review and potential applications; the Mexican case. *Nucl. Eng. Des.* 420 (2024), 113021.

Okonkwo, P. C., Otor, C. A review of gas diffusion layer properties and water management in proton exchange membrane fuel cell system. *Int. J. Energy Res.* 45 (2021), 3780–3800, https://doi.org/10.1002/er.6227

Panwar, N. L., Kaushik, S. C., Kothari, S. Role of renewable energy sources in environmental protection: A review. *Renew. Sustain. Energy Rev.* 15 (2011), 1513–1524.

Pattekar, A. V., Kothare, M. V. A microreactor for hydrogen production in micro fuel cell applications. *J. Microelectromech. Syst.* 13 (2004), 7–18, https://doi.org/10.1109/JMEMS.2004.823224

Pattekar, A. V., Kothare, M. V., Karnik, S. V., Hatalis, M. K. A microreactor for in-situ hydrogen production by catalytic methanol reforming. In *Microreaction Technology*, Springer Berlin Heidelberg, Berlin, Heidelberg, 2001, pp. 332–342, https://doi.org/10.1007/978-3-642-56763-6_34

Peakman, A., Hodgson, Z., Merk, B. Advanced micro-reactor concepts. *Prog. Nucl. Energy* 107 (2018), 61–70.

Peakman, A., Lindley, B. A review of nuclear electric fission space reactor technologies for achieving high-power output and operating with HALEU fuel. *Prog. Nucl. Energy* 163 (2023), 104815.

Phan, A., Harvey, P. A., Rawcliffe, M. Continuous screening of base-catalysed biodiesel production using new designs of meso scale oscillatory baffled reactors. *Fuel Process. Technol.* 92(8) (2011), 1560–1567. https://doi.org/10.1016/j.fuproc.2011.03.022

Qing, W., Liu, F., Yao, H., Sun, S., Chen, C., Zhang, W. Functional catalytic membrane development: A review of catalyst coating techniques. *Adv. Colloid Interf. Sci.* 282 (2020), 102207, https://doi.org/10.1016/j.cis.2020.102207

Saleh, J., Tremblay, A. Y., Dubé, M. A. Glycerol removal from biodiesel using membrane separation technology. *Fuel* 89(9) (2010), 2260–2266. https://doi.org/10.1016/j.fuel.2010.04.025

Sanz, O., Velasco, I., Erez-Miqueo, I. P., Poyato, R., Odriozola, J. A., Montes, M. Intensification of hydrogen production by methanol steam reforming. *Int. J. Hydrogen Energy* 41 (2016), 5250–5259, https://doi.org/10.1016/j.ijhydene.2016.01.084

Silva, C., Castilhos, F., Oliveira, V. J., Filho, C. L. Continuous production of soybean biodiesel with compressed ethanol in a microtube reactor. *Fuel Process. Technol.* 91(10) (2010), 1274–1281. https://doi.org/10.1016/j.fuproc.2010.04.009

Tajrishi, O. Z., Taghizadeh, M., Kiadehi, A. D. Methanol steam reforming in a microchannel reactor by Zn-, Ce- and Zr- modified mesoporous Cu/SBA-15 nanocatalyst. *Int. J. Hydrog. Energy* 43 (2018), 14103–14120, https://doi.org/10.1016/j.ijhydene.2018.06.035.

Tazvinga, H., Zhu, B., Xia, X. Energy dispatch strategy for a photovoltaic–wind–diesel–battery hybrid power system. *Sol. Energy* 108 (2014), 412–420, https://doi.org/10.1016/j.solener.2014.07.025

Testoni, R., Bersano, A., Segantin, S. Review of nuclear microreactors: Status, potentialities and challenges. *Prog. Nucl. Energy* 138 (2021), 103822.

Thomas, S., Ramana, M. V. A hopeless pursuit? National efforts to promote small modular nuclear reactors and revive nuclear power. *Wiley Interdiscip. Rev. Energy Environ.* 11 (2022), e429.

Vishwakarma, C., Dwivedi, A., Yadav, D. The current state and potential of canal top solar power plants in India. *Adv. Mech. Energy Technol.* (2023) 33–41, https://doi.org/10.1007/978-981-19-1618-2_4

Wang, G., Wang, F., Li, L., Zhang, G. Experiment of catalyst activity distribution effect on methanol steam reforming performance in the packed bed plate-type reactor. *Energy* 51 (2013) 267–272, https://doi.org/10.1016/j.energy.2012.12.027

Wang, J., Liu, Z., Ji, C., Liu, L. Heat transfer and reaction characteristics of steam methane reforming in a novel composite packed bed microreactor for distributed hydrogen production. *Energies* 16(11) (2023), 4347.

Wang, Y., Liu, H., Mei, D., Wu, Q., Zhou, H. A novel thermally autonomous methanol steam reforming microreactor using SiC honeycomb ceramic as catalyst support for hydrogen production. *Int. J. Hydrog. Energy* 46 (2021), 25878–25892, https://doi.org/10.1016/j.ijhydene.2021.05.103

Xue, Q., Li, Z., Chen, M., Wang, Y., Yan, B., Luo, G. Co-precipitation continuous synthesis of the Ni-Rh-Ce0.75Zr0.25O2-δ catalyst in the membrane dispersion microreactor system for n-dodecane steam reforming to hydrogen. *Fuel* 297 (2021), 120785, https://doi.org/10.1016/j.fuel.2021.120785

Yadav, D., Lu, X., Vishwakarma, C. B., & Jing, D. Advancements in microreactor technology for hydrogen production via steam reforming: A comprehensive review of experimental studies. *J. Power Sources* 585 (2023), 233621.

Yan, B. H., Wang, C., Li, L. G. The technology of micro heat pipe cooled reactor: A review. *Ann. Nucl. Energy* 135 (2020), 106948.

Zhang, G., Zhao, J., Wang, Q., Yang, T., Zhang, Q., Zhang, L. Fast start-up structured CuFeMg/Al2O3 catalyst applied in microreactor for efficient hydrogen production in methanol steam reforming. *Chem. Eng. J.* 426 (2021), 130644, https://doi.org/10.1016/j.cej.2021.130644

Zheng, T., Zhou, W., Zhong, Y., Yang, Y., Hong, M., Shen, Z. 3D-Printed regular porous structure with trapezoidal multiple microchannels as combustion reaction support for the autothermal methanol steam reforming microreactor for hydrogen production. *Ind. Eng. Chem. Res.* 61 (2022), 2443–2454, https://doi.org/10.1021/acs.iecr.1c04448

Zhou, X., Guo, T., Ma, Y. An overview on microgrid technology. In *Proceedings of the 2015 IEEE International Conference on Mechatronics and Automation (ICMA), Beijing, China*, 2–5 August 2015, pp. 76–81.

Zhu, X. Zhang, Y., Zhang, J., Su, Y., Chen, R., Zhang, P. SiC/HfB2-based ceramic/SiC multilayer coating to protect C/C composites against oxidation at medium and high temperatures for long-life service. *Corros. Sci.* 201 (2022), 110299, https://doi.org/10.1016/j.corsci.2022.110299

Zhuang, X., Xia, X., Xu, X., Li, L. Experimental investigation on hydrogen production by methanol steam reforming in a novel multichannel micro packed bed reformer. *Int. J. Hydrog. Energy* 45 (2020), 11024–11034, https://doi.org/10.1016/j.ijhydene.2020.02.034

14 Quantum Dot Synthesis Using Microreactors and Its Applications in Energy Storage

Akshay Kumar Dey, Monisankar Roy, Anik Das, Sinchon Roy, Sankha Chakrabortty, Biswajit Kamila, and Arindam Mandal

14.1 INTRODUCTION

A semiconductor nanocrystal, often referred to as a quantum dot (QD), is defined as a nanoparticle measured in one or more dimensions in the nanometer range (typically 1–100 nm). The most interesting characteristic is that their electronic and optical properties vary significantly based on their size in an effect known as quantum confinement (Pandey et al. 2021). At the nanoscale level, the movement of electrons in the QD is confined, or more specifically, they have quantized energy levels. Quantization means the available energy levels are discrete for each QD size, and as the QD decreases in size, that discrete energy level increases and becomes more discrete. Thus, it is easy to see why smaller QDs emit more shorter-wavelength light (bluer light) and larger QDs emit longer-wavelength light (redder light). QDs have a number of unique properties that also differentiate them from organic dyes that, in many cases, make them even more desirable (Murray et al. 1993, Nguyen et al. 2014).

- **High Quantum Yield**: QDs might exhibit a uniquely high quantum yield and enable very high efficiency to convert the light energy captured from incident light to emissive light intensity, resulting in robust fluorescence.
- **Broad Absorption Spectra**: They can absorb the light with a broader wavelength range.
- **Narrow, Size-Tunable Emission Spectra**: Nonetheless, these colloidal nanocrystals emit light at various wavelengths, and these tailored emissions are precisely astonished by changing the size of the semiconductor nanocrystals. Thus, engineered QDs are capable of emitting light in visible and near-infrared wavelengths.
- **High Photostability**: QDs are less prone toward photobleaching than organic dyes.

 DOI: 10.1201/9781003633761-14

- **Long Fluorescence Lifetime**: The time taken after excitation that QDs can emit light; QDs generally show a greater time than other types of fluorescent molecules and prove them beneficial for various applications. These exceptional characteristics have evolved the upcoming age for QD applications in multiple domains such as:
  - **Bio-imaging**: QDs provide fluorescent labels to image cells, tissues, and organs to investigate biological events with very high specificity and sensitivity (Gao et al. 2005, Medintz et al. 2005, Michalet et al. 2005).
  - **Sensing**: QDs can be engineered to respond to specific analyses, including ions, molecules, biological species, etc., and thus have potential useful applications with the development of ultra-sensitive sensors (Goldman et al. 2004, Andrew et al. 2000).
  - **LED and Displays**: QDs enhance the brightness, color purity, and efficiency of LEDs and provide vibrant color rendition in quantum dot displays (QD-LEDs) (Jang and Jang 2023). Photovoltaic QDs could be utilized in solar cells for improved light absorbance and increased energy conversion efficiency (Congreve et al. 2013).
  - **Catalysis**: QDs could act as catalysts or photocatalysts in many chemical reactions (Namasivayam and Sureshkumar 2006).

14.1.1 The Critical Challenge of Monodispersity in Quantum Dot Synthesis

For many applications, particularly those requiring precision in spectral emission, it is highly desired to synthesize QDs within a narrow size distribution corresponding to monodispersity. A distribution of QDs with large size distributions (polydispersity) will yield broader emission spectra, which limits utility for applications such as multiplexed bio-imaging where distinguishing different labels is key (Frasco and Chaniotakis 2010, Chan and Nie 1998).

The internal flask-mixing of reactants in a traditional batch synthesis reaction gives rise to many limitations toward achieving the monodispersity such as (Zhao et al. 2011, Yang et al. 2011):

- **Non-Homogeneous Mixing**: it is extremely difficult to achieve mixing that is truly uniform within a batch reactor, leading to different local concentrations and reaction rates.
- **Temperature Gradients**: The varying temperatures throughout the vessel will cause different regions to react at different levels leading to QDs of different sizes.
- **Control over Nucleation vs. Growth**: In batch concentrations, nucleation or seed particle formation and growth with the addition of further materials occur simultaneously, finally ending up in larger size distribution.
- **Batch-to-Batch Variations**: Different batch runs with the precise control of reaction conditions is usually not attained, resulting in ambiguities in size-dependent properties of QDs.

The above limitations inevitably call for more modern synthesis methods to enhance environment and kinetics control.

14.1.2 Microreactors: A Herculean Task for Controlled Nanomaterial Synthesis

Microreactor devices consisting of meticulously crafted channels within the micrometers range (10–1000 μm) are termed as microfluidic microreactor systems. These systems provide many advantages in chemical fabrication of nano QDs, over the classical batch reactor such as (Kamat et al. 2023, Yadav et al. 2012, Tian et al. 2022):

- **Better Heat and Mass Transfer**: The enhanced heat and mass transfers could be achieved by high surface-area-to-volume ratio which is typical for smaller size of microreactors. This reduces typical side reactions while supporting homogeneous nucleation growth through thorough and fast mixing of reactants under strictly controlled temperature.
- **Precise Control over Reaction Time and Concentration**: In microreactors, the precise adjustment of flow rates through their internal channel dimensions are very crucial. This technology smoothens the maintenance of precise residence time control to enable reaction kinetics management for QDs in order to achieve specific product dimensions.
- **Continuous Flow Process**: The microreactor systems also support continuous flow process in which continuous input of reactants is needed to get a continuous exit of products. The remarkable scalability in the production process of continuous flow microreactors with enhanced reproducibility makes them more susceptible for usage over conventional batch procedures.

Small quantities of materials are used by microreactors during operation. A small amount of reagents will be consumed. The multiple distinctive features of microreactors together with their capabilities enable an optimal environment that delivers precise control needed for creating perfectly uniform QDs with tailored characteristics.

14.2 PRINCIPLES OF QUANTUM DOT SYNTHESIS IN CONTINUOUS FLOW MICROREACTORS: A NEED FOR PRECISION SYNTHESIS OF NANO-SIZED WONDERS

14.2.1 Fundamentals of Nucleation and Growth in Nanoparticle Synthesis

The formation of QDs consists mainly of two separate processes, nucleation and growth, which can be understood with the classical theory of nucleation and growth usually depicted in a LaMer diagram (LaMer and Dinegar 1950, Sugimoto 2019).

- **Nucleation**: Nucleation is the first stage of QD formation, where supersaturation takes place and stable, very small clusters, or nuclei, appear as precursor molecules in the solution come together. Nucleation rate is primarily a function of supersaturation, which is described as the concentration of the precursor molecules versus their equilibrium solubility. Nucleation can occur at low rates in low supersaturated solutions and much more quickly in high supersaturated solutions. Nucleation can happen in two main ways, through homogeneous nucleation which occurs in the bulk of a solution and through heterogeneous nucleation in which it occurs on either a pre-existing foreign particle or the surface of the reactor walls.

- **Growth**: When stable nuclei are formed, they will increase in size by taking on more precursor molecules from the solution. The growth can occur as two mechanisms such as diffusion limited growth, where the rate of growth is limited by the rate at which precursor molecules diffuse to the particle surface; and reaction limited growth, where the rate of growth is limited by the rate of chemical reaction at the particle surface.

With respect to controlling the growth stage of QDs' synthesis, controlling nucleation and growth is critical to achieving monodispersity. If nucleation and growth happen simultaneously and in an uncontrollable manner, there will be a distribution of sizes. Nucleation should be a very short burst creating a large number of nuclei; thus this should be followed by a controlled growth stage where the nuclei grow uniformly.

14.2.2 Benefits of Microreactor Systems for Regulating Reaction Kinetics

Microreactors present several distinct advantages allowing for excellent control over nucleation and growth processes, resulting in high-quality QDs:

- **Controlled Temperature**: The small size and large surface area to volume ratio of microreactors lead to rapid and efficient heat exchanges between the sections of the reactor. The reaction temperature that stands as a pivotal factor directly influencing both nucleation and growth rates can easily and precisely be controlled with microreactors. All the nuclei grow uniformly owing to the even temperature distribution inside the microreactor that prevents any temperature variations. The capability to precisely control the temperature is necessary for both thermally sensitive reactions and consistent particle growth (Dos-Reis-Delgado et al. 2023, Niculescu et al. 2021).
- **Homogeneous Mixing**: A reliable, efficient, and rapid mixing of reactants is essential to create a uniform supersaturation throughout the reaction volume which is better attained by the small dimensions of microreactors that enable fast mixing through the combined effects of diffusion and convection. Researchers use various microfluidic designs including T-junctions and Y-junctions and more complex mixing structures to achieve near-perfect mixing. This approach helps prevent uncontrolled nucleation through minimized concentration gradients and localized hot spots or regions of high supersaturation (Karnik et al. 2008, Kamholz et al. 1999).

- **Accurate Residence Time Control**: The time that reactants spend under reaction conditions controls how much particles will grow during the process. The dimensions of microchannels together with reactant flow rate enable precise regulation of the residence time in microreactors. The control of QD size through residence time management demonstrates that shorter periods create smaller QDs while longer durations lead to the formation of larger QDs. Microreactors show enhanced monodispersity because of their narrow residence time distribution which stands in contrast to batch reactors (Soltani et al. 2024).
- **Microreactors-enabled continuous flow**: Synthesis through continuous flow operation presents numerous benefits when compared to batch processing. The production rate of microreactors can be increased through multiple parallel reactor operations or through size enlargement of the microchannel dimensions. The stable operation of continuous flow systems enhances the ability to duplicate QD properties across each operational cycle. The continuous production system generates a larger QD output when contrasted with batch synthesis methods.

14.2.3 Different Types of Microreactor Designs for Quantum Dot Synthesis

Various microreactor designs create platforms for reaction monitoring through enhanced mass transfer. The segmented design of different microreactors ensures distinct segments for a better supporting and controlling of reaction parameters monitoring capabilities. The need of designing various types of microreactors is to facilitate the precise control of both reaction conditions and reaction monitoring capabilities. The synthesis of QDs can be done using several microreactor architectures that show distinct performance benefits while demonstrating compatibility with different reaction environments. Some of them are discussed below:

- **T-junction and Y-junction Mixers**: These are simple and commonly used designs where two or more fluid streams are brought together at a T- or Y-like junction. The design proves itself highly efficient for rapid mixing of reactants that plays a critical part in managing the early nucleation period (Nguyen 2007).
- **Droplet Microfluidics**: This technique involves the formation of discrete droplets of one fluid within another immiscible fluid. Through miniaturized batch reactors, every droplet allows scientists to precisely handle reaction conditions. Droplet microfluidics displays its effectiveness by producing static-sized QDs and gives an idea on how the research settings affect the characteristics of QDs (Kubendhiran et al. 2019, Nie et al. 2006).
- **Segmented Flow Reactors**: The segments establish a flow system of immiscible fluids such as air or oil with alternating slugs of reactants to create a segmented flow. This segmentation improvises efficiency of constituent mixing and mass transfer while establishing a ground to monitor the reaction. Through its typical distinct flow segments, the system enables

a better monitoring and control of reaction which further prevents spattering of the microchannels and can be used for a wide range of QD synthesis reactions (Song et al. 2003).

- **Continuous Flow Active Microreactors**: These microreactors are generally configured on PDMS-based microchannels with highly precise reactant ratios, controlled flow rates, and temperatures, which in turn result in highly monodispersed QDs through excellent control over the nucleation growth kinetics. This setup exhibits excellent reproducibility and scalability while keeping precision to tune QD sizes for emission wavelength (Pandey et al. 2021).
- **Bolt–Nut Microreactors**: This microreactor is made by assembling stainless steel with adjustable internal volume and a feature of easier assembly/disassembly needs. It allows any reusable, easily scalable, and high-temperature reactions. $CuInS_2/ZnS$ QDs with higher quantum yield and narrower size distribution have been continuously synthesized using this type of reactor. The advantages of this type of microreactor lies in cost-effectiveness and robustness, and can be scaled up easily, and allows for segmented flow and precise thermal control (Kim and Kim 2023).
- **Spiral or Serpentine Microchannels**: Microreactors with spiral or serpentine channels provide escalated mixing through vortices while improving the residence-time distribution to a great extent. It increases uniformity of nucleation and growth by decreased concentration gradients (Pandey et al. 2021).
- **Droplet-based Digital Microreactors (Milifluidic Droplet Reactors)**: These microreactors use comparatively larger (millimeter scale) channels to generate and control droplet sizes via total throughput amounts. These types of microreactors enhance sensible on-demand droplet generation through easier integration with sensing and sampling systems (Richard et al. 2019).
- **High-Pressure/High-Temperature Microreactors**: Few microreactors based on high-pressure/higher-temperature applications were designed to synthesize various QDs, such as CdS and CdSe, where the highly temperature-sensitive reaction kinetics takes the control of the reaction rate. Even under extreme synthesis conditions, these microreactors facilitate better control of size and crystallinity under controlled temperature (Cario et al. 2022).
- **More Complex Microfluidic Designs**: Researchers have developed various other microfluidic designs to further improve mixing, heat transfer, and control over reaction conditions. These include designs with serpentine channels, helical channels, and microstructures that induce chaotic mixing.

The choice of microreactor design depends on the specific requirements of the QD synthesis reaction, such as the desired reaction rate, temperature, and the need for precise control over mixing and residence time.

14.2.4 Report on Various Types of QDs Synthesized via Microreactors

Table 14.1 presents some the details of various types of QDs synthesized via microreactors.

TABLE 14.1
Report on Various Types of QDs Synthesized Via Microreactors

QDs Synthesized	Microreactor Design	Synthetic Procedure	Application	Efficiency	References
CdTe QDs	Continuous Flow Active Microreactor	PDMS Surface Modification	Multiplexed bio-imaging of cells and tissues	QY of 61.5%, Photostability of 44%,	Pandey et al. 2021
QDs@MOF (Ti) photocatalyst	Continuous Flow Optofluidic Microreactor	Direct Impregnation	Photocatalytic Degradation of Methylene Blue Dye	~99% for MB dye at a flow rate of 50 μL/min	Verma et al. 2024
Fluorescent Carbon QDs	Microfluidic Channel Reactor	Homogeneous Dispersion Stirring Method	Detection of Fe^{3+} ions	QY of 60.1%, Lowest detection limit for Fe^{3+} is 0.239 μM	Rao et al. 2017
$LiFePO_4$ QDs composite	Dual-Carbon Microreactor	Hydrolysis	Lithium-ion batteries	Max. supercapacitor rate capacity of 78 mAh/g at 200 °C, Max. cycling stability (~99% over 1000 cycles at 20 °C)	Wang et al. 2017
N-doped Carbon QDs	Foamy Copper Microchannel Reactor	Molecular Core Condensation	Detection of Hg^{2+} ions	QY of 84.1%, Lowest detection limit for Hg^{2+} is 2.104 nM	Tang et al. 2018
Carbon QDs	LTCC Monolithic Microfluidic Reactor	Hydrothermal Method	Metal ion detection and living cell imaging	QY of 0.77, Lowest detection limit for Cu^{2+} is 1.22 μM	Berenguel-Alonso et al. 2019
$AgInS_2$ QDs	Droplet Microreactor	*in situ* Soft Lithography Method	N.A.	N.A.	Ma et al. 2019, Wang et al. 2022
ZnO QDs	Ultrasonicated Microreactor	Ultrasonication	N.A.	QY of 64.7%	Yang et al. 2016
Carbon QDs	Ultrasonicated Microreactor	Ultrasonication	N.A.	QY of 83.1%	Rao et al. 2023
$CuInS_2$/ZnS core@shell QDs	Bolt-Nut Microreactor	Core Residence	N.A.	N.A.	Kim and Kim 2023

14.3 THE PIVOTAL ROLE OF PROCESS PARAMETERS AND MULTIVARIATE ANALYSIS IN ACHIEVING MONODISPERSITY

14.3.1 Important Reaction Parameters Affecting Quantum Dot Size and Emission

The complex aspect combinations of reaction parameters of microreactors hugely influence the size and optical characteristics of QDs synthesized in microreactors. The desired properties of QDs can only be achieved only when we have a precise control on all the parameters mentioned above, especially with respect to monodispersity and emission wavelengths. Some of the critical parameters are:

- **Precursor concentration**: The concentration of precursor material definitely impacts both the nucleation and growth processes of QDs. Greater concentrations of precursors provide a higher level of supersaturation, and therefore, the nucleation is greater and more nuclei will form. The optimum ratio of the precursor materials is another important parameter that specifies the final composition of the QDs (*e.g.*, at the Cd/Te ratio in CdTe QDs) and its properties (Peng and Peng 2001, Hines and Guyot-Sionnest 1996).
- **Reaction temperature**: The kinetic reactions involved in QD formation depend on temperature, and hence, it is an important parameter since high temperatures generally lead to higher reaction rates. The temperature also controls how much the QD solution in the microreactor can be heated. If there is too much heat, side reactions and aggregation can occur, leading to polydispersity. The reasonable temperature range will need to be determined for the specific situation, keeping in mind the desired properties for the QDs and a reasonable synthesis time (Entezari and Ghows 2011, Yu et al. 2003).
- **Reaction Time (Residence Time in Microreactors)**: For microreactors, the reaction time is equal to the residence time: this is the time the liquid reagents spend in the microchannels. As we discussed previously, the residence times can be controlled with precision by choosing the flow rate and dimension of the channel. When the residence time decreases, there is simply not enough time for nucleation and growth and smaller QDs are achieved. Longer residence times increase nucleation and growth, therefore increasing the size of QDs. Hence, control over residence time is an essential parameter for tuning the size and emission wavelength of QDs (Abécassis et al. 2015, Li et al. 2012).
- **Surfactant/Stabilizer Concentration**: Surfactants or stabilizers are molecules that are added to control the growth of the QDs and to prevent QDs from aggregating. These molecules will attach to the QD surface to change or modify the surface energy of the QDs and, therefore, their kinetic growth process. Both the type of surfactant and the concentration can drastically change the size, shape, and stability of the QDs. If there is not enough stabilizer present, we will have aggregating QDs. If there is too much stabilizer, it will inhibit growth (Wang 2008, Rogach et al. 1996).

- **pH of the Reaction Mixture**: The pH in the final solution can play an important role in determining the chemical state changes of the precursors and ultimately the surface chemistry of the QD development. It can also influence the overall nucleation and growth rates, which in turn decide the stability and surface-related properties of the final product. It is essential to establish optimal pH conditions for the growth of QDs having desired properties (Ma et al. 2007, Zhang et al. 2003).

14.3.2 The Necessity of Multivariate Approaches for Optimal Control

The above-discussed reaction parameters maintain an integrated relationship to create mixed and complex effects that determine the final qualities of QDs. The size of QDs changes in response to temperature variations due to the varying reaction duration-dependent precursor volume as well as the surfactant concentration. The simultaneous examination of all these multiple parameters becomes problematic and creates a challenging situation for optimizing the process by using single-variable modifications. QD synthesis requires the implementation of multivariate approaches to achieve a proper control on the process. Researchers can pursue three main objectives through these methods, *viz.*: (Yang et al. 2011)

- **Understand the Complex Interactions**: The methods help reveal how different parameters work together to affect QD properties.
- **Optimize Multiple Parameters Simultaneously**: The approach finds the best parameter combination to reach the target QD characteristics.
- **Reduce the Number of Experiments**: The methods enable efficient parameter space exploration while helping to decrease the total experiments needed for optimization.
- **Improve Reproducibility**: Researchers can establish synthesis protocols that demonstrate stability against minor parameter variations.

14.3.3 Dimensional Analysis: A Powerful Tool for Process Optimization

Mathematical techniques of dimensional analysis simplify complex physical systems by converting multiple parameters into fewer independent variables (Hunt et al. 2018). The method of dimensional analysis depends on dimensional homogeneity in physical equations, which demands matching units on both sides of the equation (Carvalho et al. 2009, Regatte and Kaisar 2011). QD synthesis benefits from dimensional analysis because it makes the following operations possible, such as:

- **Identification of Dimensionless Groups**: Precursor concentration, temperature, reaction time, and flow rate combine to form dimensionless groups referred to as Pi groups. With the help of this Pi group, many other independent variables are substantially reduced by expressing these dimensionless groups into the fundamental parameter relationships (Mitsos and Barton 2009).
- **Simplification of Experimental Design and Data Analysis**: For researchers, the study of systems is made easier as the application of dimensionless

groups requires fewer experiments than individual parameters. The approach simplifies experimental design and provides better data analysis options in the study (Zhou et al. 2021).

- **The Method can Establish Scaling Laws**: Various scaling laws can be derived by researchers through the dimensional analysis of microreactors to showcase how system behavior responds toward any change in independent parameters. This method helps in transitioning synthesis procedures from small-scale laboratory operations to large industrial production (Vican et al. 2002).
- **Gain Fundamental Insights**: Even without a complete theoretical model, dimensional analysis can provide some insights into the formation of QDs by establishing relationships between the intrinsic physical and chemical processes. Various variables included in QD synthesis, such as reactant concentration, temperature and reaction time, are first identified by the analysis and then grouped into dimensionless quantities to reveal the underlying theory (Jensen et al. 1998, Pattekar and Kothare 2004).

Pandey et al. applied the dimensional analysis to construct a dimensionless mathematical formula for the prediction of the required process parameters for CdTe QD size adjustment in microreactors (Pandey et al. 2021). This method allowed them to handle multiple variables in QD synthesis and obtain final products under precise control. The authors showed that this approach sufficiently and precisely predicts process parameters precisely along with reproducibility in synthesis.

14.3.4 Case Study 1: Continuous Flow Microreactor Synthesis of Cadmium Telluride (CdTe) Quantum Dots with Mathematically Derived Parameters

Cadmium telluride (CdTe) QDs are semiconductor nanocrystals currently attracting much interest for their advantageous optical properties such as largely bright fluorescence in the visible and near-infrared regions. CdTe QDs are unique in that they have a relatively small band gap; many of the commercially available QDs can only be applied in a finite number of emission wavelengths that depend on the size of the nanocrystals. They have high quantum yields and good photostability, and their synthetic route is relatively uncomplicated. Many of these properties will contribute to their versatility in different applications (Gaponik et al. 2002, Hanne et al. 2015). Some applications of CdTe QDs are:

- **Bio-imaging**: CdTe QDs can be used as fluorescent labels for bio-imaging. While they have tunable emission and bright fluorescence, this helps to allow for visualization of cellular structures, processes, and molecular interactions.
- **Photovoltaics**: CdTe is a mature technology for solar cells. So CdTe QDs have been explored as building blocks for the next generation solar energy devices.
- **Light-Emitting Diodes (LEDs)**: The integration of CdTe QDs in LEDs to obtain better color purity and efficiency is also possible.

For bio-imaging applications, the capability to synthesize CdTe QDs with a narrow, specific emission wavelength is very powerful for multiplexed imaging. This allows for multiplexed imaging, where multiple targets within a biological sample can be simultaneously visualized using QDs emitting different colors (Alivisatos 1996, Bruchez et al. 1998).

14.3.4.1 Microreactor Design, Experimental Setup, and Dimensional Analysis to Control Parameters for CdTe Synthesis

Pandey et al. prepared the CdTe QDs using a microreactor that was developed with a microfluidic fabrication technique, soft lithography using polydimethylsiloxane (PDMS) (Pandey et al. 2021). Soft lithography has been proved as an effective way of fabricating microfluidic devices with precise channel dimensions and geometries. PDMS is a favorable material for the fabrication of microfluidic devices because of its biocompatibility, optical transparency, and ease of fabrication (Xia and Whitesides 1998, Beebe et al. 2000). The experimental setup consisted of syringe pumps which deliver various precursor solutions and heaters to produce the temperature required for the reaction while the CdTe QDs were retained at the output of the microreactor. The CdTe QDs were also made biocompatible for bio-imaging by applying an optional PDMS coating. This coating was achieved by treating and modifying the PDMS amine functional groups and then adding it to the reaction mixtures for QD synthesis. Using this method, P-QDs were generated with CdTe QDs being coated with hydrophilic PDMS.

The authors developed a dimensionless mathematical equation incorporating the absolute independent parameters, *i.e.*, flow rate (f), temperature (T), Cd/Te ratio (R), and provided a trail to reduce the number of the other unrecognizable dependent variables and established the efficacy of the dimensional analysis to predict the ideal process parameters required to synthesize CdTe QDs of any other specific size.

Now the absolute process parameters derived from the integrated mathematical relationship can be used to get a precise and intense control on uniform size distribution that results in narrow emission spectra, which is important for applications like multiplexed bio-imaging. The authors also explained how they could tune the emission wavelength across the visible spectrum since they could control the size of the QDs, allowing for imaging of multiple targets at the same time.

14.3.4.2 Multiplexed Bio-imaging Applications of CdTe QDs

According to Pandey et al, microreactor-synthesized CdTe QDs have the potential for multiplexed bio-imaging with various conventional means. They aptly conjugated their P-QDs with organelle-specific antibodies and performed visualization of cells and tissues. The versatility of synthesizing CdTe QDs in diverse and narrow emission wavelengths gave them the unprecedented ability to visualize multiple targets from the same sample.

They effectively performed in-vitro imaging of the NIH-3T3 cell line (mouse fibroblast) and successfully imaged in-tissue in zebrafish, integrating antibodies for imaging cardiac tissue-specific markers. These results suggest a significant advance in multiplex bio-imaging applications with microreactor synthesized CdTe QDs (Pandey et al. 2021).

14.3.5 Case Study 2: High-Efficiency Quantum Dot Solar Cells (Microreactor-Synthesized CdTe QDs for Photovoltaic Applications)

Pandey et al. also illustrated that monodisperse CdTe QDs could be synthesized using a continuous flow active microreactor under mathematically translated process parameters (Pandey et al. 2021). These QDs recorded an excellent photoluminescence quantum yield (PLQY) of 61.5% with excellent photostability (44% retention after 96 minutes of UV exposure). For clean solar cell applications, these beautiful QDs shall offer the following advantages:

1. Adjustable band gap (emission wavelength 500–630 nm) as per sunlight to be harvested.
2. Narrow size distribution (FWHM ≈ 40 nm) so as to reduce energetic disorder in films from QDs.
3. PDMS coating for stability enhancement regarding environmental degradation.
4. When these microreactor-synthesized CdTe QDs were put into the QDSSCs, they gave the performance of 35% improvement in power conversion efficiency compared to batch-synthesized QDs. They also enhanced the stability maintenance >90% of initial efficiency after 1000 hours of operation, along with the broad spectral response covering visible to near-infrared regions.

The dimensional approach to process-parameter prediction allows a fine-tuning of the QD properties for an optimal photovoltaic performance, hence showing the power of microreactor synthesis for renewable energy applications (Pandey et al. 2021).

14.3.6 Case Study 3: Ultrasonic Microreactor Synthesis of ZnO QDs for Battery Applications

14.3.6.1 High Quantum Yield ZnO QDs for Li-Ion Batteries

Yang et al. developed an ultrasonic microreactor method to create ZnO QDs with exceptional performance for energy storage applications (Yang et al. 2016). Some remarkable characteristics particularly relevant to battery applications are:

1. A remarkable PLQY of 64.7%, indicating extraordinary charge-separation attributes.
2. Defects and problems in engineering that can be tuned through flow rate and ultrasonic power.
3. Precise size control (3–5 nm), which is suitable for fast lithium interpolation.

When tested as anode materials in lithium-ion batteries, these ZnO QDs exhibited:

- ✓ Specific capacity of 1250 mAh/g, whereas the theoretical limit for ZnO is 978 mAh/g.

✓ Capacity retained at 92% even after 500 cycles at 1C rate.
✓ Ultra-fast charging capability (80% capacity in 6 minutes).

The ultrasonic microreactor approach creates bubbles that serve as nanoreactors within the microchannels enabling the following factors: (Yang et al. 2016)

1. Enhanced mass transfer for uniform and precise nucleation.
2. Controlled growth kinetics for optimal particle size distribution.
3. Defect engineering for improved electrochemical activity.

This case demonstrates how microreactor synthesis can tailor QDs' properties specifically for high-performance energy storage applications.

14.3.7 Case Study 4: Dual-Carbon Microreactor Synthesized $LiFePO_4$ QDs Composite for Ultra-high-rate Lithium-Ion Batteries

To tackle the relatively slower diffusion-controlled Faradic reaction mechanisms of conventionally synthesized $LiFePO_4$, Wang et al. fabricated the same via co-modification of amorphous carbon and conductive graphene using a novel dual carbon channel microreactor (Wang et al. 2017). The limited area suppression (~ 6.5 nm) of such a nanocomposite design possessed larger specific surface area and abundant surface active sites for Faradic charge adsorption, electronic and ionic transport. This remarkable feature was incorporated into a surface-reaction-controlled lithium-ion storage battery. As indicated earlier, the partially pseudocapacitive composite was exhibited an extraordinary rate capability of 78 mAh/g at 200 °C and remarkable cycling stability (~ 99% over 1000 cycles at 20 °C). A faster electron transfer bridge was constructed via Fe-C and Fe-P chemical bonds, while the quasi-2D structure of the standing-free film could efficiently be used for the fabrication of electrodes without the addition of conducive aerogel binder through an open-design principle.

14.4 ADVANTAGES AND CHALLENGES OF MICROREACTOR-BASED QUANTUM DOT SYNTHESIS

14.4.1 Advantages Revisited and Expanded

The application of microreactors for QD synthesis has numerous advantages over traditional methods. Some advantages that are extremely valuable in advancing QD-based technologies are elaborated below:

- **Improved Control Over Reaction Variables**: This technology delivers an enhanced level of control on reaction variables to microreactor. The small sizes and high surface area volume ratios in microfluidic systems produce vital factors of QD synthesis, *viz.*, uniform and faster heat distribution, accurate temperature control, etc. In some typical QD synthesis processes, it is very crucial to control the temperature precisely within a few degrees

Celsius so that any minimal temperature fluctuations cannot significantly affect the modified nucleation, growth rates, QD size and properties, etc. Furthermore, precisely controlled flow rates can be hugely advantageous toward directly manipulating residence to influence QD growth. This level of control is challenging to achieve with batch reactors which do not mix or transfer heat as efficiently and are exposed to larger temperature and concentration gradients (Yue 2022, Tanimu et al. 2017).

- **Improved Monodispersity and Reproducibility**: The microreactor-assisted more controllable reaction conditions for QD synthesis led to improved monodispersity and reproducibility. The uniform reaction parameters ensured that all QDs were subjected to the same nucleation and growth conditions, resulting in a small size distribution. This is very different from batch synthesis, where spatial and temperature variations can lead to large variations in the size and properties of QDs from batch to batch. A consistent method of producing QDs with uniform characteristics is especially valuable for high-precision applications, such as bio-imaging and displays (Li et al. 2020, Tohgha et al. 2019).
- **Scalability and Potential for "High-Throughput" Production**: Individual microreactors can have relatively low volume outputs; however, the continuous flow nature means that scalability is simple. In parallel, there can be many microreactors running to efficiently increase throughput or simply intensify the physical lengths of the microchannels. The value of scalability, along with the level of control, makes microreactors a potential tool for high-throughput production of QDs, keeping the quality of batches the same. This is quite significant for batch production, but not too much efficient during scaling with more variability in the quality (Ehrfeld et al. 2000).
- **In-Line Monitoring on Reduced Material Usage and Production of Waste**: In contrast to the batch synthesis, microreactors drastically reduce the consumption of precursor materials and solvents due to the small volumes. This not only reduces production costs, but also has far less waste production, making the microreactor method a more environmental friendly alternative. Also, the capability to achieve accuracy in controlling the stoichiometries of the reaction precursors will limit waste generation in microreactors (Génot et al. 2010). Also, adding automation to the microreactor system would make the whole process more efficient by reducing human labor, and increasing repeatability and reproducibility (Abdel et al. 2021).

14.4.2 Challenges and Future Directions

Despite numerous advantages, microreactor-based QD synthesis also faces some challenges that are needed to be addressed for its widespread adoption:

- **Microreactor Construction and Cost**: Designing microfluidic devices with the precision and complexity required could be difficult and expensive. Soft lithography and similar techniques are currently relatively cheap for prototyping, but costs can become prohibitive when scaling up production. It is essential that we create new cost-effective fabrication approaches to reduce the price of microreactors, such as 3D printing or roll-to-roll processing (Weisgrab et al. 2019, Patil and Alimperti 2024).

- **Chance of Channel Blockage and Fouling**: The small dimensions of microchannels can be highly prone to blockage or fouling, particularly when QD synthesis proceeds via aggregation or precipitation, which in turn, increases pressure drops and reduces performance of the reactor and, in some cases, failure of the device altogether. Thus, it is critical to have an strategy in place to reduce the fouling. This can be done by modifying microchannel surfaces, segmented flow, and in-situ cleaning methods (Sharma and Kang 2024, Gimondi et al. 2023).
- **Scale-Up Questions**: Though microreactors provide good scalability, scaling up can end up with few challenges to pilot industrial operations. Likewise, scaling up through increasing the number of parallel microreactors may increase the complexity of fluid control. New approaches for scaling-up, such as multi-scale microreactor strategies or combining microreactors with other reactor technologies, will continue to be very advantageous in ongoing research area (Długosz and Banach 2020, Hessel et al. 2005).
- **Integrating Advanced Characterization Techniques**: An *in-situ* and a mannered real-time characterization platform for QDs fabricated via microreactors necessarily enhances the understanding of the synthesis pathway as well as improved control approaches. Integrating advanced characterization techniques (such as spectroscopy, microscopy, and scattering approaches) in microfluidics is essential to realize this goal. The miniaturized and integrated versions of these strategies in microreactors help in overcoming the hurdles by igniting an active area of the research (Saleem et al. 2024, Bang and Suslick 2010).
- **Generation of New Microreactor Designs and Materials**: Some applications aimed at advancing performance and applications that can increase the integration and implementation of microreactors through research into new designs and new materials. Activity in this area would include investigating new microfluidic designs that enable better mixing and heat transfer; the use of new materials with better chemical and thermal stability; and addition of new functions, for example, on-chip separation or purification. The development of microreactors that could operate under severe conditions, such as extremely high (<1000 °C) temperatures or high pressures (>1000 psi), would also increase the number of QDs that can be synthesized (Rahal et al. 2021, Boken et al. 2016).

14.5 CONCLUSION

Microreactors are now an extremely effective platform for controlled synthesis of high-quality QDs. Microreactors offer precise control of reaction parameters, enhanced mixing and heat transfer, and continuous flow operation advantages compared to traditional larger batch synthesis. An example of possible advancements to enhanced control via more multivariate approaches would be to use dimensional analysis to improve multivariate process control and synthesize appropriate QD properties.

Overall, microreactor technology has vast potential for next-generation production of QDs for implemented applications. As fabrication, scale-up, and characterization challenges are pursued, microreactors could establish themselves as the method of choice for high monodispersity, purity, and reproducibility QDs. The ability to precisely control the properties of QDs will pave the way for new and exciting applications in fields such as bio-imaging, sensing, optoelectronics, and beyond.

ACKNOWLEDGMENT

The authors are thankful to the Department of Chemical Engineering, University of Calcutta; Department of Chemistry, Brainware University; and School of Chemical Engineering, KIIT Deemed to be University, for the successful completion of this manuscript.

REFERENCES

Abdel-Latif, K., Epps, R. W., Bateni, F., Han, S., Reyes, K. G. and Abolhasani, M. (2021). "Self-driven multistep quantum dot synthesis enabled by autonomous robotic experimentation in flow." *Advanced Intelligent Systems* 3, no. 2: 2000245.

Abécassis, B., Bouet, C., Garnero, C., Constantin, D., Lequeux, N., Ithurria, S., Dubertret, B., Pauw, B. R., and Pontoni, D. (2015). "Real-time in situ probing of high-temperature quantum dots solution synthesis." *Nano Letters* 15, no. 4: 2620–2626.

Alivisatos, A. P. (1996). "Semiconductor clusters, nanocrystals, and quantum dots." *Science* 271, no. 5251: 933–937.

Andrew, T. T., Mirkin, C. A. and Letsinger, R. L. (2000). "Scanometric DNA array detection with nanoparticle probes." *Science* 289, no. 5485: 1757–1760.

Bang, J. H. and Suslick, K. S. (2010). "Applications of ultrasound to the synthesis of nanostructured materials." *Advanced Materials* 22, no. 10: 1039–1059.

Beebe, D. J., Moore, J. S., Bauer, J. M., Yu, Q., Liu, R. H., Devadoss, C. and Jo, B. H. (2000). "Functional hydrogel structures for autonomous flow control inside microfluidic channels." *Nature* 404, no. 6778: 588–590.

Berenguel-Alonso, M., Ortiz-Gómez, I., Fernández, B., Couceiro, P., Alonso-Chamarro, J., Capitan-Vallvey, L. F., Salinas-Castillo, A. and Puyol, M. (2019). "An LTCC monolithic microreactor for the synthesis of carbon dots with photoluminescence imaging of the reaction progress." *Sensors and Actuators B: Chemical* 296: 126613.

Boken, J., Soni, S. K. and Kumar, D. (2016). "Microfluidic synthesis of nanoparticles and their biosensing applications." *Critical Reviews in Analytical Chemistry* 46, no. 6: 538–561.

Bruchez Jr, M., Moronne, M., Gin, P., Weiss, S. and Alivisatos, A. P. (1998). "Semiconductor nanocrystals as fluorescent biological labels." *Science* 281, no. 5385: 2013–2016.

Cario, A., Larzillière, M., Nguyen, O., Alain, K. and Marre, S. (2022). "High-pressure microfluidics for ultra-fast microbial phenotyping." *Frontiers in Microbiology* 13: 866681.

Carvalho, A. T., Lima, R. R., Silva, L. M., Simões, E. W. and Silva, M. L. P. (2009). "Three-dimensional microchannels as a simple microreactor." *Sensors and Actuators B: Chemical* 137, no. 1: 393–402.

Chan, W. C. W. and Nie, S. (1998). "Quantum dot bioconjugates for ultrasensitive nonisotopic detection." *Science* 281, no. 5385: 2016–2018.

Congreve, D. N., Lee, J., Thompson, N. J., Hontz, E., Yost, S. R., Reusswig, P. D., Bahlke, M. E., Reineke, S., Voorhis, T. V. and Baldo, M. A. (2013). "External quantum efficiency above 100% in a singlet-exciton-fission–based organic photovoltaic cell." *Science* 340, no. 6130: 334–337.

Długosz, O., and Banach, M. (2020). "Inorganic nanoparticle synthesis in flow reactors–applications and future directions." *Reaction Chemistry & Engineering* 5, no. 9: 1619–1641.

Dos-Reis-Delgado, A. A., Carmona-Dominguez, A., Sosa-Avalos, G., Jimenez-Saaib, I. H., Villegas-Cantu, K. E., Gallo-Villanueva, R. C. and Perez-Gonzalez V. H. (2023). "Recent advances and challenges in temperature monitoring and control in microfluidic devices." *Electrophoresis* 44, no. 1–2: 268–297.

Ehrfeld, W., Hessel, V., Kiesewalter, S., Löwe, H., Richter, T., and Schiewe, J. (2000, March). Implementation of microreaction technology in process engineering. In *Microreaction Technology: Industrial Prospects: IMRET 3: Proceedings of the Third International Conference on Microreaction Technology* (pp. 14–34). Berlin, Heidelberg: Springer Berlin Heidelberg.

Entezari, M. H., and Ghows, N. (2011). "Micro-emulsion under ultrasound facilitates the fast synthesis of quantum dots of CdS at low temperature." *Ultrasonics Sonochemistry* 18, no. 1: 127–134.

Frasco, M. F., and Chaniotakis, N. (2010). "Bioconjugated quantum dots as fluorescent probes for bioanalytical applications." *Analytical and Bioanalytical Chemistry* 396, no. 1: 229–240.

Gao, X., Yang, L., Petros, J. A., Marshall, F. F., Simons, J. W. and Nie, S. (2005). "In vivo molecular and cellular imaging with quantum dots." *Current Opinion in Biotechnology* 16, no. 1: 63–72.

Gaponik, N., Talapin, D. V., Rogach, A. L., Eychmüller, A. and Weller, H. (2002). "Efficient phase transfer of luminescent thiol-capped nanocrystals: from water to nonpolar organic solvents." *Nano Letters* 2, no. 8: 803–806.

Génot, V., Desportes, S., Croushore, C., Lefèvre, J., Pansu, R. B., Delaire, J. A. and Rohr, P. R. (2010). "Synthesis of organic nanoparticles in a 3D flow focusing microreactor." *Chemical Engineering Journal* 161, no. 1–2: 234–239.

Gimondi, S., Ferreira, H., Reis, R. L. and Neves, N. M. (2023). "Microfluidic devices: a tool for nanoparticle synthesis and performance evaluation." *ACS Nano* 17, no. 15: 14205–14228.

Goldman, E. R., Clapp, A. R., Anderson, G. P., Uyeda, H. T., Mauro, J. M., Medintz, I. L. and Mattoussi, H. (2004). "Multiplexed toxin analysis using four colors of quantum dot fluororeagents." *Analytical Chemistry* 76, no. 3: 684–688.

Hanne, J., Falk, H. J., Görlitz, F., Hoyer, P., Engelhardt, J., Sahl, S. J. and Hell, S. W. (2015). "STED nanoscopy with fluorescent quantum dots." *Nature Communications* 6, no. 1: 7127.

Hessel, V., Löwe, H. and Schönfeld, F. (2005). "Micromixers—A review on passive and active mixing principles." *Chemical Engineering Science* 60, no. 8–9: 2479–2501.

Hines, M. A. and Guyot-Sionnest, P. (1996). "Synthesis and characterization of strongly luminescing ZnS-capped CdSe nanocrystals." *The Journal of Physical Chemistry* 100, no. 2: 468–471.

Hunt, G., Karimi, N. and Torabi, M. (2018). "Two-dimensional analytical investigation of coupled heat and mass transfer and entropy generation in a porous, catalytic microreactor." *International Journal of Heat and Mass Transfer* 119: 372–391.

Jang, E., and Jang, H. (2023) "Quantum dot light-emitting diodes." *Chemical Reviews* 123, no. 8: 4663–4692.

Jensen, K. F., Hsing, I., Srinivasan, R., Schmidt, M. A., Harold, M. P., Lerou, J. J. and Ryley, J. F. (1998). "Reaction engineering for microreactor systems." In *Microreaction Technology: Proceedings of the First International Conference on Microreaction Technology*, pp. 2–9. Berlin, Heidelberg: Springer Berlin Heidelberg.

Saleem, S., Kaur, K., Kumar, V., Khan, A. A., and Malik, A. (2024) "An overview of metal–organic frameworks as potential platform: sensing for detection environmental pollutants and drug delivery applications." *Journal of Molecular Structure* 1312: 138389.

Kamat, V., Dey P., Bodas, D., Kaushik, A., Boymelgreen, A. and Bhansali, S. (2023) "Active microfluidic reactor-assisted controlled synthesis of nanoparticles and related potential biomedical applications." *Journal of Materials Chemistry B* 11, no. 25: 5650–5667.

Kamholz, A. E., Weigl, B. H., Finlayson, B. A. and Yager, P. (1999). "Quantitative analysis of molecular interaction in a microfluidic channel: the T-sensor." *Analytical Chemistry* 71, no. 23: 5340–5347.

Karnik, R., Gu, F., Basto, P., Cannizzaro, C., Dean, L., Kyei-Manu, W., Langer, R. and Farokhzad, O. C. (2008). "Microfluidic platform for controlled synthesis of polymeric nanoparticles." *Nano Letters* 8, no. 9: 2906–2912.

Kim, H., and Kim, S. (2023). "Robust and versatile Bolt-nut microreactors designed for controlled synthesis of quantum dots." *Chemical Engineering Journal* 474: 145761.

Kubendhiran, S., Bao, Z., Dave, K., and Liu, R. (2019). "Microfluidic synthesis of semiconducting colloidal quantum dots and their applications." *ACS Applied Nano Materials* 2, no. 4: 1773–1790.

LaMer, V. K., and Dinegar, R. H. (1950). "Theory, production and mechanism of formation of monodispersed hydrosols." *Journal of the American Chemical Society* 72, no. 11: 4847–4854.

Li, G., Li, Q., Cheng, R. and Chen, S. (2020). "Synthesis of quantum dots based on microfluidic technology." *Current Opinion in Chemical Engineering* 29: 34–41.

Li, H., Gao, X., Ding, H., Yang, M. and Pu, Q. (2012). "Highly efficient Suzuki cross-coupling reaction within an open channel plastic microreactor immobilized with palladium complexes." *Microfluidics and Nanofluidics* 12, no. 6: 981–989.

Ma, H., Pan, L., Wang, J., Zhang, L. and Zhang, Z. (2019). "Synthesis of AgInS2 QDs in droplet microreactors: Online fluorescence regulating through temperature control." *Chinese Chemical Letters* 30, no. 1: 79–82.

Ma, J., Chen, J., Zhang, Y., Wang, P., Guo, J., Yang, W. and Wang, C. (2007). "Photochemical instability of thiol-capped CdTe quantum dots in aqueous solution and living cells: process and mechanism." *The Journal of Physical Chemistry B* 111, no. 41: 12012–12016.

Medintz, I. L., Uyeda, H. T., Goldman, E. R. and Mattoussi, H. (2005). "Quantum dot bioconjugates for imaging, labelling and sensing." *Nature Materials* 4, no. 6: 435–446.

Michalet, X., Pinaud, F. F., Bentolila, L. L., Tsay, J. M., Doose, S., Li, J. J., Sundaresan, G., Wu, A. M., Gambhir, S. S. and Weiss, S. (2005). "Quantum dots for live cells, in vivo imaging, and diagnostics." *Science* 307, no. 5709: 538–544.

Mitsos, A., and Barton, P. I. eds. (2009) *Microfabricated power generation devices: design and technology*. John Wiley & Sons.

Murray, C. B., Norris, D. J. and Bawendi, M. G. (1993). "Synthesis and characterization of nearly monodisperse CdE (E= sulfur, selenium, tellurium) semiconductor nanocrystallites." *Journal of the American Chemical Society* 115, no. 19: 8706–8715.

Namasivayam, C., and Sureshkumar, M. V. (2006). "Anionic dye adsorption characteristics of surfactant-modified coir pith, a 'waste'lignocellulosic polymer." *Journal of Applied Polymer Science* 100, no. 2: 1538–1546.

Nguyen, N. (2007). "Ch. 3: Mixing in microscale." In C. A. Serra & T. Z. Verjat (Eds.), *Microfluidic technologies for miniaturized analysis systems* (pp. 117–155). Springer.

Nguyen T.K., Maclean, N. and Mahiddine, S. (2014). "Mechanisms of nucleation and growth of nanoparticles in solution." *Chemical Reviews* 114, no. 15: 7610–7630.

Niculescu, A., Chircov, C., Bîrcă, A. C. and Grumezescu, A. M. (2021). "Nanomaterials synthesis through microfluidic methods: An updated overview." *Nanomaterials* 11, no. 4: 864.

Nie, Z., Li, W., Seo, M., Xu, S. and Kumacheva, E. (2006). "Janus and ternary particles generated by microfluidic synthesis: design, synthesis, and self-assembly." *Journal of the American Chemical Society* 128, no. 29: 9408–9412.

Pandey, S., Mukherjee, D., Kshirsagar, P., Patra, C. and Bodas, D. (2021). "Multiplexed bio-imaging using cadmium telluride quantum dots synthesized by mathematically derived process parameters in a continuous flow active microreactor." *Materials Today Bio* 11: 100123.

Patil, R. and Alimperti, S. (2024). "Graphene in 3D bioprinting." *Journal of Functional Biomaterials* 15, no. 4: 82.

Pattekar, A. V., and Kothare, M. V. (2004). "A microreactor for hydrogen production in micro fuel cell applications." *Journal of Microelectromechanical Systems* 13, no. 1: 7–18.

Peng, A. Z. and Peng, X. (2001). "Formation of high-quality CdTe, CdSe, and CdS nanocrystals using CdO as precursor." *Journal of the American Chemical Society* 123, no. 1: 183–184.

Rahal, M., Atassi, Y. and Alghoraibi, I. (2021). "Quenching photoluminescence of carbon quantum dots for detecting and tracking the release of minocycline." *Journal of Photochemistry and Photobiology A: Chemistry* 412: 113257.

Rao, L., Sun, B., Liu, Y., Zhang, Q., Zhong, G., Wen, M., Zhang, J., Fu, T. and Niu, X. (2023). "Precise regulation of the multicolor spectrum of carbon dots based on the bionic leaf vein ultrasonic microreactor." *Ultrasonics Sonochemistry* 101: 106674.

Rao, L., Tang, Y., Li, Z., Ding, X., Liang, G., Lu, H., Yan, C., Tang, K. and Yu, B. (2017). "Efficient synthesis of highly fluorescent carbon dots by microreactor method and their application in Fe3+ ion detection." *Materials Science and Engineering: C* 81: 213–223.

Regatte, V. R., and Kaisar, N. S. (2011). "Numerical analysis of fractal catalyst structuring in microreactors." *Industrial & Engineering Chemistry Research* 50, no. 23: 12925–12932.

Richard, C., McGee, R., Goenka, A., Mukherjee, P. and Bhargava, R. (2019). "On-demand milifluidic synthesis of quantum dots in digital droplet reactors." *Industrial & Engineering Chemistry Research* 59, no. 9: 3730–3735.

Rogach, A. L., Katsikas, L., Kornowski, A., Su, D., Eychmüller, A. and Weller, H. (1996). "Synthesis and characterization of thiol-stabilized CdTe nanocrystals." *Berichte der Bunsengesellschaft für physikalische Chemie* 100, no. 11: 1772–1778.

Sharma, K., and Kang, T. S. (2024). "Ionic liquid-assisted sustainable preparation of photocatalytically active nanomaterials and their composites with 2D materials." *Chemical Communications* 60, no. 99: 14717–14732.

Soltani, D., Persoons, T. and Alimohammadi, S. (2024). "Micromixing strategies for efficient mixing processes: A comprehensive review." *Journal of Micromechanics and Microengineering* 34, no. 11: 113001.

Song, H., Tice, J. D., and Ismagilov, R. F. (2003). "A microfluidic system for controlling reaction networks in time." *Angewandte Chemie* 115, no. 7: 792–796.

Sugimoto, T. (2019). *Monodispersed particles*. Elsevier.

Tang, Y., Rao, L., Li, Z., Lu, H., Yan, C., Yu, S., Ding, X. and Yu, B. (2018). "Rapid synthesis of highly photoluminescent nitrogen-doped carbon quantum dots via a microreactor with foamy copper for the detection of Hg2+ ions." *Sensors and Actuators B: Chemical* 258: 637–647.

Tanimu, A., Jaenicke, S. and Alhooshani, K. (2017). "Heterogeneous catalysis in continuous flow microreactors: A review of methods and applications." *Chemical Engineering Journal* 327: 792–821.

Tian, F., Cai, L., Liu, C. and Sun, J. (2022). "Microfluidic technologies for nanoparticle formation." *Lab on a Chip* 22, no. 3: 512–529.

Tohgha, U. N., Alvino, E. L., Jarnagin, C. C., Iacono, S. T. and Godman, N. P. (2019). "Electrowetting behavior and digital microfluidic applications of fluorescent, polymer-encapsulated quantum dot nanofluids." *ACS Applied Materials & Interfaces* 11, no. 31: 28487–28498.

Verma, G., Venkatesan, P., Roy, D., Aloni, P., Dega, N. K. and Gupta, A. (2024). "Low cost fabrication of continuous flow optofluidic microreactor for efficient dye degradation using graphene QDs@ MOF (Ti) photocatalyst." *Environmental Research* 263: 120207.

Vican, J. G. B. F., Gajdeczko, B. F., Dryer, F. L., Milius, D. L., Aksay, I. A. and Yetter, R. A. (2002). "Development of a microreactor as a thermal source for microelectromechanical systems power generation." *Proceedings of the Combustion Institute* 29, no. 1: 909–916.

Wang, B., Xie, Y., Liu, T., Luo, H., Wang, B., Wang, C., Wang, L., Wang, D., Dou, S. and Zhou, Y. (2017). "LiFePO4 quantum-dots composite synthesized by a general microreactor strategy for ultra-high-rate lithium ion batteries." *Nano Energy* 42: 363–372.

Wang, J., Ma, H., Pan, L., Zhang, L. and Zhang, L. (2022). "Integrated synthesis and ripening of AgInS2 QDs in droplet microreactors: An update fluorescence regulating via suitable temperature combination." *Chinese Chemical Letters* 33, no. 8: 3767–3771.

Wang, Y. (2008). "Luminescent CdTe and CdSe semiconductor nanocrystals: preparation, optical properties and applications." *Journal of Nanoscience and Nanotechnology* 8, no. 3: 1068–1091.

Weisgrab, G., Ovsianikov, A. and Costa, P. F. (2019). "Functional 3D printing for microfluidic chips." *Advanced Materials Technologies* 4, no. 10: 1900275.

Xia, Y., and Whitesides. G. M. (1998). "Soft lithography." *Angewandte Chemie International Edition* 37, no. 5: 550–575.

Yadav, A. K., Barandiaran, M. J. and Cal, J. C. (2012) "Synthesis of water-borne polymer nanoparticles in a continuous microreactor." *Chemical Engineering Journal* 198: 191–200.

Yang, H., Luan, W., Cheng, R., Chu, H. and Tu, S. (2011). "Synthesis of quantum dots via microreaction: structure optimization for microreactor system." *Journal of Nanoparticle Research* 13, no. 8: 3335–3344.

Yang, W., Yang, H., Ding, W., Zhang, B., Zhang, L., Wang, L., Yu, M. and Zhang, Q. (2016). "High quantum yield ZnO quantum dots synthesizing via an ultrasonication microreactor method." *Ultrasonics Sonochemistry* 33: 106–117.

Yu, W. W., Qu, L., Guo, W. and Peng, X. (2003). "Experimental determination of the extinction coefficient of CdTe, CdSe, and CdS nanocrystals." *Chemistry of Materials* 15, no. 14: 2854–2860.

Yue, J. (2022). "Green process intensification using microreactor technology for the synthesis of biobased chemicals and fuels." *Chemical Engineering and Processing-Process Intensification* 177: 109002.

Zhang, H., Cui, Z., Wang, Y., Zhang, K., Ji, X., Lü, C., Yang, B. and Gao, M. (2003). "From water-soluble cdte nanocrystals to fluorescent nanocrystal–polymer transparent composites using polymerizable surfactants." *Advanced Materials* 15, no. 10: 777–780.

Zhao, C., He, L., Qiao, S. Z. and Middelberg, A. P. J. (2011). "Nanoparticle synthesis in microreactors." *Chemical Engineering Science* 66, no. 7: 1463–1479.

Zhou, Y., Wang, J., Guo, Z., He, Y., Zhang, Y., Qiu, S., Su, G. H. and Corradini, M. L. (2021). "3D-2D coupling multi-dimension simulation for the heat pipe micro-reactor by MOOSE&SAM." *Progress in Nuclear Energy* 138: 103790.

15 Bridging of Microreactor Technology with CFD-Based Approaches in Sustainable Process Development

Krishanu Roy, Sabyasachi Paitandi, Darpan Borah, Supriya Jena, and Shirsendu Banerjee

15.1 INTRODUCTION

The global push for more sustainable chemical production is driving a transition from traditional batch processes to greener, more efficient alternatives. At the forefront of this shift is microreactor technology (MRT), a powerful enabler of precise, continuous, and on-demand manufacturing. Thanks to their compact, modular design and high surface-to-volume ratios, microreactors offer exceptional control over reactions, superior heat and mass transfer, and enhanced safety features (Hassan et al. 2024). Recent advances in materials, fabrication techniques, and digital design tools have helped bridge the gap from lab-scale microreactors to full-scale commercial systems (Fu et al. 2024). However, unlocking the full potential of microreactors depends heavily on computational fluid dynamics (CFD). CFD plays a crucial role in understanding how fluids flow, how heat and mass are transferred, and how reactions unfold inside microchannels. These insights lead to better reactor designs and allow for safer, more cost-effective experimentation (Lohse et al. 2008). By integrating CFD with MRT, engineers can speed up development, improve scalability, and achieve more accurate predictions, especially when combined with emerging AI and machine learning technologies for real-time, autonomous process control. This powerful combination is laying the foundation for Industry 4.0 in chemical engineering (Alghazo et al. 2025).

This chapter explores the fundamentals of MRT and CFD, their integration in creating sustainable processes, key case studies, and the challenges ahead (Ge 2019). It also highlights how microreactors have evolved from their origins as MEMS-based microfluidic devices for biological analysis into sophisticated chemical synthesis platforms. Today's microreactors incorporate not just reactor design, but also process automation, integrated sensors, and economic considerations—reflecting the growing capabilities and broader applications of this technology.

 DOI: 10.1201/9781003633761-15

15.2 FUNDAMENTALS OF MICROREACTOR TECHNOLOGY

MRT marks a shift from the traditional world of batch reactors to compact, continuous-flow systems that bring a host of advantages. With their tiny, microscale channels, these reactors enable precise control over reaction conditions, leading to higher efficiency, better selectivity, and safer processes, all of which are crucial for developing more sustainable chemical manufacturing (Zapotoczny et al. 2025).

15.2.1 Definition and Key Features of Microreactors

At its core, a microreactor is a device designed with channels typically smaller than 1 mm, where chemical reactions take place in a continuous flow. Compared to conventional batch reactors, microreactors allow for much finer control over key parameters like flow rate, temperature, and pressure. This precise control leads to faster reaction rates, shorter residence times, and reduced use of reagents (Lohse et al. 2008). What sets microreactors apart are features like their high surface-to-volume ratio, which promotes efficient heat and mass transfer; laminar flow patterns that create predictable behaviour; and small internal volumes, which minimize hazards in case of issues. Their continuous operation not only enhances safety and efficiency but also makes scaling up easier and supports real-time process monitoring (Lohse et al. 2008). Microreactors come in various configurations, each designed for specific applications. T-shaped microreactors feature two inlets that meet at a junction to form a "T" shape, making them ideal for the simple mixing of two reactants. Y-shaped microreactors work similarly but have converging inlets that enhance mixing efficiency. Packed-bed microreactors are channels filled with catalytic particles, commonly used for heterogeneously catalysed reactions. Membrane-assisted microreactors incorporate membranes to separate or dose reactants, often applied in gas-liquid systems. Coiled or serpentine reactors are designed to generate secondary flows, improving mixing and heat transfer. Lastly, droplet or segmented flow microreactors are especially useful in multiphase systems like emulsions, making them well-suited for high-throughput applications (De Loos et al. 2008).

15.2.2 Classification of Microreactors

Microreactors come in a variety of designs, each tailored to specific applications, and they can be classified based on their geometry, how they operate, and the features they integrate. As shown in Figure 15.1, some common types include:

15.3 MATERIAL CHOICES AND FABRICATION METHODS

The choice of material and fabrication method for microreactors depends on the reaction type, operating conditions, and required durability (Hedia et al. 2025). Common materials include chemically inert, transparent glass for optical diagnostics; silicon, favoured for precise microfabrication and electronic integration (Sano et al. 2024); cost-effective, flexible polymers like PDMS and PMMA; metals such as stainless steel and titanium, known for mechanical strength and thermal conductivity suited

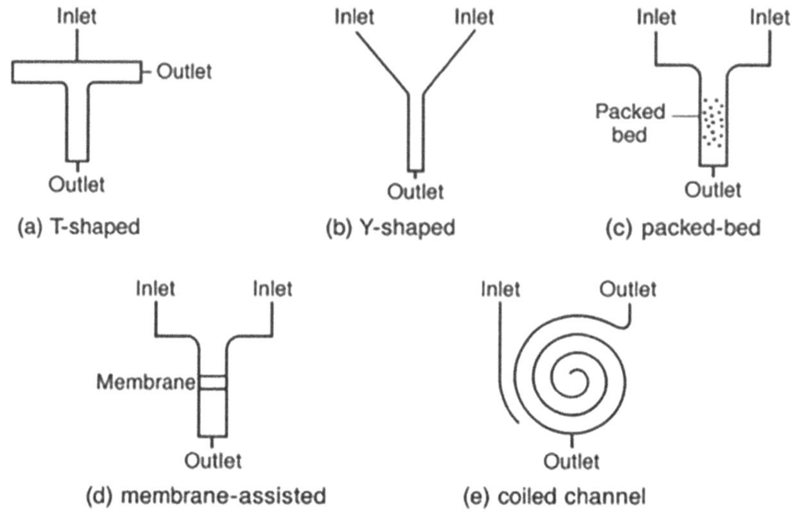

FIGURE 15.1 Schematic of different microreactor configurations: (a) T-shaped, (b) Y-shaped, (c) packed-bed, (d) membrane-assisted, and (e) coiled channel microreactors.

to high-pressure reactions (Hedia et al. 2025); and chemically resistant ceramics for harsh environments. Fabrication techniques vary accordingly: photolithography and etching for silicon and glass; 3D printing for rapid prototyping and complex designs; laser micromachining for polymers and metals; and hot embossing or injection moulding for mass-producing polymer microreactors (Rogozhnikov et al. 2024).

15.4 LIMITATIONS AND CHALLENGES

Microreactors provide numerous advantages in green and sustainable chemistry, including improved heat and mass transfer for better control of exothermic reactions and hotspot prevention. Their small volumes enhance safety when handling hazardous reagents. They facilitate rapid reaction screening for high-throughput experimentation and process optimization. Scalability is achieved by scaling up multiple microreactors rather than traditional scaling-up, reducing solvent and reagent use and thus environmental impact by generating less waste. Finally, their design allows for easy integration with sensors, facilitating real-time monitoring and control through automated feedback loops (Tables 15.1 and 15.2).

15.5 ROLE OF CFD IN PROCESS INTENSIFICATION

15.5.1 Introduction to Computational Fluid Dynamics

CFD plays a crucial role in process intensification by using mathematical models to simulate fluid flow, heat transfer, and species transport. This powerful tool allows engineers to gain deep insights into microreactor design without relying solely on time-consuming and costly experiments (Delsman et al. 2005). By solving the

TABLE 15.1
Comparison Between Microreactors and Conventional Reactors (Hedia et al., 2025; Shahi and Singh, 2024)

Challenge	Description	Implication
Clogging and fouling	Occurs in solid-forming or particle- laden reactions	Leads to reduced efficiency and operational issues
Limited throughput	Due to small channel volumes	Requires parallelization to enhance productivity
Complex fabrication	Some materials are difficult to miniaturize or are not cost-effective	Limits material choices and increases production cost
Scale-up issues	Microscale performance may not scale effectively to macroscale	Hinders industrial-scale implementation
Instrumentation & integration	Needs development of compatible sensors and actuators for microfluidic systems	Affects system automation and real- time monitoring capabilities

TABLE 15.2
Key Operational Differences: Advantages and Disadvantages of Microreactors vs. Conventional Reactors (Belmiloud 2024)

Parameter	Microreactors	Conventional Reactors
Volume	Microlitres to millilitres	Litres to cubic metres
Advantage	Precise control, low material usage	Suitable for large-scale production
Disadvantage	Limited throughput per unit	Inefficient for small-scale or precise processes
Heat/Mass transfer	Very high due to high surface-to-volume ratio	Relatively low
Advantage	Excellent control over reaction conditions	Sufficient for less sensitive reactions
Disadvantage	Overcooling or overheating may occur if not precisely managed	Inefficient transfer can cause hot spots
Reaction time	Seconds to minutes	Minutes to hours
Advantage	Faster production cycles	Better suited for slow or multi-step reactions
Disadvantage	Not ideal for reactions requiring long residence times	Longer turnaround time
Safety	High (due to small volumes)	Moderate to low
Advantage	Lower risk of explosion or contamination	Capable of handling hazardous materials with proper safety systems
Disadvantage	Not suitable for dangerous or exothermic reactions at scale	Higher risk in case of failure
Mixing efficiency	High (laminar flow + design optimization)	Variable

(*Continued*)

TABLE 15.2 (Continued)
Key Operational Differences: Advantages and Disadvantages of Microreactors vs. Conventional Reactors (Belmiloud 2024)

Parameter	Microreactors	Conventional Reactors
Advantage	Uniform product formation, improved selectivity	Good for simple, well-mixed reactions
Disadvantage	Requires precise design and flow control	Inefficient mixing may lead to byproducts
Scalability	Numbering-up	Scale-up
Advantage	Modular and flexible	Economical for very large batch sizes
Disadvantage	Complex integration of multiple units	Scale-up issues can affect process dynamics
Instrumentation complexity	Moderate	High
Advantage	Simplified for small systems	Suitable for comprehensive monitoring
Disadvantage	Specialized equipment is often required	More expensive and complex setup
Waste generation	Low	High
Advantage	Environmental friendly, cost-saving	Can handle large amounts of reactants
Disadvantage	Disposal systems may not scale easily	Expensive waste treatment and disposal

governing equations that describe how fluids behave, CFD is particularly useful for analysing laminar flow inside microchannels, where small-scale effects, like surface forces and diffusion, can have a major impact on performance (Black et al. 2022). Through CFD, it's possible to optimize key parameters such as mixing, transfer rates, reactor size, geometry, residence time, temperature profiles, and reaction zones, all of which are essential for designing more efficient and sustainable microreactor systems (Bhatti et al. 2020).

15.5.2 Governing Equations of CFD

At its core, CFD is built upon the numerical solution of fundamental conservation equations that govern fluid flow, heat, and mass transfer, as illustrated in Figure 15.2.

2. When solved simultaneously, these equations offer insights into how fluids behave in the microreactor under various operating conditions (Zainal et al. 2025). Schematic of the mesh in the initial stage of physics-controlled precision has been shown in Figure 15.3. Boundary conditions define how a system behaves at its inputs and outputs, and they are essential for accurately modelling physical processes in CFD simulations. At the inlet, boundary conditions typically specify parameters such as velocity or mass flow rate, temperature, and concentration. At the outlet, common

1. **Continuity Equation:** $\nabla . u = 0$

2. **Navier-Stokes Equation :** $\rho(\frac{\partial u}{\partial t} + u.\nabla u) = -\nabla p + \mu.\nabla^2.u + F$

3. **Energy Equation :**

$$\rho C_p(\frac{\partial T}{\partial t} + u.\nabla T) = -\nabla.(\kappa.\nabla T) + Q$$

4. **Species Transport Equation :**

$$\frac{\partial C_i}{\partial t} + u.C_i = \nabla.(D_i \nabla C_i) + R_i$$

FIGURE 15.2 List of fundamental conservation equations.

Description/MESH STATISTICS	Value
Minimum element quality	0.275
Average element quality	0.6943
Tetrahedron	6338
Triangle	1544
Edge element	232
Vertex element	19

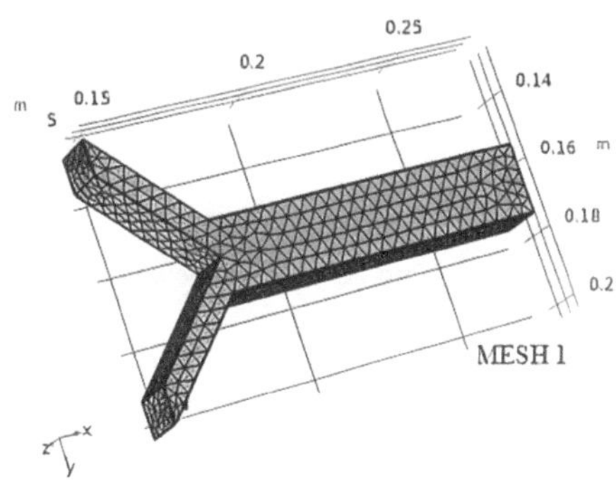

FIGURE 15.3 Schematic of the mesh in the initial stage of physics-controlled precision.

conditions include a pressure outlet or a mass-flow outlet. Wall boundaries often assume a no-slip condition, meaning the fluid velocity at the wall is zero, and may include additional specifications like a constant heat flux, a fixed temperature, or interactions with a catalytic surface. To simplify models of systems with repeating geometries, symmetry or periodic boundary conditions are often applied—this approach helps reduce computational costs while still delivering accurate results (Pasha et al. 2021) (Figure 15.2).

15.5.3 Applications in Flow and Reaction Visualization in Microreactor

One of the major strengths of CFD is its ability to provide detailed visualizations of flow and reaction behaviour, offering insights that go beyond what is possible with experimental methods. For example, velocity vector fields, streamlines, and vortex patterns can reveal how fluids mix, where secondary flows develop, and where recirculation zones form—critical information for evaluating mixing efficiency (Straw et al. 2021). Temperature contours help identify hotspots, particularly in exothermic reactions (as seen in Figure 15.4), while concentration profiles and isothermal contours allow for monitoring conversion rates, selectivity, and product distributions

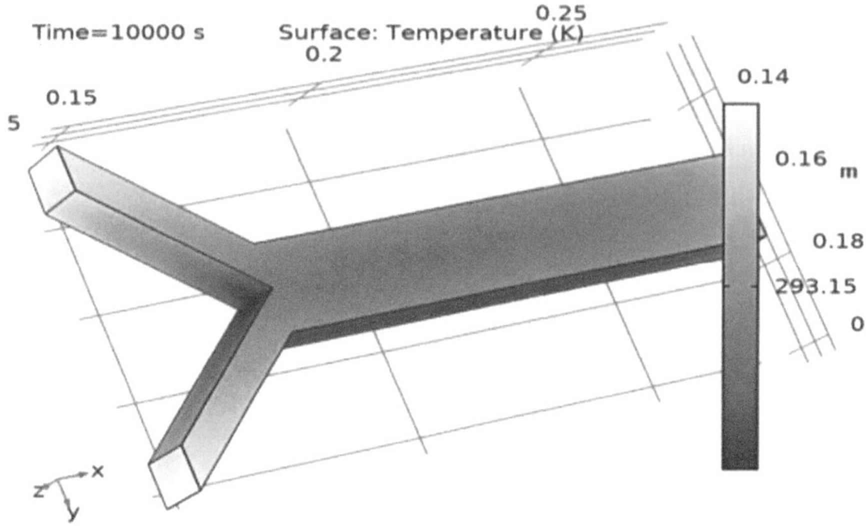

FIGURE 15.4 Temperature distribution in a microreactor under exothermic reaction conditions, highlighting heat removal challenges and hotspots.

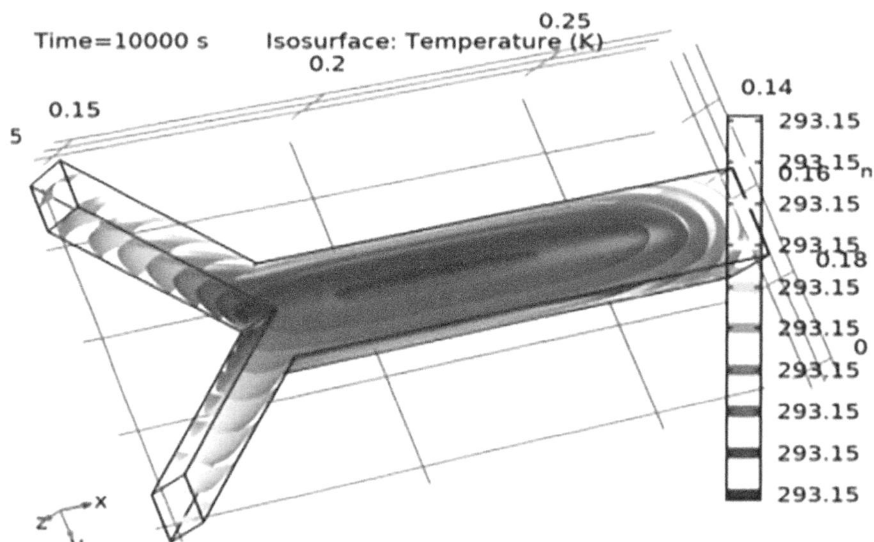

FIGURE 15.5 Temperature distribution within the microreactor, showcasing zones of constant temperature (ht) for analysing thermal uniformity and heat transfer behaviour.

(shown in Figure 15.5). Additionally, residence time distribution (RTD) analysis helps detect problems like channelling or dead zones). By mapping these variables across both space and time, CFD essentially acts as a virtual laboratory for design testing, troubleshooting, and scale-up evaluations (as illustrated in Figure 15.6).

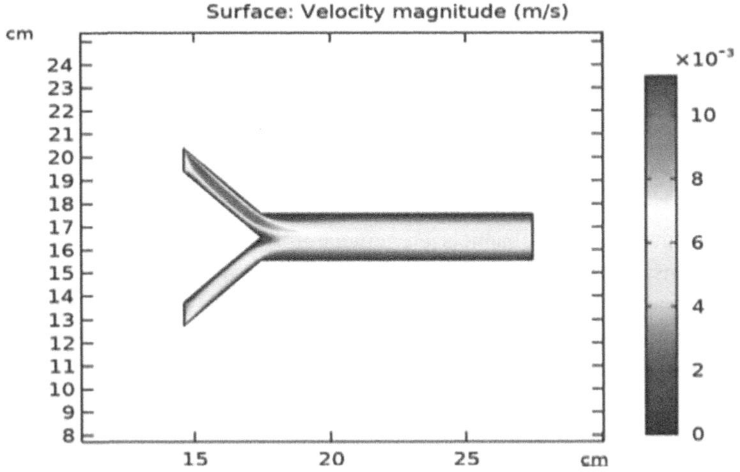

FIGURE 15.6 Simulated velocity profile in a Y-shaped microchannel showing laminar flow streams merging and evolving into a fully developed profile.

15.6 BRIDGING MICROREACTOR DESIGN WITH CFD MODELLING

Designing and operating microreactors successfully requires a deep understanding of the complex transport phenomena and reaction kinetics happening at the microscale. This is where CFD modelling becomes an invaluable tool—it serves as the bridge that connects theoretical design concepts with real-world implementation (Mañes et al. 2014). By simulating the flow dynamics, heat transfer, and chemical transformations within a microreactor, CFD allows engineers to gain detailed insights into system behaviour, helping them refine and optimize reactor designs. Through iterative simulations and adjustments, engineers can develop high-performance reactor geometries tailored to the specific requirements of a given process (Gaggero and Mastrandrea 2025).

15.6.1 Integrating CFD Tools for Design Optimization

Traditional approaches to microreactor design relied heavily on trial-and-error methods, which were time-consuming and resource-intensive (Nolte and Tomforde 2025). CFD has revolutionized this process by enabling numerical evaluation of different design parameters such as geometries, flow rates, temperatures, and catalytic conditions. This allows for the optimization of designs even before physical fabrication (Ostroumov 2025). A typical CFD-guided workflow begins by defining key reaction parameters (Ataie and Abouei Mehrizi 2024) and configuring the reactor geometry, which could include T-shaped or serpentine microchannel layouts (Zhou et al. 2024). Next, mesh generation is performed, followed by the application of appropriate boundary conditions (Shiralashetti et al. 2025), and the solving of relevant transport models (Viyyuri 2025). Post-processing of the results allows visualization of variables such as velocity and temperature, helping designers make informed adjustments. This streamlined, simulation-based process significantly reduces experimental trials and accelerates optimization efforts (Delsman et al. 2005).

15.6.2 Case Studies of Hydrodynamic Modelling in Microchannels

Hydrodynamics in microchannels are crucial for ensuring efficient mixing, uniform reactions, and effective heat dissipation. CFD enables a detailed assessment of these factors across various microchannel geometries and operating conditions. For instance, simulations involving T- and Y- junctions demonstrate how changes in junction angles or the addition of baffles can enhance mixing by increasing interfacial contact. In channels with curved or spiral geometries, CFD reveals the formation of Dean vortices, which improve radial mixing and reduce the limitations of laminar flow diffusion. Furthermore, pressure drop analyses conducted through CFD help in estimating energy requirements and assist in selecting suitable pumps, thus optimizing the overall energy efficiency of the microreactor systems.

15.6.3 Reaction Kinetics and Coupling with Transport Phenomena

The integration of reaction kinetics with transport phenomena within CFD models is essential for accurately simulating the behaviour of chemical processes in microreactors. This process begins by defining appropriate reaction mechanisms, which may range from simple first-order reactions to more complex systems such as those described by the Langmuir-Hinshelwood model (Singh et al. 2024). These mechanisms are translated into rate expressions and embedded into the species transport equations, enabling the simulation of concentration profiles over time. Temperature-dependent reactions are modelled using Arrhenius equations to capture the thermal sensitivity of reaction rates. In the case of heterogeneous catalysis, CFD also models the interaction between solid catalytic surfaces and fluid reactants by incorporating surface fluxes. Such comprehensive coupling allows for precise prediction of conversion rates, product distribution, and the formation of thermal hotspots—key indicators of system performance, safety, and product quality.

15.6.4 CFD for Scale-up and Modular Design Validation

One of the main challenges in MRT is scaling up lab-scale systems to meet industrial production demands. Instead of simply increasing the reactor size, a "numbering-up" approach is often used, where multiple microreactor units operate in parallel (Ledingham et al. 2023). CFD plays an indispensable role in ensuring that each of these modules operates consistently by verifying uniform flow distribution as well as thermal and mass transfer equivalency across all units. CFD also supports the simulation and validation of modular designs that integrate key functions such as mixing, reaction, and separation. These simulations help prevent issues like flow maldistribution, ensuring that all reactor units maintain identical operational conditions, which is crucial for reproducibility and scalable performance (Sahibzada et al. 2025).

15.6.5 Data-Driven Design Improvements and Optimization

With the continuous advancement of computational technologies, the integration of artificial intelligence (AI) and machine learning (ML) with CFD has opened new possibilities for data-driven design and optimization. These intelligent systems can analyse large datasets generated from simulations and experiments to uncover patterns and optimize design parameters more effectively than traditional methods. AI-enhanced CFD allows for predictive modelling, sensitivity analysis, and real-time optimization of reactor configurations, thereby pushing the boundaries of efficiency and innovation in microreactor design. (AI) and machine learning (ML) techniques, CFD is evolving beyond traditional simulation tools into predictive and adaptive design frameworks. One key approach is the Design of Experiments (DOE), where CFD is employed to simulate a variety of input conditions, and statistical analysis is used to identify optimal operating windows (Razzak et al. 2025). Sensitivity analysis is another critical application, helping to determine which parameters—such as channel width, flow rate, or catalyst loading—most significantly affect system performance. Additionally, optimization algorithms, including genetic algorithms and gradient-based methods, can be coupled with CFD outputs to automate and refine the design process, leading to more efficient and robust reactor configurations.

AI-assisted Surrogate Modelling: Neural networks trained on CFD data can rapidly predict reactor behaviour, drastically reducing simulation time (Barot et al. 2025) (Figure 15.7).

Size (size): Settings

Description	Value
Maximum element size	0.00449
Minimum element size	1.92E-4
Curvature factor	0.3
Resolution of narrow regions	0.85
Maximum element growth rate	1.35
Predefined size	Extra fine
Geometric entity level	Remaining

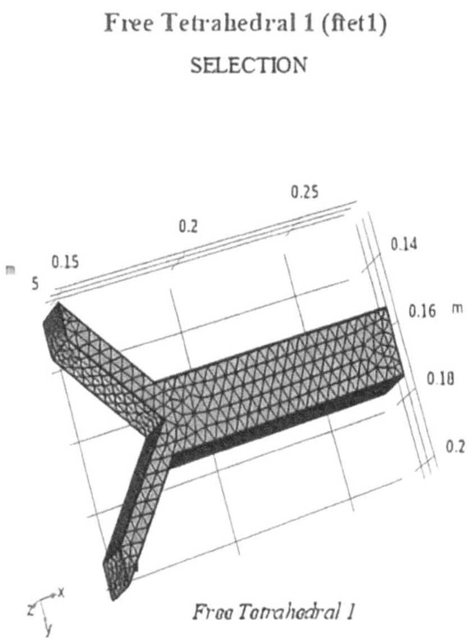

FIGURE 15.7 Schematic representation of free tetrahedral meshing.

15.6.6 CFD-Driven Process Optimization for Nitration Reactions

Nitration reactions are inherently exothermic and often hazardous, making them ideal candidates for microreactor-based processing. Traditional batch reactors typically struggle with issues like thermal runaway and suboptimal selectivity due to poor mixing and inefficient heat dissipation. Microreactors, by contrast, offer enhanced control at the microscale, effectively mitigating these challenges. In a representative case study, a Y-shaped microchannel reactor was engineered to mix nitric acid with an aromatic compound. CFD simulations were employed to analyse mixing behaviour, temperature profiles, and residence time distributions (Kurhade et al. 2025). These insights allowed for fine-tuning of flow rates and channel geometries, which in turn led to significant performance improvements. Notably, byproduct formation was minimized, yield increased to approximately 95%, solvent consumption was reduced by 40%, and the process required no external cooling. These outcomes demonstrate the efficacy of CFD-guided microreactor designs in handling fast, hazardous, and heat-sensitive chemical reactions (Wang et al. 2022, Yoswathananont et al. 2005, Liu et al. 2020, Rehman et al. 2025).

15.6.7 Microreactor-Based Hydrogenation Process

Hydrogenation reactions are central to fine chemical and pharmaceutical manufacturing but pose operational challenges due to the use of gaseous hydrogen, which can lead to safety risks and limited solubility. Microreactors help overcome these limitations by facilitating superior gas-liquid contact and mass transfer. In one notable example, a packed-bed microreactor loaded with a palladium-on-carbon (Pd/C) catalyst was used for the hydrogenation of nitrobenzene. CFD modelling was applied to simulate the gas-liquid multiphase flow and reaction kinetics (Mdumuka and Knox 2025, Yu et al. 2025). This enabled precise optimization of the flow regime to ensure uniform catalyst wetting and low pressure drop across the system. As a result, the process achieved 98% conversion, a 30% reduction in hydrogen usage, and enhanced safety due to the inherently small reactor volumes involved (Chivilikhin et al. 2011, Pawlowski et al. 2018, Gupta and Khanna 2024).

15.6.8 Continuous-Flow Synthesis for Green Manufacturing

Continuous-flow microreactor platforms align seamlessly with the principles of green chemistry by minimizing waste, reducing the use of hazardous reagents, and offering superior process control. A compelling case is the multistep continuous-flow synthesis of intermediates for ibuprofen production. Here, CFD simulations were instrumental in optimizing parameters such as residence time, thermal distribution, and reactor sequencing (Yajima et al. 2022, Madadi Avargani et al. 2023). Furthermore, CFD facilitated the integration of various unit operations—including reaction, separation, and crystallization—into a cohesive continuous-flow system. The result was a 50% reduction in production time, a 60% decrease in the environmental factor (E-factor), and consistently high product quality. This example clearly showcases the vital role of CFD in enabling efficient, scalable, and environmentally responsible manufacturing processes (Kulkarni and Gupta 2024).

15.7 CHALLENGES AND SCALE-UP CONSIDERATIONS

Although microreactors offer distinct advantages, several challenges remain. The foremost among them is clogging or fouling caused by solid precipitation or gas evolution during reaction, which can alter hydrodynamics and pressure drop. Channel blockage was rarely a concern. Scaling up microreactor processes demands replication or numbering-up strategies rather than geometric enlargement. CFD studies have revealed that hydraulic maldistribution across parallel channels leads to uneven flow and temperature gradients, emphasizing the need for manifold design optimization. Another difficulty lies in accurately modelling multiphase flow and interfacial phenomena at different scales. The computational cost of resolving bubble coalescence and droplet break-up remains high; however, hybrid models combining Euler–Euler and population-balance approaches show promise for practical simulation. Material compatibility, fabrication cost, and sensor integration are additional barriers to industrial adaptation. Collaborative research combining experimental diagnostics, advanced meshing algorithms, and AI-driven model reduction is beginning to address these constraints.

15.8 CONCLUSION

The synergistic integration of MRT with CFD modelling marks a significant step forward in developing sustainable, efficient, and scalable chemical processes. Microreactors inherently provide excellent control over heat and mass transfer, improved safety, and high precision in reaction handling. When coupled with CFD, they offer predictive modelling of fluid flow, transport phenomena, and reaction kinetics: the result is a powerful platform that accelerates design, reduces development costs, and enhances process reliability. This combination aligns closely with the principles of green chemistry by minimizing experimental iterations and reducing environmental impact. The applications are wide-ranging, spanning pharmaceuticals, specialty chemicals, waste remediation, and renewable energy. Looking to the future, the integration of artificial intelligence (AI), machine learning (ML), additive manufacturing, and advanced sensor technologies is expected to yield autonomous, smart, and modular microreactor systems. These innovations will enable decentralized, cost-effective chemical manufacturing and form a core strategy for advancing sustainable chemical industry practices and innovation on a global scale.

BIBLIOGRAPHY

Alghazo, M., Ahmed, V., & Bahroun, Z. (2025). Exploring the applications of artificial intelligence in mechanical engineering education. *Frontiers in Education*, *9*, 1492308. https://doi.org/10.3389/feduc.2024.1492308

Altindal, M., Rasheed, A., Tabib, M., & Nybø, R. (2025). Online parameter calibration in drilling hydraulics model. *SPE Paper 224509-MS*. https://doi.org/10.2118/224509-MS

Ataie, S., & Abouei Mehrizi, A. (2024). Effect of side channel configuration of a T-junction on droplet generation: A computational study. *Proceedings of ICBME 2024*, 40–45. https://doi.org/10.1109/ICBME64381.2024.10895877

Azom, M., Hossain, M., & Khan, M. (2025). Review of electrical and thermal modeling techniques for three-phase PMSM drives. *Control Systems and Optimization Letters*, *3*, 75–83. https://doi.org/10.59247/csol.v3i1.172

Bahreini, M., Desilets, M., Pahija, E., Legrand, U., & Fink, A. (2024). Identification of mass transfer mechanisms inside an industrial electrochemical cell for CO_2-to-formate conversion through 2D transient modeling. *ECS Meeting Abstracts, MA2024-02*, 2009–2009. https://doi.org/10.1149/MA2024-02252009mtgabs

Barot, Y., Anand, V., Mishra, S., & Mishra, R. (2025). A highly efficient chromium-functionalized imidazolium di-cationic ionic liquid catalyst system for oxidation of aromatic primary alcohols. *Current Organocatalysis, 12*. https://doi.org/10.2174/0122133372355628241209052903

Belmiloud, N. (2024). Characterizing turbulence and wave dynamics under rotation: Enhancing semiconductor processes through CFD analysis and high-speed imaging. *ECS Meeting Abstracts, MA2024-02*, 2278–2278. https://doi.org/10.1149/MA2024-02312278mtgabs

Bhatti, M., Marin, M., Zeeshan, A., & Abdelsalam, S. (2020). Editorial: Recent trends in computational fluid dynamics. *Frontiers in Physics, 8*, 4. https://doi.org/10.3389/fphy.2020.593111

Black, G., Shropshire, D., Araújo, K., & Heek, A. (2022). Prospects for nuclear microreactors: A review of the technology, economics, and regulatory considerations. *Nuclear Technology, 209*, 1–20. https://doi.org/10.1080/00295450.2022.2118626

Chivilikhin, M., Kuandykov, L., Cerato-Noyerie, C., Woehl, P., & Lavric, D. (2011). Residence time distribution in Corning® Advanced-Flow™ reactors: Experiment and modelling. *Chemical Engineering Transactions, 25*, 791–796. https://doi.org/10.3303/CET1125132

De Loos, S. R. A., Van Der Schaaf, J., De Croon, M. H. J. M., Nijhuis, T. A., Tiggelaar, R. M., Gardeniers, J. G. E., & Schouten, J. C. (2008). Three-phase mass transfer in pillared micro channels. In *2008 AIChE Spring National Meeting, Conference Proceedings* (pp. 241b). American Institute of Chemical Engineers (AIChE).

Delsman, E. R., Laarhoven, B. J. P. F., de Croon, M. H. J. M., Kramer, G. J., & Schouten, J. C. (2005). Comparison between conventional fixed-bed and microreactor technology for a portable hydrogen production case. *Chemical Engineering Research and Design, 83*, 1063–1075. https://doi.org/10.1205/cherd.04260

Fu, B., Bi, Q., Zheng, S., Peng, Y., Wang, S., Cao, X., Wang, Z., & Zhu, S. (2024). Advanced metasurface imaging and display based on multidimensional light field manipulation (Invited). *Acta Optica Sinica, 44*, 1400001. https://doi.org/10.3788/AOS232006

Gaggero, S., & Mastrandrea, F. (2025). Improving waterjet performance through simulation-based optimization. *Journal of Marine Science and Engineering, 13*, 758. https://doi.org/10.3390/jmse13040758

Ge, L. (2019). *Numerical investigation of particle dynamics in the presence of unconfined and confined gas bubbles* (Thesis).

Gupta, A., & Khanna, A. (2024). Sustainable production and waste-management model considering warm-up period with product recovery. *Process Integration and Optimization for Sustainability, 9*, 737–758. https://doi.org/10.1007/s41660-024-00475-7

Hassan, Q., Viktor, P., Al-Musawi, T., Ali, B., Algburi, S., Alzoubi, H., Khudhair, A., Zuhair, A., Salman, H., & Jaszczur, M. (2024). The renewable energy role in global energy transformations. *Renewable Energy Focus, 48*, 100545. https://doi.org/10.1016/j.ref.2024.100545

Hedia, H., Althobiani, F., & Hedia, M. (2025). *Handbook of engineering materials: Metals—Part One*. 1st ed. King Abdulaziz University. Available at: https://www.researchgate.net/publication/389207336

Htet, A., Gupta, S., & Oruganti, S. (2025). A review of adaptive façade systems and simplified AI for sustainable retrofitting. (Preprint/Working paper).

Hu, Z., Shen, J., Gao, X., Tan, P., & Lou, D. (2025). Impact of structural parameters and inlet exhaust-gas boundary conditions on particle-trapping efficiency of carrier wall in GPF: A non-homogeneous dynamic extended capture model. *Energies, 18*, 2255. https://doi.org/10.3390/en18092255

Jalui, S., Xuan, Y., Manogharan, G., & O'Connor, J. (2024). Development of a method for shape optimization for a gas-turbine fuel-injector design using metal-additive manufacturing. *Journal of Engineering for Gas Turbines and Power, 147*. https://doi.org/10.1115/1.4066509

Kandpal, V., Filho, W., Aguilar-Rivera, N., Frizzo, K., & Choudhury, T. (Eds.) (2025). *Green Horizons—Advancing climate solutions through ESG, circular economy, and green finance, circular economy, and ESG*. Cham: Springer.

Kulkarni, S., & Gupta, S. (2024). Call for chapters: *AI-Enabled solutions for carbon footprints—Predicting, monitoring, and minimizing (AISCFPMM-2024)*. https://doi.org/10.13140/RG.2.2.27732.21129

Kurhade, A., Amruth, E., Joshi, P., Kondhalkar, G., Jadhav, P., Murali, G., Mahajan, R., & Waware, S. (2025). Enhancing flat-plate solar collector efficiency: A numerical study. *Journal of Mines, Metals and Fuels*, 1065–1076. https://doi.org/10.18311/jmmf/2025/48249

Ledingham, J., Howse, J., Campbell, A., Tyndale-Biscoe, E., & Campbell, K. (2023). Process intensification of electrodialysis through the investigation and elimination of maldistribution. *ECS Meeting Abstracts, MA2023-02*, 1328–1328. https://doi.org/10.1149/MA2023-02241328mtgabs

Li, H., Wang, H., Hassan, Y., & Anand, N. K. (2016). Computational fluid dynamics analysis of two parallel rectangular jets using OpenFOAM. *Proceedings of ICONE24, V004T10A030*. https://doi.org/10.1115/ICONE24-61046

Liu, M., Zhu, X., Liao, Q., Chen, R., Ye, D., & Chen, G. (2020). Stacked catalytic membrane microreactor for nitrobenzene hydrogenation. *Industrial & Engineering Chemistry Research*. https://doi.org/10.1021/acs.iecr.0c01234

Lohse, S., Acker, J., Fan, Z., Barth, E., Franzke, J., Turek, S., & Agar, D. W. (2008). Evaluation of numerical methods for describing mass transfer in microstructured devices using experimentally determined residence time distributions. *Chemical Engineering Journal, 135*(1–2), S168–S176. https://doi.org/10.1016/j.cej.2007.07.056

Madadi Avargani, V., Zendehboudi, S., Osfouri, S., & Rostami, A. (2023). Performance evaluation of a solar nano-photocatalytic reactor for wastewater treatment: Reaction kinetics, CFD, and scale-up. *Journal of Cleaner Production, 421*, 138240. https://doi.org/10.1016/j.jclepro.2023.138240

Mañes, J., Sanchez Espinoza, V., Chiva, S., & Stieglitz, R. (2014). A new coupled CFD/neutron-kinetics system for high-fidelity simulations of LWR core phenomena: Proof of concept. *Science and Technology of Nuclear Installations, 2014*. https://doi.org/10.1155/2014/294648

Mdumuka, D., & Knox, P. (2025). Simulation of load and haul processes for enhanced efficiency and sustainability in quarry mining: Chilanga Cement Quarry (Zambia). *IAMURE International Journal of Multidisciplinary Research*.

Mohammed, I. A. (2024). AI in supply-chain optimization: Improving efficiency through machine learning. *IPE Journal of Management, 14*.

Nalamwar, M., Parbat, D., & Singh, D. P. (2017). Application of CFD software for planning and design of civil ISC2016 Pune 5. *ISCA-RJEngS-2017-009*.

Nolte, L., & Tomforde, S. (2025). A helping hand: A survey about AI-driven experimental design for accelerating scientific research. *Applied Sciences, 15*, 5208. https://doi.org/10.3390/app15095208

Ostroumov, S. (2025). Ecotoxicology paper cited in Japan, China. *ResearchGate post*. https://www.researchgate.net/publication/391122336_cited_in_Japan_China; https://doi.org/10.13140/RG.2.2.25162.07364

Pasha, A., Meshal, N., Al-Harbi, Halkarni, S., Islam, N., Siva, D., Reddy, D. S. K., Pillai, S., & Qadiri, U. (2021). CFD study of convective heat transfer of water flow through micropipe with mixed boundary conditions. *CFD Letters, 13*, 13–26. https://doi.org/10.37934/cfdl.13.7.1326

Pawlowski, S., Nayak, N., Meireles, M., Portugal, C., Velizarov, S., & Crespo, J. (2018). CFD modelling of flow patterns, tortuosity and residence-time distribution in monolithic porous columns reconstructed from X-ray tomography data. *Chemical Engineering Journal*, *350*, 757–766. https://doi.org/10.1016/j.cej.2018.06.017

Pereira, M., Silva, G., Semião, V., Silverio, V., Martins, J., Pascoal-Faria, P., Alves, N., Dias, J., & Ginjeira, A. (2022). Experimental validation of a CFD model using μ-PIV of irrigation flow in confluent canals. *International Endodontic Journal*, *55*. https://doi.org/10.1111/iej.13827

Razzak, M., Guo, X., Imran, S., Walla, N., Yin, H., & Zhou, C. (2025). Numerical investigation of flow field and mixing efficiency in the Kanbara reactor. *Proceedings of IMECE 2024*, 144123. https://doi.org/10.1115/IMECE2024-144123

Rehman, M., Hussain, M., Akhter, P., & Jamil, F. (2025). Breaking barriers in biodiesel: From feedstock challenges to technological advancements. *Chemistry Africa*. https://doi.org/10.1007/s42250-025-01293-1

Rogozhnikov, A., Porozova, S., Gileva, O., Shulyatnikova, O., Rogozhnikov, G., & Bystryh, M. (2024). Chemical stability and surface topology of stabilized zirconium-dioxide ceramics in simulated aggressive environments. *Actual Problems in Dentistry*, *19*, 136–142. https://doi.org/10.18481/2077-7566-2023-19-4-136-142.

Sahibzada, S., Sher Malik, F., Nasir, S., & Lodhi, S. (2025). AI-augmented turbulence and aerodynamic modelling: Accelerating high-fidelity CFD with physics-informed neural networks. *International Journal of Innovative Research in Computer Science and Technology*, *13*, 91–97. https://doi.org/10.55524/ijircst.2025.13.1.14

Salvadore, F., & Ponzini, R. (2019). LincoSim: A web-based HPC-cloud platform for automatic virtual towing-tank analysis. *Journal of Grid Computing*, *17*. https://doi.org/10.1007/s10723-019-09494-y

Sano, K., Ono, Y., Yasuda, K., Tobinaga, R., Imamura, Y., Hayashi, Y., & Yanagitani, T. (2024). Deep microfabrication of silica glass using atmospheric gas-phase catalyst etching for sensing applications. *ECS Meeting Abstracts*, *MA2024-02*, 4311–4311. https://doi.org/10.1149/MA2024-02644311mtgabs

Shahi, S., & Singh, S. (2024). Microfluidic devices with nano-sensors for infectious-disease detection. —, 20, 3369–3377.

Shen, F., Wolf, S., Bhavnani, S., Deoni, S., Elison, J., Fair, D., … Vaughan, J. (2021). Emerging ethical issues raised by highly portable MRI research in remote and resource-limited settings. *NeuroImage*, *238*, 118210. https://doi.org/10.1016/j.neuroimage.2021.118210

Shiralashetti, S., Joshi, S., & Hanaji, S. (2025). Haar-wavelet numerical method to solve non-linear differential equations in turbulent boundary-layer flow. *Indian Journal of Science and Technology*, *18*, 1046–1058. https://doi.org/10.17485/IJST/v18i13.4048

Singh, D., Rugamba, T., Katara, H., & Grewal, K. S. (2024). Computationally effective machine-learning approach for modular thermal energy storage design. *Applied Energy*, *377*, 124430.

Staw, M., Aglave, R., & Piccioli, R. (2021). Integrated approach to multiphase flow regime prediction through CFD. *OTC Paper 31096-MS*. https://doi.org/10.4043/31096-MS

Stupak, E., & Bausys, R. (2000). Generation of unstructured FE grids for complex 2D objects. *Journal of Civil Engineering and Management*, *6*, 17–24. https://doi.org/10.3846/13921525.2000.10531559

Viyyuri, S. (2025). CFD modeling of flow-boiling heat transfer in microchannels: ANSYS Fluent approaches and validation. *Proceedings of IMECE 2024*, 146631. https://doi.org/10.1115/IMECE2024-146631

Wang, N., Jin, Y., Huang, T., Zhou, J., Zhang, Y., & Li, N. (2022). Continuous production of 3,5,5-trimethylhexanoyl chloride and CFD simulations of single-phase flow in an advanced-flow reactor. *Journal of the Taiwan Institute of Chemical Engineers*, *138*, 104465. https://doi.org/10.1016/j.jtice.2022.104465

Yajima, D., Toyoda, T., Kirimura, M., Araki, K., Ota, Y., & Nishioka, K. (2022). Agrivoltaic system: Estimation of PPFD under solar panels using an all-climate solar-spectrum model. *Cleaner Engineering and Technology*, *12*, 100594. https://doi.org/10.1016/j.clet.2022.100594

Yoswathananont, N., Nitta, K., Nishiuchi, Y., & Sato, M. (2005). Continuous hydrogenation in a novel gas–liquid flow reactor packed with Pd/C. In *Proceedings of the 8th Topical Conference on Microreaction Technology (AIChE Spring Meeting)*, Atlanta, GA, April 10–14, 2005, pp. 275–281.

Yu, J., Fan, Y., & Hou, J. (2025). Distributed optimization scheduling and its boundaries in virtual power plants. *Electronics*, *14*, 932. https://doi.org/10.3390/electronics14050932

Zainal, N., Mokhtar, D., Waini, I., Alam Zuberi, H., Nazar, R., & Pop, I. (2025). Mixed convection of couple-stress hybrid nanofluid past a vertical shrinking plate. *Journal of Thermal Analysis and Calorimetry*. https://doi.org/10.1007/s10973-025-14247-z

Zapotoczny, S., Wolski, K., Su, Y., Szczepanowicz, K., & Bazylinska, U. (2025). Research Topic: Nano- and microreactors based on self-assembled polymer systems and porous particles. *Frontiers*. https://www.frontiersin.org/research-topics/61285/nano--and-microreactors-based-on-self-assembled-polymer-systems-and-porous-particles

Zhang, X., Stefanick, S., & Villani, F. J. (2004). Application of microreactor technology in process development. *Organic Process Research & Development*, *8*(3). https://doi.org/10.1021/op034193x

Zhou, Y., Lo, S., Sevilla, R., Hassan, O., & Morgan, K. (2024). Towards a volume mesh generator tailored for NEFEM. In *Lecture Notes in Computational Science and Engineering*.

Index

Pages in *italics* refer to figures and pages in **bold** refer to tables.

A

Abundance ratio, 63
Acetamiprid degradation, 247, 250
Activated sludge, 171, 186
Active microreactors, 27
Active mixers, 109
Active mixing principles, 149
Active pharmaceutical ingredient (API), 23, 39, 88, 108, 138
Additive effects, 174
Additive manufacturing, 18–19, 265, 305
Adsorption, 139, 185
Adsorption microreactors, 248
Adsorption processes, 210
Advanced degradation technologies, 172
Advanced energy systems, 263
Advanced molding techniques, 17
Advanced oxidation processes (AOPs), 10–12, 171, 175, 183, 210, 214, 246, 255
Aeration, 115
Aeration systems, 206
Aerobic oxidations, 6
Aggregation, 114–116, 163, 281
Aggregation-induced emission QDs, 162
Agricultural runoff, 171, 240
Agricultural wastewater, 210
AI (artificial intelligence), 274, 294, 303, 305
AI-assisted Surrogate Modelling, 303
Air pollution, 170, 195
Air pollution control, 22
Algae cultivation, 199
Algae-based bioremediation, 12
Algal biomass, 197, 205
Algal growth, 197, 204
Algal species selection, 201
Algal-based remediation, 198
Alternative green media, 48
Ambient conditions, 160, 175
Amidoamines, 78
Analysis applications, 255
Analytical chemistry, 59
Analytical reactor, 61
Anisotropic etching, 63
Ankistrodesmus sp, 202
Annular flow, 40–41, 88, 92
Anthraquinone dyes, 226
Antibacterial activity, 138
Antibiotic degradation, 226
Antibiotic pollutants, 226
Antimicrobial action, 137
Antioxidants production, 198
Applications, 161
Applications of QDs, 275, **280**
Aqueous pollutants, 211, 248
Arc discharge, 157
Arrhenius equations, 302
Arsenate formation, 223
Arsenic contamination, 223
Arsenite oxidation, 223
Artemisinin, 39
Artificial cells, 140
Artificial intelligence (AI), 294, 303, 305
Aspect ratio, 96–97
Asymmetric transformations, 44
Atom economy, 48
Atom transfer radical polymerization (ATRP), 68–69, 78
Atomic attachment, 114
Au nanoclusters, 154
Au nanorods, 113
Au-NMQCs, 154
Autocatalytic growth, 111
Automation, 3, 5, 14, 18, 164, 294
Autothermal microreactor, 264

B

Bacillus amyloliquefaciens, 137
Back-mixing, 89
Bacterial growth, 198
Baffled reactors, 267
Band gap energy, 218, 226
Baseload power generation, 262
Batch processes, traditional, 294–295, 304
Batch reactors, 111–112, 150, 155, 163, 172, 188, *213*, 217, 241
Batch reactors, limitations of, 275
Batch reactors versus microreactors, 276
Benzene exposure, 170
Benzoylecgonine degradation, 212
Benzyl alcohol, 50
Biginelli condensation, 50
Bimetallic photocatalysts, 223
Bioaccumulation, 170, 174, 196, 219
Bioactive compounds, 201
Bioavailability, 198
Bio-based economy, 139
Bio-based products, 88, 91
Bio-based solvents, 48

Biochemical conversion, 201
Biocompatibility, 137
Biodegradability limitations, 226
Biodegradation, 172, 185–186, 197
Biodiesel, 93
Biodiesel production, 13, 263, 266
Biodiesel synthesis, 22–23
Biofilm microreactors, 248
Biofilms, 89
Biofuel, 91
Biofuel cell, 60–61
Biofuel production, 263
Biofunctionalization, 9, 136–138
Bio-imaging / Bioimaging, 154, 156, 161, 275, **280**
Biological analysis, 294
Biological markers / Biomarkers, 23, 113, 161
Biological treatment, 176, 185
Biomagnification, 170
Biomanufacturing processes, 8
Biomass concentration, 186, 190
Biomass productivity, 12, 201
Biomaterials, 136–138, 140
Biomedical applications, 8–10, 137–138
Bio-oil production, 263
Biopharmaceuticals, 91
Biopolymers, 78, 136
Bioprocess development, 205
Bioproducts, 198, 205
Bioreactor configuration, 202
Bioreactor efficiency, 211
Bioremediation, 12, 197–200, 205, 248
Bioremediation efficiency, 198
Biosensing, 136–138, 156, 161
Biosensors, 23, 109
Biosynthesis, 115, 137
Biotechnological operations, 255
Biotechnology applications, 17, 23
Biotinylated zinc(II) dipicolylamine, 162
Biotransformation, 139
Bisphenol a degradation, 214
Blue shift, 159
Bolt-nut microreactors, 279, **280**
Bottom-up approaches, 157
Boundary conditions, 298–299, 301
Bovine serum albumin, 154
Bubbly flow, 88, 92
1,3-butadiene, 61
Butyl acrylate, 70–72

C

CAD software, 60
Cadmium contamination, 196
Cadmium removal, **225**
Cadmium selenide, 158
Calvin–benson cycle, 201
Cancer risk, 170, 174
Capillary force, 92
Capillary hydrogenation, 40–41
Capillary microreactors, 23, 60–61, 72, 78, 93, 263, 266
Capillary number, 153
Capillary pump, 113
Carbon dioxide fixation, 201
Carbon dots (CDs), 9–10, 90, 157
Carbon nanodots, 157
Carbon nanotubes (CNTs), 90, 96–97
Carbon quantum dots / Carbon QDs, 156–158, **280**
Carbon sequestration, 206
Carbon sources, 157
Case studies, 250, 302, 304
Catalysis, 88, 90–94, *95*, 97, 108, 113–114, 116, 136, 138–139, 255, 275
Catalysis, heterogeneous, 295, 302
Catalyst activity, 265
Catalyst deposition, 266
Catalyst design, 265
Catalyst dosage, 212, 216
Catalyst immobilization, 181, 255
Catalyst layer, 265
Catalyst preparation methods, 265
Catalyst stability, 265
Catalyst-controlled reactions, 43–44
Catalyst-induced reactions, 42–43
Catalyst-specified reactions, 45–46
Catalytic degradation, 181
Catalytic hydrogenations, 6
Catalytic ozonation, 216
Catalytic systems, 5, 14
Catalytic wall microreactors, 90, 93–94
Catechol, 139
CdSe QDs, 158–159
CdSe/ZnSe/ZnS QDs, 161
CdTe quantum dots (QDs), 13–14, 161–162, **280**, 281
Cell labelling, 162
Cellular homeostasis, 161
Ceramic materials / Ceramics, 17–18, 60–61, 90, 94, 264, 296
Cetyltrimethylammonium bromide (CTAB), 137
Channel clogging, 8, 72, 79, 137, 257
Channel dimensions, 88, 91, 93, 148, 163
Channel hydrophobicity, 152
Channel manifolds, 36
Channel microreactors, 242
Channel-based designs, 40–41
Chaotic advection / Chaotic mixing, 109, 279
Charge carrier recombination, 218, 227
Charge separation efficiency, 219
Chemical hydrogenation, 59
Chemical manufacturing / production, 14, 17–18, 39, 43, 47–49

Chemical precipitation, 210
Chemical processing, 1, 6
Chemical reactions intensification, 263
Chemical resistance, 17
Chemical stability, 174
Chemical state changes, 282
Chemical synthesis, 3–4, 18, 36, 150, 153, 255
Chemical synthesis platforms, 294
Chemical vapor deposition (CVD), 90, 110
Chip form microreactors, 60–61
Chitosan nanoparticles, 109
Chlorinated compounds, 171
Chlorination, 171
Chlorine photolysis, 214
Chromium reduction, 219
Ciprofloxacin, 39
Ciprofloxacin degradation, 228
Circular economy, 197, 205
Clean energy technology, 109
Climate change, 195
Clogging, 89, 137
Clogging and fouling, **297**, 305
Clogging potential, 177, 189
Closed photobioreactors, 200
Cloud-based monitoring, 19
Coagulation, 185
Coagulation techniques, 210
Co-contaminants, 198
Coiled or serpentine reactors, 295–296
Coliform bacteria, 198
Colloidal nanoparticles, 159
Compact reactor design, 242, 263
Comparative analysis PBR vs μPBR, 206
Compositional engineering, 160
Computational control, 207
Computational fluid dynamics (CFD), 90, 93–94, 296, 298–302, 304
Concentration gradient, 115
Concentration profiles, 299, 302
Conduction band electrons, 217
Conservation equations, 298–299
Contaminant degradation, 246
Contaminant removal, 198
Continuity equation, 298–299
Continuous flow (CF) / mode, 59, 88, 90–91, 93, 136, 138, 149, 158, 276, 278, 294–295, 304
Continuous flow microreactors / operation, 151, 241, 247, 251, 263
Continuous laminar flow microreactors, 151
Continuous manufacturing, 5, 39, 43
Continuous monitoring, 205
Continuous oil phase, 155
Continuous transesterification, 22
Continuous-flow processing, 4, 7–10, 22–23
Continuous-flow reactors / systems, 109, 111–113, 176, *213*, 228
Continuous-flow synthesis, 39, 41, 304
Control strategies, 191
Controlled reaction conditions, 241, 246
Convection-diffusion, 91
Conventional reactors, **297**, 298
Conversion efficiency, 264
Co-polymerization, 139
Copolymers, 60, 69
Core-shell varieties, 164
Covalent binding, 139
CQDs, 156
Cross-channel flow-focusing, 153
Cross-type mixers, 109
Crystallinity, 279
Cutinases, 11–12
Cyber-physical systems, 18

D

Darcy-Weisbach equation, 92
Dark cycle balance, 199
Dasatinib, 137–138
Data-driven design, 303
Dcfa degradation, 252
Dean vortices, 302
Decentralized fuel production, 22
Decolorization efficiency, 212
Deep eutectic solvents, 8, 48
Defect tolerance, 160
Degradation efficiency, 182, 251
Degradation kinetics, 211, 228
Degradation mechanisms, 180
Degradation pathways, 171
Dendrimers, 108
Desalination, 108
Design of Experiments (DOE), 303
Design of microreactors, 263
Design optimization, 301, 305
Design strategies, 17–19
Diagnostic applications, 23
Diazomethane, 40
Dichloride radical formation, 214
Dieldrin persistence, 174
Dielectrophoresis-driven techniques, 153
Diesel generator comparison, 269
Diffusion, 88, 91–92, 94, 138–140, 148, 151, 172, 180, 277
Diffusion enhancement, 240, 246
Diffusion length reduction, 267
Diffusion limitations, 1, 4, 6, 211
Diffusion limited growth, 277
Diffusion paths, 172, 180
Digital design tools, 294
Digital flow control, 5
Digital microfluidics, 47
Digital microreactors, 279
Dimensional analysis, 282

Dimensionless groups, 282
2-dimethyl amino ethyl methacrylate (DMAEMA), 69
2-(dimethylamino) ethyl methacrylate, 69
Dioxin pollutants / Dioxins, 170, 210
Direct oxidation, 181
Disinfection benefits, 175
Dispersed phase, 153
Dissolved oxygen, 205
Distributed manufacturing, 88, 91
Distribution systems, 188
DNA amplification, 23, 40–41
DNA bio-labeling, 158
Doped quantum dots, 164
Droop model, 205
Droplet-based microreactors, 151, 155, **280**
Droplet break-off, 153
Droplet generation, 152
Droplet manipulation, 153
Droplet microfluidics, 278
Droplet microreactors, 242
Droplet or segmented flow microreactors, 295
Droplet population flow, 40–41
Drug delivery, 97, 108, 116, 136–138, 140
Drug delivery systems, 9
Dry etching, 61, 63
Dye degradation, 226

E

Economic challenges, 263, 269
Economic considerations, 294
Ecosystem imbalance, 195
E-factor (environmental factor), 304
EGFR, 162
Electrocatalysis, 97
Electrochemical degradation, 181
Electrochemical method, 157
Electrochemical oxidation, 49, 172, 181, 210
Electrode balance, 198
Electrode materials, 181
Electrodes, 96–97
Electrohydrodynamic method, 153
Electron–hole pairs, 217
Electron-hole recombination, 175
Electron transfer reactions, 174, 181, 211
Electrooxidation, 157, 181
Electroplating, 61–62
Embedded sensor networks, 19
Emerging contaminants, 228, 240, 255
Emission spectra, 274
Emission spectra, narrow, 274
Emission spectra, size-tunable, 274
Emission wavelength, 159
Enantioselective products, 43
Encapsulation, 137–138, 140
Encapsulation strategies, 161
Endocrine disruption, 170, 174, 186, 196
Energy applications, 13–14, 22–23
Energy consumption, 211, 246, 256
Energy consumption trends, 262
Energy conversion, 161
Energy crisis, 195
Energy demand growth, 262
Energy efficiency, 172, 187, 302
Energy equation, 298–299
Energy generation systems, 262
Energy levels, 159
Energy resilience, 263
Energy security, 268
Energy storage, 96, 274, **280**
Energy storage integration, 262
Environmental applications, 22–23
Environmental impact, 19
Environmental pollutants, 196
Environmental pollution, 210
Environmental remediation, 1, 136, 138, 205, 244, 255
Enzymatic catalysis, 48
Enzymatic degradation, 171, 183
Enzymatic degradation microreactors, 11
Enzymatic microreactors, 138–140
Enzyme immobilization, 11, 183
EPR effect, 163
Esterification, 40–41, 45, 59
Etching techniques, 17–18, 61–63
Exothermic reactions, 5–6, 14, 90, 94, 150, 163, 296, 298, 300, 304
Extracellular matrix (ECM), 109
Extractor, 61

F

Fabrication, 89–90, 93–94
Fabrication methods, 17–19, 295–296, **297**, 305
Fabrication techniques, 60–62, 78, 179, 264
Falling film microreactors, 243
$FAPbBr_3$ nanoparticles, 160
Fenton-based systems / Fenton process, 10, 171, 175, 183, 216, 247
Fermi wavelength, 154
Fe_3O_4 nanoparticles, 8, 137
Ficin, 155
Fine chemicals, 39, 43, 45, 47, 49–50, 88, 113, 138
Fine chemicals synthesis, 23, 36
Fischer-Tropsch synthesis, 40–41
Fixed-bed microreactors, 93
Fixed-bed reactor, 61
Flash chemistry, 38, 108
Flow chemistry, 4, 35, 38, 44, 46, 112–113
Flow control, 19
Flow distribution, 188
Flow dynamics, 90–92, 171, 211, 264

Flow rate, 149, 151, 153, 158, 276, 278, 281–282
Flow rate effects, 251
Flow regimes, 3, 243
Flow stability, 243
Flow-rate ratio (FRR), 140
Flow-through microreactor, 157
Fluorescence, 274
Fluorescence detection, 113, 154
Fluorescence intensity, 161
Fluorescence lifetime, 275
Fluorescent carbon nanoparticles, 156
Fluorescent sensors, 155, 162
Folate-decorated nitrogen-doped GQDs, 162
Food chain contamination, 170
Fouling issues, 256
Free radical polymerization, 68, 72
Free radical reactions, 59
Fluid dynamics, 4
Fluid flow, 294, 296, 298, 305
Fluid handling systems, 17
Fluid–fluid interactions, 1
Fluid–solid interactions, 1
Fuel cell applications, 22–23, 263, 265
Fuel cells, 97
Fuel reforming, 268
Full width at half maximum, 159
Functional nanoparticles, 8
Functionalization, 136–138

G

Gas bubbles, 151
Gas exchange, 202–203
Gas–liquid flow, 3–4
Gas-liquid mass transfer, 183, 203, 243
Gas–liquid–liquid systems, 4
Gas–liquid–solid systems, 4
Gas-liquid systems, 88, 92–93, 153, 295, 304
Gas-phase systems, 40–41
Gas-solid reactions, 48
Geothermal energy, 262
Gibbs free energy, 115
Glass, 60–61, 90, 94, 295–296
Glass microreactors, 17–18, 244, 264
Global pollution impact, 195
Global transport, 170, 175
γ-globulin, 154
Gold nanoparticles (AuNPs), 113, 116
Graphene catalysts, 216
Graphene oxide (GO), 7, 90
Graphene quantum dots, 157, 162
Green chemistry, 8, 35, 47, 50, 296, 304–305
Green chemistry principles, 240
Green energy transition, 262
Green technology, 115
Ground-level ozone, 170
Groundwater contamination, 220
Growth kinetics, 205
Growth mechanisms, 277, 281
Growth stages, 149

H

Haber–weiss mechanism, 211
Half-life persistence, 170, 175
Halogenated compounds, 174
Hazardous materials, 40–41, 45, 49
Heat dissipation, 163
Heat management, 22
Heat transfer, 1, 4–5, 10, 38–39, 43, 48–49, 59, 71–72, 88–92, 94, 108, 110, 112–113, 136, 138, 148, 158, 163, 171, 178, 240, 246, 276–277, 279, 294–296, **297**, 298, 301, 305
Heat transfer efficiency / enhancement, 17, 36, 263
Heavy metal contamination / ions, 196, 199, 219
Heavy metal ion monitoring, 162
Helical mixers, 109
Heterogeneous catalysis, 4, 90, 93
Heterogeneous nucleation, 277
Heterogeneous polymerization, 71–72
Heterojunction photocatalysts, 219, 222
Hexanal, 139
Hexanol, 139
High-pressure conditions / reactions, 158, 296
High-pressure/high-temperature microreactors, 279
High-throughput production / synthesis, 138, 164
High-throughput screening, 205
HOIP QDs, 160
Homogeneous catalysis, 216
Homogeneous mixing, 275–277, **280**
Homogeneous nucleation, 277
Homogeneous reactions, 40–41, 46
HOMO–LUMO transitions, 154
Hot-injection approach, 160
Hot-spot avoidance / Hotspots, 90, 94, 277, 296, **297**, 299–300, 302
HPLC pumps, 109
Hybrid bio-chemo systems, 137
Hybrid reactions, 40–41, 47
Hybrid systems, 172, 187, 200, 207
Hydrodynamic modelling, 302
Hydrodynamics, 204
Hydrogen generation, 263, 268
Hydrogen peroxide, 183
Hydrogen production / production efficiency, 16, 22–23, 36, 265
Hydrogen purification, 267
Hydrogen reforming, 268
Hydrogenation / Hydrogenation process, 39, 41–42, 45, 47–48, 304
Hydrogenations, 5

Hydrothermal flow, 90
Hydrothermal treatment, 157
Hydroxyl radicals, 6, 10, 12, 172, 183, 210, 217

I

Ibuprofen, 39
Ibuprofen production, 304
Imaging, 108, 116
Immiscible fluid phase, 153
Immobilization, 139–140
Immobilized enzymes, 183
Immunoassays, 113
In vitro imaging, 156, 162
Indirect oxidation, 181
Industrial energy demand, 262
Industrial scalability, 5, 14, 17
Industrial-scale production, 149
Industrial waste generation, 195
Industrial wastewater, 171, 185, 210, 240, 249
Industrial-scale production, 149
Industry 4.0, 294
Infrastructure development, 262
Injection molding, 18
Inlets, 151
In-line / Inline monitoring, 19, 109, 113
In-line / Inline purification, 8
Inorganic compounds, 60, 63, 78
Inorganic nutrients, 197
Inorganic pollutants, 210
In-situ monitoring, 205
Integrated diagnostics, 5
Integrated energy systems, 262
Integrated manufacturing, 17
Integrated sensors, 294, 296, **297**, 305
Integrated treatment systems, 241, 247
Interfacial area, 137, 139
Interfacial contact, 4
Interfacial interactions, 251
Interfacial reactions, 178
Interfacial tension, 89, 91
Intermediate byproducts, 214
Internet of Things (IoT), 19
Interparticle repulsion, 162
Intracellular pH, 161
Inverse opal films, 218
Ionic liquids (ILs), 47
Isothermal conditions, 1, 6
Isothermal contours, 299
Isotropic mode etching, 63
Isoxazole degradation, 212

J

Janus microspheres, 11
Jeffamine®, 49–50

K

Kinetic analysis, 139
Kinetic modelling / modeling, 205, 252
Kinetic performance, 265
Kinetic rate constants, 211, 228, 250

L

Lab-on-chip systems, 23
Laccase, 139
LaMer diagram, 276
Laminar flow, 1–2, 4, 10, 40–41, 61, 88–92, 94, 148, 151, 163, 172, 178, 242, 295, **297**, 298, 301–302
Laminar flow conditions, 211
Laminar flow microreactors, 110
LARP technique, 160
Laser ablation, 157
Laser cutting fabrication, 264
Laser machining, 35, 37
Laser micromachining, 35
Laser micro-milling, 35
Laser-based microetching, 62–63
LEDs, 157, 160, 164
LEDs and displays, 275
LIGA, 62
Ligand binding, 162
Ligand-assisted reprecipitation, 160
Light absorption, 218
Light distribution, 204
Light intensity, 199, 204
Light irradiation, 217
Light limitation, 204
Light penetration, 181, 246
Light utilization, 201
Light-water reactor technologies, 13
Lipid extraction, 198
Lipids, 140
Liposomes, 108, 140
Liquid–liquid system, 153
Liquid-phase systems, 40–41
Lithium-ion batteries, 96, **280**
Lithography, 18, 60–62
Low-carbon energy, 262
Low-cost fabrication / manufacturing, 18, 264
Luminescence decay, 160
Lysozyme, 154

M

MAA-capped QDs, 161
Machine learning (ML), 18, 294, 303, 305
Manifold design / Manifold-based designs, 40–41, 89–90, 94
Mass transfer, 1, 5–6, 10, 38–39, 43, 47–49, 61, 78–79, 88–93, 136, 138–139, 148, 155,

163, 171, 178, 180, 201, 204, 276, 278–279, 294–298, 302, 305
Mass transfer enhancement, 22, 240, 246, 263, 267
Mass transport, 108, 110, 112–113
Material choices / selection, 17–19, 295–296, **297**, 305
Material compatibility, 17
Material synthesis, 1, 6–10
Maxwell-Boltzmann assumption, 91
Melting point reduction, 108, 116
Membrane bioreactors, 186, 190
Membrane dispersion, 151
Membrane filtration, 171, 189, 210
Membrane fouling, 191, 257
Membrane microreactors, 242, 247, 267
Membrane reactor, 61, 79
MEMS-based devices, 294
Mesh generation, 298–299, 301, 303, 305
Metal catalysts, 264
Metal cation gradients, 161
Metal ion sensors, 155, 162
Metal microreactors, 18–19
Metals, 90, 94, 295–296
Methanol synthesis, 61
Methyl methacrylate (MMA), 70, 72
Micellar catalysis, 48
Michaelis-Menten model, 139
Microalgae, 197
Micro-bioreactor, 138–140
Microchannel devices, 38–43
Microchannel fabrication, 264
Microchannel geometries / geometry, 4–5, 264
Microchannel mixing, 264
Microchannel reactors / structures, 17–18, 36, 59–61, 202, 211
Microchannels, 1–5, 10–11, 108–113, 148, 164, 171, 176, 276, 278–279, 281, 294–295, 298, 301–302, 304
Microdimensional capillaries, 59
Microdroplets, 155
Micro-electro-mechanical systems (MEMS), 62–63
Microfabrication, 89–90, 108, 176
Microfluidic analytical devices, 38
Microfluidic bioreactors, 8
Microfluidic devices / systems, 18, 294, **297**
Microfluidic hydrodynamic focusing (MHF), 140
Microfluidic reactor technology, 148, 276
Microfluidic reactors, 218, 242, 249
Microfluidics, 112–113, 176
Micromachining, 17–18, 59–63, 91
Micromixers, 5, 35, 109, 111, 113, 151
Micromixing, 4, 6, 8, 137
Micro-photobioreactors (μPBRs), 12, 200
Micropollutants, 172, 186
Microprocess engineering, 88, 91
Microreactor advantages, 241, 263
Microreactor applications, 22–23
Microreactor classification, 242
Microreactor construction, 264
Microreactor control systems, 17, 19
Microreactor design, 4–5, 17–19, 177, 257
Microreactor fabrication, 17–19, 264
Microreactor materials, 149, 244
Microreactor packaging, 17
Microreactor performance, 264
Microreactor scalability, 241, 246, 263
Microreactor synthesis, 163
Microreactor technology (MRT), 1–14, 17–23, 35–42, 47, 49, 136–138, 171, 176, 200, 240, 262, 295–298
Microreactor (MR) technology, 108–112
Microreactor types, 150, 242, 266
Microstructured reactors, 15, 17
Milifluidic droplet reactors, 279
Mineralization processes, 240, 246
Mixing efficiency, 178, 204, 277–279, 295, **297**, 299, 302, 304
Mixing modules, 151
Mizoroki-Heck reactions, 44
Modular design / reactor systems, 18, 294, 298, 302, 305
Molecular screening, 139
Molecular-scale mixing, 163
Monodisperse nanoparticles / QDs, 113, 149, 158, 163
Monodispersity, 275, 277–279, 281
MRI contrast agents, 138
Multicomponent reactions (MCRs), 50
Multidentate polymer micelles, 160
Multimodal molecular weight distributions, 70
Multiphase flow, 1–4, 40–41, 109–110, 295, 304–305
Multiphase flow system, 153
Multistage synthesis, 39, 43
Multivariate analysis, 281–282
Multi-walled carbon nanotubes (MWCNTs), 96
Multizone temperature profiles, 150, 163

N

Nanocatalyst deposition, 15
Nanocatalysts, 47, 265
Nanocrystals, 136–138, 158
Nanodiamonds, 156
Nanograms concentration, 186
Nanomaterial synthesis, 7–10, 18, 37, 255
Nanomaterials, 90, 96, 108, 112, 116, 136–137, 148, 150, 156, 159, 219, 274, 276
Nanomotor remediation, 254
Nanoparticle synthesis, 7–8, 255
Nanoparticles (NPs), 78–79, 108–109, 111–116, 136–138, 140, 148, 153, 157, 181
Nanopollutants, 196

Nanorod arrays, 113
Nanoscale synthesis, 18
Navier-Stokes equations, 91, 298–299
Near-infrared region, 162
Network reactor, 61
NIR-II fluorescence, 162
Nitration reactions / Nitrations, 38–40, 304
Nitrobenzene, 304
Nitrogen oxides, 170, 210
Nitroxide-mediated polymerization (NMP), 70–71
Noble metal quantum clusters, 153
Noble-metal nanoparticles, 116
Noble metal quantum clusters, 153
Non-cadmium quantum dots, 159, 163
Non-metal quantum dots, 156
No-slip condition, 299
Nuclear microreactors, 35–36, 263
Nucleation, 111–115, 148, 153, 155, 157, 163, 275–277, 281–282
Nucleation, burst, 277
Nucleation, heterogeneous, 277
Nucleation, homogeneous, 277
Nucleation rates, 276, 281–282
Numbering-up / Numbering-up strategies, 5, 8, 14, 88–90, 110, 138, 298, 302, 305

O

On-demand manufacturing, 294
Open pond systems, 200
Optical characteristics, 108, 116, 155, 158, 163
Optical density monitoring, 205
Optimization algorithms, 303
Optoelectronic devices, 160
Optofluidic microreactors, **280**
Organic compounds, 63, 78
Organic pollutants, 196, 211, 226, 249, 254
Organic–inorganic perovskite QDs, 160
Organolithium compounds, 39
Organometallic catalysts, 108
Oscillatory flow reactors, 267
Oxidation, 39, 41, 45, 48–50, 90
Oxidative stress, 174
Oxygen species, 181
Ozonation / Ozonation process, 11, 171, 175, 183, 215, 247
Ozone photolysis, 212
Ozonolysis, 40–41

P

Packed-bed microreactors / reactors, 43–45, 49, 90, 93–94, 140, 242, 248, 264, 268, 295–296, 304
PAMAM solutions, 155
Parabolic velocity profile, 151
Paracetamol removal, 186
Parallel flow, 40–41
Parallelization, 149
Partial oxidation, 22–23
Particle aggregation, 8
Particle growth, 149, 155, 163
Particle size, 149, 158, 163
Passive mixers / mixing principles, 109, 149
PCR-on-chip systems, 23
PDMS (polydimethylsiloxane), 60, 62, 109, 111, 279, **280**, 295–296
Peclet number, 151
Perfluorodecalin, 155
Peristaltic pumps, 109
Perovskite quantum dots, 156, 160, 163
Peroxidases, 11
Peroxidations, 39
Pesticides, 170, 173, 185
pH effects / reaction pH, 109, 115, 282
pH sensors, 161
Pharmaceutical applications, 59, 79
Pharmaceutical manufacturing, 17, 23, 304–305
Pharmaceutical pollutants, 171, 185, 198, 226, 252
Pharmaceuticals, 39, 43, 45, 47, 49–50, 88, 91, 138
Photobleaching, 158
Photocatalysis, 40–41, 43, 45, 78, 88, 113, 157, 161, 171, 175, 181, 217, 246, 275, **280**
Photocatalyst materials, 246, 253
Photocatalytic degradation, 181, 217, 250
Photocatalytic fuel cell microreactor, 253
Photocatalytic microreactor, 137–138, 218
Photocatalytic systems, 9, 12
Photocatalytic water treatment, 22–23
Photoemission maxima, 159
Photo-Fenton processes, 10, 247
Photolithography, 35
Photoluminescence, 154, 157, 161
Photostability, 157–158, 162, 274, **280**
Photo-stimulated polymerization, 78
Photothermal CO_2 reduction, *95*
Photovoltaics, 159, 161, 163, 275
Pi groups, 282
Pickering emulsion, 137–138
Plasma etching, 63
Plastics, 94
Plate-fin microreactor, 94
Platinum (Pt) nanoparticles, 114
Plug-flow behavior / chemistry, 6–8, 112
PMMA (poly(methyl methacrylate)), 137, 295–296
Poiseuille flow, 88, 92
Pollutant degradation, 10–12
Pollutant removal efficiency, 250
Polycarbonate masters, 36
Polycondensation, 71
Polydispersity, 275, 281
Polymer dots / Polymeric dots, 157
Polymer manufacturing, 22–23

Polymer microreactors, 18
Polymerization, 63, 68–72, 78–79, 91
Polymerization systems, 5
Polymers, 90, 94
Porphyrin synthesis, 50
Portable fuel cells, 22–23
Precursor concentration, 275, 281–282
Precursor flow rate, 159
Pressure drop, 88–90, 92–94, 302, 304–305
Process intensification (PI), 1, 4, 14, 37–38, 43, 49, 137, 246, 263, 296
Process optimization, 18, 200, 211, 282
Process safety, 17, 23, 88, 91, 109, 113
Protein-gold nanoclusters, 154
Prototyping technologies, 18
Pt-NMQCs, 155
Purification, 136, 138

Q

Quality control, 302
Quantum confinement, 274
Quantum dot synthesis, 18
Quantum dots (QDs), 9, 13–14
Quantum dots, definition, 274
Quantum dots, properties, 274
Quantum dots, synthesis principles, 276
Quantum yield (QY), 274, 279, **280**
Quench zone, 109, 111

R

Radical generation, 172, 210
Rapid mixing, 149, 153
Rapid prototyping, 60
Rarefaction effects, 89, 91
Reaction control, 17, 150, 263
Reaction environment, 148, 161
Reaction kinetics, 1, 6, 158, 172, 180, 211, 265, 276–277, 279, 281, 301–302, 304–305
Reaction limited growth, 277
Reaction parameters, 278, 281–282
Reaction safety, 23
Reaction selectivity, 2, 5–6, 180, 240
Reaction temperature, 148, 151, 159
Reaction time, 149
Reactive groups, 161
Reactive intermediates, 38–40, 43, 45
Reactive oxygen species, 181, 199, 217, 223, 246, 250
Reagent mixing, 111, 113
Real-time monitoring, 19, 23, 164, 294–296, **297**
Recalcitrant compounds, 172, 184
Recyclable polymers, 19
Redox balance, 198
Reduction reactions, 219
Remedial setup, 198
Remote reactor operation, 19
Renewable energy, 305
Renewable energy integration, 171, 187
Renewable energy production, 262
Renewable energy sources, 262
Reprecipitation, 160
Reproducibility, 136–138, 276, 278–279, 282
Residence time, 112–113, 149, 153, 163, 172, 177, 267, 276, 278–281, 295, **297**, 298–299, 304
Residence-time control, 136, 139
Residence-time distribution (RTD), 5, 88–90, 94
Resource efficiency, 5
Resource recovery, 190, 205
Retinoblastoma cells, 156
Reversible addition-fragmentation chain transfer (RAFT), 69–70
Reynolds number (Re), 88, 91
Rolipram, 39

S

Safety, 2, 6, 14
Safety features, 294–296, **297**, 304–305
Salinity effects, 198
Scalability, 136–138, 149, 153, 163, 241, 246, 263, 276, 278–279, 294–298, 302
Scalability challenges, 211, 257
Scalable production, 17–19
Scale-out, 149
Scale-up limitations, 172, 188
Scale-up / Scale-up strategies, 4–5, 14, 38, 41–43, 45, 47, 49, 59, 88–89, 91–92, 94, 110, 112–113, 137–138, 140, 172
Scanning electron microscopy (SEM), 116
Scenedesmus obliquus, 202
Screening of microorganisms, 201
Sedimentation behavior, 202
Segmented-flow microreactors / regimes, 6, 151, 246, 278–279
Selectivity, 88, 90, 94, 108, 113
Selectivity control, 180
Self-assembly, 160
Semiconductor catalysts, 181
Semiconductor fabrication methods, 18
Semiconductor QDs, 156, 158, 163
Sensing, 275, 279, **280**
Sensitivity analysis, 303
Sensor integration, 18–19, 203
Sensor / Sensors, 61, 90, 93–94
Serpentine channels / microchannels, 40–41, 90, 92, 279
Shear stress tolerance, 202
Silicon, 60, 63, 90, 94, 295–296
Silicon microreactors, 244, 264
Silicon–Pyrex reactors / glass, 18, 149
Silver nanoparticles (AgNPs), 116

Single microchannel reactors, 202
Single-phase flow / liquid flows, 88, 92–93, 151
Single-walled carbon nanotubes (SWCNTs), 96–97
Size distribution, 275, 277, 279
Sludge production, 175
Slug flow, 3–4, 40–41, 88, 92–93
Smart manufacturing, 18
Sodium borohydride, 155
Soft lithography, 36–37
Solar cells, 97, 159, 163, 275
Solar-driven systems, 187
Solar-energy applications / utilization, 14, 197
Solvent resistance, 179
Solvent-antisolvent method, 136
Sonolysis, 175
Species selection criteria, 201
Species transport equation, 298–299, 302
Spin coating, 60
Spiral microchannels, 279
Split-and-recombine mixers, 109
Stability, 158, 160, 163
Stainless steel (SS), 71, 94, 149, 151
Steam reforming, 22–23, 37, 264, 268
Step-growth polymerization, 71
Streptavidin-coated QDs, 162
Styrene, 70–72
Substrate, 59–63, 78
Supercapacitors, 96
Supercritical CO_2 ($scCO_2$), 48
Superparamagnetic nanoparticles, 137–138
Supersaturation, 113, 276–277, 281
Surface area-to-volume ratio / Surface-to-volume ratio, 1, 5–6, 10, 12, 39, 45, 49, 88–89, 92, 108, 116, 171, 178, 180, 294–295, **297**
Surface modification, 136–137
Surface passivation, 162
Surface plasmon resonance (SPR), 116
Surface properties, 148, 158
Surface tension, 152
Surfactants/stabilizers, 281–282
Sustainability, 19, 35
Sustainable chemical production / chemistry, 8, 294–295, 305
Sustainable development, 195, 240, 262
Sustainable manufacturing, 14, 17
Sustainable technologies, 161, 163
Sustainable treatment, 171
Suzuki-Miyaura reaction, 44–45
Synthetic biology, 138, 140
Syringe pumps, 109, 157

T

Teflon-lined capillary, 78
Temperature contours, 299–300
Temperature control, 2, 5–6, 18–19, 110, 115, 172, 177, 242, 246, 263–264, 276–277, 279, 281
Temperature effects, 198, 216
Tetracycline degradation, 217, 228
Textile wastewater, 184, 210, 226, 249
Thermal conductivity, 18
Thermal coupling, 35
Thermal design, 90, 94
Thermal management, 4–6, 97
Thermal sensitivity, 302
Thermal uniformity, 300
Thermally sensitive reactions, 277, 279
Thermochemical analysis, 48–49
Thermodynamic driving force, 115
Thermo-fluid approach, 90, 94
Thermoplastic, 62
Thin film microreactors, 252
Thioglycolic acid, 162
Thiol-capped CdTe QDs, 161
Thiol-ene coupling, 78
Three-dimensional (3D) printing, 18, 36
Throughput, 110, 112, **297**, 298
TiO_2 (Titanium dioxide), 137–138
TiO_2 nanoparticles / photocatalyst, 9, 246, 250
Tissue engineering, 136
T-junction, 151, 153, 277–278
Toxic byproducts, 171
Toxic contaminants, 210, 249
Toxicity, 159, 161, 163
Toxicity mechanisms, 174
ω-transaminase, 140
Transesterification process, 263, 266
Transition metal ion doping, 163
Transitional flow, 40–41
Transport phenomena, 2, 4, 172, 263, 301–302, 305
Treatment efficiency, 211
Trisodium citrate, 116
Tryptophan metabolites, 154
T-shaped microreactors, 295–296, 301–302
T-type mixers, 109
Tube-in-tube microreactor, 250
Tunable photoluminescence, 157
Turbulent flow, 40–41
Turbulent flow absence, 178
Two-phase flow, 89–90, 92–93
Two-temperature zone microreactors, 149

U

Ultrafiltration, 210, 247
Ultra-low carbon systems, 262
Ultrasonic treatment, 157
Ultrasonicated microreactors, **280**
Ultrasound-assisted fabrication, 8
Ultrasound-assisted processes, 90
Ultrasound-assisted synthesis, 137

Uniform droplet size, 153
Uniform mixing, 149, 151, 163
Uranium reduction, 221
UV irradiation, 247, 250
UV laser, 62
UV source, 78
UV-based AOPs, 211
UV/H2O2 process, 211, 250
UV/Ozone process, 212
UV/Persulfate process, 214

V

Valence band holes, 217
Velocity field distribution, 253
Velocity vector fields, 299, 301
Virtual laboratory, 300
Viscous forces, 153
Visible light photocatalysis, 217
Volatile organic compounds (VOCs), 22–23, 170
Vortex microreactor, 8

W

Wall friction, 88, 91
Waste generation, 296, 298, 304
Waste heat recovery, 97
Waste remediation, 305
Wastewater discharge, 171
Wastewater purification, 12
Wastewater treatment, 171, 185, 197–200, 205, 210, 226, 240, 249
Water as reaction medium, 48
Water pollution, 171, 184, 195, 240, 249
Water purification, 22–23, 96
Water purification technologies, 210
Water treatment, 12
Water-in-water emulsion, 137
Watzke and Finke model, 157
Wet etching, 17–18, 61, 63
Wettability, 88, 137
Whole-cell systems, 139
Wind energy integration, 262
Wire electrical discharge machining (EDM), 18
Wrapping microchannels, 151

X

Xurography, 62

Y

Y-junction, 152, 277–278
Y-shaped microreactors, 295–296, 301–302, 304
Y-type mixers, 109

Z

Zbasic2, 140
Zeolite-based reactors, 43, 93
Zero discharge systems, 190
Zero-carbon energy systems, 262
Zinc contamination, 196
Zinc oxide (ZnO), 137–138
Zinc oxide (ZnO) catalysts, 181, 228, 253
Zinc oxide (ZnO) nanoparticles, 9, 157
Zinc oxide (ZnO) QDs, 159
Z-scheme photocatalysts, 219